MW01348613

THE PERMACULTURE BOOK OF

FERMENT

AND

HUMAN NUTRITION

Image from original cover artwork printed in 1993. Fay Plampka freely contributed the cover drawing from her original work.

THE PERMACULTURE BOOK OF FERMENT AND HUMAN NUTRITION

by

Bill Mollison

Tagari Publications ✧ Tasmania, Australia

Tagari Publications, Sisters Creek, Tasmania, Australia 7325
www.tagari.com

© 2011, Bill Mollison

All rights reserved. Published 2011

16 15 14 13 12 11 3 4 5 6 7
First published in 1993, reprinted in 1993 and 2011

ISBN 978-0-908228-06-5
ISBN 0-908228-06-6

The contents of this book and the word Permaculture © are copyright. Apart from any fair dealing for the purpose of private study, research, criticism or review as permitted under the Copyright Act, no part of this book may be reproduced by any process without written permission from Tagari Publications, 31 Rulla Road, Sisters Creek, Tasmania, 7325, Australia.

National Library of Australia Cataloguing-in-publication
Mollison, Bill, 1928—
The Permaculture Book of Ferment and Human Nutrition
Includes Index.
Bibliography.
631.58 M726
Subjects: Cooking, Permaculture, Nutrition.

Printed by Print Applied Technology, Hobart, Tasmania, Australia using vegetable based inks, on paper certified as originating from responsibly managed forests.

To Lisa, Sammy and Chiquita
for love everyday

TABLE OF CONTENTS

CHAPTER SIX FRUITS—FLOWERS—NUTS—OILS—OLIVES

CHAPTER SEVEN LEAF—STEM—AGUAMIELS

CHAPTER EIGHT MARINE AND FRESHWATER PRODUCTS—FISH—MOLLUSCS—ALGAE

CHAPTER NINE MEATS—BIRDS—INSECTS

CHAPTER TEN DAIRY PRODUCTS

CHAPTER ELEVEN BEERS—WINES—BEVERAGES

CHAPTER TWELVE CONDIMENTS—SPICES—SAUCES

CHAPTER THIRTEEN AGRICULTURAL COMPOSTS—SILAGES—LIQUID MANURES

CHAPTER FOURTEEN NUTRITION AND ENVIRONMENTAL HEALTH

ABOUT THE AUTHOR

Bill Mollison has been vitally concerned about the environment since the 1970s, first as a scientist and naturalist, later as a vigorous campaigner against environmental exploitation. In 1974 he shifted his emphasis from protest to positive solutions with the development of the Permaculture concept, which focuses on positive ecological design for urban and rural properties.

His work with Permaculture brought him into close contact with village and traditional societies worldwide. It became apparent that wherever people still retained their traditional methods of food preparation, great attention had been paid to the nutritional basis of agriculture. However, this knowledge was being rapidly destroyed as mass-marketed foods and modern agriculture began to replace older systems. Nutrition is rarely mentioned now in terms of crop production and processing.

Bill saw that it was vital to record these elegant and often very sophisticated food preparation systems and to use modern research to record and confirm the value of the traditional recipes presented here.

In recent years, Bill has devoted all his energies towards designing sustainable systems, writing books and articles on Permaculture and, most importantly, teaching. He has given hundreds of public lectures and courses. Bill, along with subsequent teachers, has trained many thousands of people worldwide, resulting in the expediential growth of independent associations involved in the areas of sustainable agriculture, re-afforestation, education and village economics.

Born in 1928 in the small fishing village of Stanley, Tasmania, Bill Mollison left school at the age of fifteen to help run the family bakery. He soon went to sea as a shark fisherman and until 1954 filled a variety of jobs as a forester, mill-worker, trapper, snarer, tractor-driver, and naturalist.

"As a child I lived in a sort of dream and I didn't really awaken until I was about twenty eight years old. I spent most of my working life in the bush or on the sea. I fished and hunted for my living. It wasn't until the 1950s that I noticed large parts of the system in which I lived were disappearing. First, fish stocks became extinct. Then I noticed the seaweed around the shorelines had gone. Large patches of once ubiquitous forest began to die. I hadn't realised until those things were gone that I'd become very fond of them, that I was in love with my country".

Bill joined the CSIRO (Commonwealth Scientific and Industrial Research Organisation—Wildlife Survey Section) in 1954 and for the next nine years worked in many remote locations in Australia as a biologist. In 1963 he spent a year at the Tasmanian Museum in curatorial duties, then returned to fieldwork with the Inland Fisheries Commission surveying the fauna of inland water and estuaries.

Returning to formal studies in 1966, he lived on his wits, running cattle, bouncing at dances, shark fishing, and teaching part-time at an exclusive girls' school. Upon receiving his degree in bio-geography, he was appointed to the University of Tasmania, where he later developed the unit of Environmental Psychology.

While at the University, Bill independently researched and wrote the *Three Volume Tasmanian Aborigines and their Descendants* as a response to the Tasmanian government's refusal to give schooling grants to Aboriginal children on the basis of their supposed extinction. With his genealogies, Bill was able to prove the lineage and existence of many Aboriginal families, and succeeded in forcing the government to issue schooling grants to the children of many needy families.

In 1981, Bill Mollison received the Right Livelihood Award in Stockholm "for vision and work contributing to making life more whole, healing our planet and uplifting humanity". The Right Livelihood Award, often called the Alternative Nobel Prize, honours people working on practical and exemplary solutions to the social, cultural, and environmental problems facing us today.

Following his keynote address to the Schumacher Society in the UK (1989), Bill was made an honorary Fellow of the Society, joining such luminaries as Ivan Illich, Helena Norberg-Hodge, and Dr James Lovelock.

The ABC documentary series "Visionaries", shown in 1990, featured Bill Mollison and Permaculture. This was followed in 1991 by a four-part ABC series "The Global Gardener". The enormous public response generated by this program demonstrated a desire by people in all walks of life to find safe, sustainable solutions to our global environmental problems.

Of all the achievements attributed to Bill's undying commitment to Permaculture education, the one he is most proud of is receiving the Vavilov Medal for contributions to sustainable agriculture in Russia and his invitation and acceptance into the Russian Academy of Agricultural Science in 1991, the first foreigner to be so honoured.

AWARDS

- Nominated for Australian of the Year 2011
- 2010 Senior Australian of the Year for the State of Tasmania
  Awarded by the National Australia Day Council
- 2010 Senior Australian of the Year—National Finalist
  Awarded by the National Australia Day Council
- 1999 Australian Icon of the Millennium in the field of Ecology
  Awarded by the Australian Broadcasting Corporation in recognition of a visionary, and for contribution to his field of expertise.
- 1996 Steward of Sustainable Agriculture, USA
  Eco-Farm Conference, Asilomar, California, USA
- 1994 The Banksia Environmental Award, Australia
  Awarded for the promotion of sustainable systems worldwide.
- 1993 Outstanding Australian Achiever Award
  Awarded by the National Australia Day Council.
- 1991 Vavilov Medal, Moscow, Russia
  Inaugural award by the Russian Academy for contributions to sustainable agricultural and community systems.
- 1991 Member of the Russian Academy of Agricultural Science
  The first foreigner to be so honoured.
- 1989 Reconocimiento Mexico
  Awarded by the Governor of Sonora, Mexico for work with poor urban, and rural campesinos.
- 1989 Honorary Fellow of the Schumacher Society, Surrey, UK
  Delivered the Annual Schumacher Lecture.
- 1988 Tree Tax Award, Holland
  Given by De Twaalf Ambachten for initiating a tree tax on all Permaculture publications.
- 1981 Right Livelihood Award
  This award honours people working on practical and exemplary solutions to the social and environmental problems facing the world (referred to as the Alternative Nobel Prize).

Bill's publications include: *Translations available

- *Three Volume Tasmanian Aborigines and their Descendants (Chronology, Genealogies and Social Data)*
- *Permaculture One* with David Holmgren
- *Permaculture Two: Practical Design for Town and Country in Permanent Agriculture*
- *PERMACULTURE: A Designers' Manual**
- *Introduction to Permaculture* with Reny Mia Slay*
- *The Permaculture Book of Ferment and Human Nutrition*
- *Travels in Dreams: An Autobiography*

ABOUT THE PERMACULTURE INSTITUTE

The Permaculture Institute was established in 1979, by Bill Mollison, its founding director. Its purposes are to provide education in Permaculture design, promote careful thought and understanding and the positive effects of its only three ethics: 1. Care of the Earth 2. Care of People 3. Return of surplus to the first two ethics.

Bill Mollison devoted the years following its establishment to travelling the world, where no internet existed, educating people via the popular The Permaculture Design Certificate Course, in 120 countries. The success of these courses is founded on the fact that they all followed an effective and carefully crafted curriculum which is found in the book, *PERMACULTURE: A Designers' Manual.*

The effect of Permaculture Design has grown exponentially due to Bill's foresight in making sure that each course must produce new teachers of the curriculum. Few system expansions can compare to this.

Bill has had a profound influence on people all over the world, from small gardeners to royalty. His gift is one of knowledge and inspiration.

PREFACE

As McDonalds and fast food chains proliferate and supersede the foods of different world cultures, there is a danger that traditional processes and recipes will be lost, very much as traditional seeds, customs, dress, and behaviour have been lost.

Thus, I think a very cogent reason for collecting every possible traditional process in detail is imperative to preserve traditional seeds and recipes—they work. Just as we need to bring root inoculants with trees to ensure their nutrition, I believe that we need to bring traditional ferment and cooking methods with any new foods, for our own nutrition.

While many people make noises about biotechnology providing food for the poor, the actual facts are that the poor already use a great variety of sophisticated applied microbiological skills, and academic biotechnologists, for the main part, work for large multinational companies, providing inessential beers or spirits, or monopolising the condiment, bread and cheese markets.

Increasingly, through varietal selection and gene recombination, and using very large continuous or column fermentation, large firms are patenting organisms and processes and producing a wide range of medical and industrial products (and the associated ferment organisms). Very little attention is paid to assisting poor people to make the most of their foods, and this is what this book is about.

No book is ever complete, and I appeal to my readers to send in additional material, corrections, or traditional recipes for future editions.

BM

ACKNOWLEDGMENTS

Many friends have already contributed to this book, and have been patient with my research in their kitchen libraries, and I thank them all. They suffered my presence with good cheer, and provided ferments for my sustenance.

Janet Lane converted my roughs to legible line drawings and flow charts, and the photographs are my own, or in part from the library of the Permaculture Institute.

Other friends or contributors are acknowledged in the text.

PUBLISHER'S ACKNOWLEDGMENTS

As the popularity of this book continues to grow, it is with great pleasure that we bring to the market a new print run, which has undergone some minor updates including:

Consultation with the author to clarify recipe and method instructions; re-formatted imagery and text for the front matter, back matter and chapter contents; reviewed text for uniformity; updated references; updated resources appendix; new pictures of the author; photo section re-styled and expanded; updated illustrations; a glossary; updated index; botanical and Latin names for clarity; updated layout; and a new cover.

Publishing Team—Lisa Mollison, Nicki Trewin, Michelle Walker, Cathy Shliapnikoff and Carolyn Watson-Paul.

Agricultural Terms

There are many terms used throughout the book which are colloquial. Some may not make sense to readers unfamiliar with rural practices or the various fermenting processes and food preservation techniques. We have endeavoured to place these into a glossary to assist the reader, but some terms may not have been included. Thanks to the internet, a quick search ought to provide illumination to the phrase or term in question.

Recipes Collected

Many of the recipes and methods presented in this book have come from interviews conducted by Bill, and often with the help of a local translator or taxi driver. Therefore the reader can expect to encounter measurements and turns of phrase that are as unique as the many individuals who shared them.

Personal View Points

Every author has a view point and opinion on a variety of matters. Bill is no exception. Some people may find some of the content of this book offensive.

THE PERMACULTURE BOOK OF FERMENT AND HUMAN NUTRITION

INTRODUCTION

All the recipes given herein are traditional; they belong to humanity, even though they have been collected or tried by various authors, they have all been used for centuries by thousands of human beings. Only a few recipes are my own inventions (you may guess at these) but even these derive from my family or friends in their main ingredients or procedures.

The book grew out of my interest in gardening and teaching gardening. Gardeners are always interested in food preparation and storage, especially when dealing with bumper harvest and limited room in the freezer! I am a lazy gardener, using mulch for growing food, rather than digging too much, and growing perennial foods. At my farm I maintain a varied framework of perennials and small patches of annuals in my gardens, over the years I have milked goats and cows, made fermented milk products, preserved meats, vegetables and fruits, and so did my mother and grandmother, and their mothers before them.

So this is a book about making the preserves, sauces, pastes, and preserved or dry foods needed for many recipes, or used as side dishes that add vital nutrition and culinary interest to starchy foods. It is also about developing some additional essential nutrients in food by fermentation techniques.

I have added a small section on eating the earth (geophagy) and small vertebrate, invertebrate, and insect foods, as these interesting foods are often neglected by gourmets (with the exception of frogs' legs and snails).

This treatment does not include many recipes for sweets, unfermented beverages, or normal cooked or raw fresh foods (fish, meat, vegetables). It does include the normal condiments, sauces, and some special products (gelatine, pectin) that we can make in the kitchen, and notes on the preparation, preservation, and storage of common classes of foods and/or ingredients.

None of the recipes within this book are to be regarded as forming any but a small part of a diet. I believe that the very concept of a staple food is foolish; with twenty to thirty foods to choose from, no staple is actually needed (a staple food comprises 50% or more of the diet). A good garden duplicates and extends the hunter-gatherer diet. We originated as broad-spectrum omnivores and remain so today. We can all eat well if we garden.

A stimulus to writing this work came from field data collected in Africa, India, amongst Native Americans, and with tribal peoples. It should be read and well digested by anyone encountering other cultures where traditional processes are still preserved. We should always try to describe, evaluate, and understand why traditional foods are so frequently subject to ferment or microbiological processes, and why they are cooked in special ways.

As a professional master baker (when I was a young man), and in my travels among villages and tribal peoples, I have tried many of the foods herein, including various meats, insects, ferments and organic acid products, and made many of the pickles and ferments listed. I have found that practise in ferment is needed to be certain of a good product until the process becomes routine.

I have also tried to fill in some important data on trace elements for gardens, to explain the role of enzymes, and have appended a note on very alkaline sites. Not only are trace elements critical to good health in plants, but in the microbial world, and in our bodies, they join with enzymes from micro-organisms in a catalytic role to both break down complex foods into amino acids, and to assemble those special proteins needed by the body as enzymes, muscle tissues, hormonal, connective, or nerve tissue substances.

The book may also assist field workers and paramedical staff working in areas of poor nutrition in preserving, valuing, describing, improving and trying out many traditional food ferments and food storage systems that would cover a wide range of foods. By using ferment methods we can create many more valuable foods from otherwise low-quality food bases using readily available micro-organisms and enzymes.

TRADITIONAL

Many societies, and especially tribal societies, have carefully apportioned food to suit age and sex needs (testicles for zinc to breeding males, liver for iron to menstruating women) while others, often male-dominated, make it difficult for women and children to obtain full nutrition.

There is a profound difference between cultures derived from true nomadic herders and settled village peoples. Nomads often classify ferment as "magic" and may shun or forbid fermented foods, thus giving rise to religious prohibitions. Some food fads (vegetarianism in upper caste Hindus, and in some modern peoples) are adopted just to differentiate themselves, or to assert superiority. This also is the basis of the European aristocracy and their punishments for poaching salmon, deer, and gamebirds. Special diets and special foods are seen to be part of power and privilege.

In many of the existing traditional societies, we get clues as to how good practices, leading to successful cultures, have arisen. The Lapps (Saami) still use reindeer stomachs (for rennin) to curdle and store milk. The Tibetans roast pigeon manure (to sterilise it) and add this to their sour roast-millet porridges, thus providing major microbial nutrients and trace elements. Many Native Americans add ash to flours and ferments for the same reason. The Japanese use seawater to curdle legume milks, thus again adding a wide range of minerals and mineral salts for specific microbial nutrition. Ash and lime provide the alkaline base for beneficial moulds in foods.

How to derive, or create "feasts of good food in the middle of want" (Couffignal, 1949) has called on all our resourcefulness, talent, good science (or observation plus memory), courageous exploration of unfamiliar foods, and perhaps has arisen from the sheer desperation of people over thousands of years. We commonly risk famine from climatic or economic change, war, abandonment, banishment, cultural prohibitions, spoilage of stored foods, and the loss of forests and soils.

Perhaps only the spur of poverty and deprivation has initially driven us to try eating "spoilt" food, insects, or earth, but from subsequent survival and observation we have learned lessons that have indicated where relatively inedible, unfamiliar, or insufficient foods were made edible or more nutritious by ferment; lessons on the selection, preparation, and particular food value of invertebrate foods, and on the value of certain earths and clays in food preparation. We have often, in the last resort or as religious or mystical beliefs dictate, tried cannibalism, or even institutionalised cannibalism (symbolised in Catholic ceremonies as partaking of "flesh and blood").

In difficult climates or hostile environments, people have been forced to turn to unusual foods, sometimes as a source of those trace elements lacking in soils, sometimes to supply sufficient food, minerals or vitamins for good health. Often, the rationale for traditional diets and for habits of food preparation or food combinations has been lost, or never known in the sense that it could not be explained. We are still explaining the role of trace elements such as chromium in the diet, and vitamins were virtually unknown less than a century ago. Enzymatic processes (although used) could not be explained before the development of sophisticated science and technology.

TRADITIONAL FERMENT

Most of this book is about fermentation, for it is an excellent way to prolong the life of many foods, and to build proteins and vitamins into starchy or low-grade foods. Most western peoples are familiar with yeast breads, sourdoughs, cheese, and beers. But few of us realise how skilfully traditional peoples enhance the flavours in their diet, or make simple carbohydrates more nutritious through fermentation.

Not that ferment is used only for food production, it is also part of silage (fodder) and compost making, a method of producing fibres and dyes by retting (rotting) plants, a way to free seeds from awkward capsules, or hair from hides, and an old method of preparing seeds for germination by the removal of inhibiting substances, such as saponins (soap) from the seed coat.

For almost everybody in the west who is not a scientist or a food technologist, ferment processes are a mystery. Beer, bread, cheese and yoghurt are the most familiar products of ferment and also a fairly large part of rural diets. In the eastern world however, fermented fish and vegetables are the normal foods of the villager, and in Asia take many forms, based on many substrates (rice, wheat, legumes, vegetables).

Curiously, almost all micro-organisms used in cultures to inoculate foods are usually found in quite specific foods, in very specific places (e.g. *Lactobacilli* on milk or cream in dairies, rarely in nature). It is probable that we have co-evolved these useful species over thousands of years, as much as we have domesticated larger organisms like dogs or cattle. How else could a milk-specific organism have persisted? How also could brewer's yeasts have evolved, when we know that it was ourselves who make bread and beer?

Such comments do not apply to spontaneous ferments, e.g. vinegars in open containers, radish in pits, grain in compost, or fish in salt; many materials carry their own enzymes and ferment organisms, able to operate in these artificial environments, and so are dependable cultures of themselves. But how did we discover these things except by building pits, composts, and salt packs? It seems probable that many of our attempts to store food actually resulted in beneficial ferment, and that we proceeded to refine the process over thousands of years.

However ferments originated, once established they were marvellously refined to thousands of products, and form an important part of the totality of human nutrition.

Milk, bread or beer may very well "create" some microbiological species, and will certainly generate varieties suited to specific processes or flavours. This would explain why we can't readily find many useful species outside human created environments, and also why our near ancestors fermented in wood, earth pits, unglazed pottery, gourds, and skins; they preserved their micro-organisms in the containers which they carried with them, or used every year (as in pits). Thus, the substrate creates varieties, and we create the substrate and perpetuate successful cultures.

Although we in the west can afford to eat rice as a simple starchy grain (in an already over-sufficient diet), very few people in the second and third world can afford to omit some form of starch enrichment by ferment, and to create foods consumed as sour drinks, beers, sour porridges and sprouted seeds. Many field biologists now point out that indigenous ferment methods apply to homes and villages, and use common low-quality starches or cellulose wastes to create protein enrichment.

A BRIEF HISTORY OF MICROBIOLOGY AND FOOD

The spread of cultivated plants has strong associations with armies (as hard rations), slaving (as crops and rations), missionary work (people going to and returning from distant lands), and of course the development of ancient trade routes. But vegetatively-grown plants lagged behind, as seeds were easier to carry and propagate, so that bananas, breadfruit, yams, taro, and so on spread more slowly and often depended for their distribution on the development of large trading vessels, and of sheet glass for deck hothouses.

Today, several thousand botanical centres exchange seed and seed catalogues, and many private plant collectors assemble large and unusual collections. The very recent emphasis is not on decorative or rare plants (cacti, succulents, flowers, foliage plants) as it was in the eighteenth and nineteenth centuries, but on useful plants for Permaculture assemblies of food, firewood, building, thatch and windbreak (Mollison, 1988, 1991). The potential is as yet hardly realised for complex regional "forests with clearings" composed of the thousands of useful trees, shrubs, vines, ground covers, bulbs, rhizomes, fungi, and grasses available to us in modern times, although home gardens of two hundred to five hundred species are now in fact common, as are village forests of up to four hundred species of useful trees and vines.

As we assemble these modern and productive soil-creating systems, we enter a new era of human resource development and conservation of the earth. We learn to value the total genetic resource of the earth, now rapidly being impoverished by deliberate supplanting of cultivated local grain and vegetable races by patented crops, and a reckless destruction of forests for wastepaper; and shrublands, desert border floras, or heaths for the grazing of beef and growing of wheat.

Fermented foods are almost certainly coincidental with the development of mankind from earliest times. Many fruits and aguamiels (sugary saps and nectars) ferment of themselves, and were always used by people. Food gatherers and cooks would therefore have had many of their foods ferment naturally, and eventually would deliberately come to ferment products for flavour, storage, or for improved nutrition. Only modern food introductions (soybeans from 1000 AD on) brought to new areas by sea, would be brought as seed for gardens, and the ferment organisms left behind.

By 1800 AD mycorrhiza were brought in soils, with their host plants in Ward boxes (small greenhouses on the decks of boats) as plants in living soils, and ferment technology is slowly catching up as yoghurts and sour porridges are developed in new countries, or immigrants bring their "seeds" of ferment by sea and air to their host nations (many Dutch immigrants to Australia in the 1950s brought koumiss and cheese starters). The Japanese and Chinese set up miso and soy manufacture where they settle, together with pungent leaf ferments, and the northern Europeans their sauerkrauts.

The bland diets of many western Europeans kept pungent and fermented products (chilli, garlic, kim chee, sour porridges) from their tables until recent years, and it may be decades before these methods are fully accepted by many conservative peoples.

Thus, the degree of ferment in the diet is changing slowly. My trust is that this book will help many people take over their own food processing and improve regional and local nutrition.

HOW TO USE THIS BOOK CREATIVELY

First, read all the recipes and get a general idea of the field. It is then possible to plan for several benefits.

For example:

- Reading in nutrition enables us to first plan a balanced diet, and from that to plan family menus over a year.
- Planning diets and menus points to the essential garden and livestock species that we need to achieve health; so we plan a garden.
- Try to invent substitutes for ingredients to suit your climate or garden; just as long as they are in the same plant or animal groups they should work.
- If you belong to a garden club or grower cooperative, suggest group processing to save energy and to create viable small industries; many grow, a few need to process to "value-added" products.

CHAPTER ONE

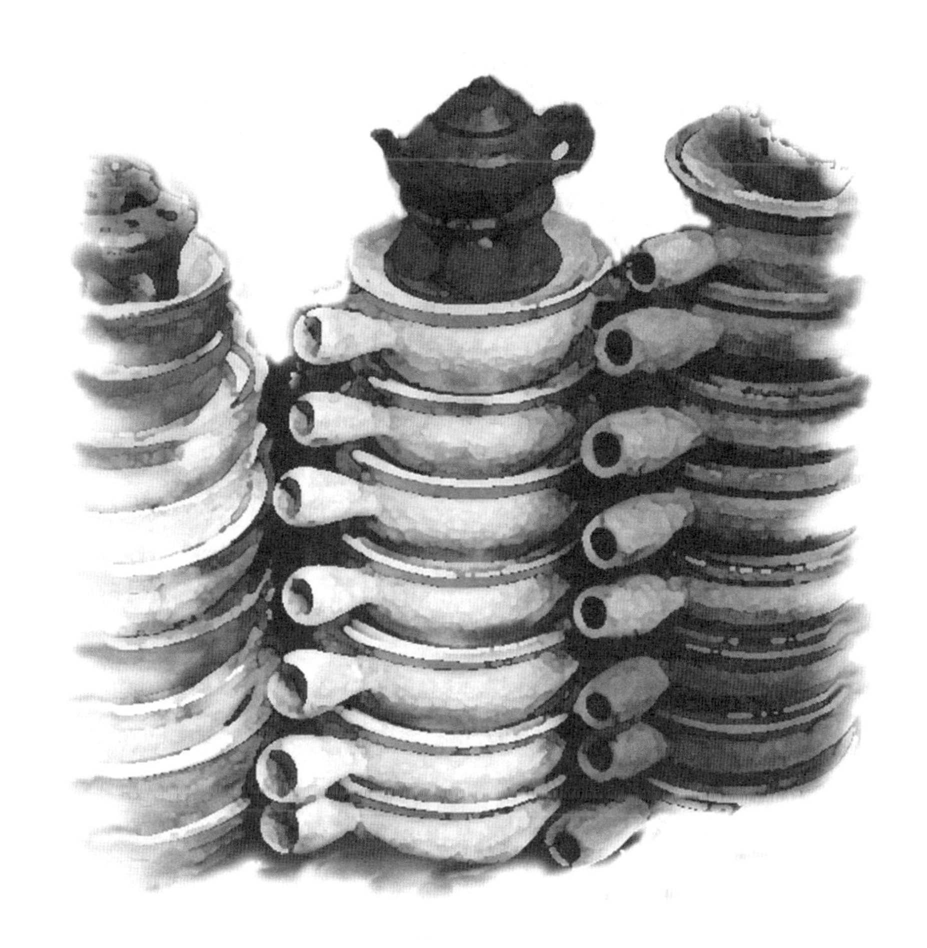

Storing • Preserving • Cooking Foods

1.1 INTRODUCTION

In this section we will briefly review some preparation, preserving, and storage methods. Section 1.10 covers specific fermentation processes, which is the main subject of this book.

In western countries, before the advent of refrigeration, excess food was always "put up" in the pantry or in the stillroom. These foods were dried, salted, pickled and fermented, and what was produced and stored away were aromatic cheeses, sausages, wines and beers, pickles, cured hams, salted fish, and even dyes and soaps. Pantries and stillrooms are not now built into modern houses, but there is no reason not to add one to the kitchen, even if it is only a small corner of the kitchen or basement.

A small storage shed, perhaps insulated, would be ideal for curing the cheeses and sausages, and storage of wines and spirits. No fancy equipment is needed and once you get a routine down, it takes very little time to produce favourite cheeses, beverages, tofu, pickles, and other fermented products. The main thing to remember is cleanliness, so that the possibility of contamination is reduced.

Types of Storage and Preservation Techniques

- Refrigeration and freezing—lowers temperatures and slows decay. Refrigeration is useful only for short-term use, while frozen food can last from several months to a year. Freezing is a high-energy storage method, except in permafrost.
- Canning—storage in glass jars or (commercially) tins, most suitable as a home method when using sugar as a preservative. Requires heat, jars with special lids to seal contents, or paraffin for jams and jellies.
- Preservatives—using sugar, acids, spices (all natural preservatives); or either sulphur dioxide and benzoic acid (artificial preservatives).
- Gas storage—storing grains in carbon dioxide or fruits in ethylene. Also food packed under anaerobic conditions, in sealed jars.
- Irradiation—ultraviolet or alpha radiation, now commercially used on fruits and vegetables in some countries, is said to prolong shelf-life, but is still suspect for breakdown products; such systems need clear labelling for public consumption. Also commercial use of fast, high heat for such products as long-life milk.
- Special packing—storing products in such substances as waterglass, cold sand, oils and peat.
- Filtration and pasteurisation—removing sediments from fruit juices and alcoholic drinks. Also heating to temperatures high enough to kill spoilage bacteria.
- Dehydration—drying products via the sun or ovens, and storing in jars or in sand. Suitable for fruits, vegetables, and meats.
- Smoking—the coating of bactericides on meats through smoke. Usually cured first by salting.
- Pickling—curing with salts, vinegar, acids, oils or a combination of these.
- Fermentation—really a sub-unit of pickling, it employs microbes (yeast, bacteria, moulds) to preserve the product, impart a distinctive flavour, and, in many cases, provides vitamin or protein enrichment.

Many of these techniques are used in combination. For example, partial drying, pressing, or withering often comes before ferment processes (those for tea, nuka ferment, cheese, fish pastes), while salting or sugaring is used to draw out liquids before processing (sauerkraut, kim chee). For this reason, even though the focus of this book is on ferment techniques, I include some of the other preserving methods such as drying, sugaring, and salting with a brief mention of smoking.

1.2 STORAGE METHODS

Many cultures have used foods buried in earth material as a storage method. Some of the ways this has been useful are:

Earth Storage

- Peat storage—swamp peats and bogs preserve both by excluding air (anaerobic) and by the presence of humic acids, sulphurous gases, and carbon dioxide. In older times, barrels of butter, beech-nut oil, and hoards of acorns were stored in swamps by European and Amerindian peoples. Long-term mummification of bodies is commonly recorded for people buried in peats or bogs, and no doubt hides and leather were buried for tanning in acid peats. Bottles, jars and cans packed in sawdust keep cool on hot days, and do not freeze in cold spells. Many pickled goods can be so kept in cellars or storerooms.
- Earth curing—game meats, and whole game such as hares, pheasants, quail and venison have been traditionally hung to tenderise them in cool seasons, but in summer periods they are wrapped in paper or bark and buried 20–30 cm (8–12") deep in soil to undergo controlled rotting for about a week to ten days. The traditional jugged hare is prepared in this way before cooking. Such game is more easily cooked and digested, is insect-free, and is tenderised. "Thousand-year eggs", beloved by the Chinese, are buried in volcanic sulphurous mud or anaerobic mud below mangroves. Here (over one to six years) they undergo controlled rot, to a jelly-like green-black semi-solid, then eaten as is. Faster methods are given herein.
- Ash storages—bacon, eggs, root crop, and (wrapped) dry fruit or vegetables can be ash-stored in cellars or cool rooms. Ground nuts and seeds are stored in wood ash to prevent weevil attack. Fine grass ash, and the ash of specially selected plants may be used.
- Sand storages—fresh sound fruits, wrapped in silk paper or fine paper, are buried in cool dry sand, and kept for weeks beyond normal periods. A few sand boxes enable different fruits to be stored for longer keeping times using this method. Citrus kept cool and dry becomes well ripened, thin-skinned, and sweeter.

Figure 1.1—Staddle stones, cut to support grain bins or barns used to store roots and grains, prevent entry of rodents or large ground insects like termites.

Sand has traditionally been used in the following ways to assist food preservation:

- Roasting and drying
- Cooling and storing
- Fermenting and infusing

Throughout South East Asia, barley, chickpeas, groundnuts, and nuts have been roasted by stirring them in hot sand. This is called parching. The sand is first carefully washed, sieved, and graded so that it is easily sieved from the grains to be roasted. Then it is heated in a shallow iron pan or wok, and the grains or nuts stirred in until evenly browned and roasted. This prevents burning by contact with the pan. The roasted material is then easily sieved from the sand. This process is found in the many wayside stalls or cafes in India.

Volcanic (black) sands reach very high surface temperatures in sunlight 45–72°C (112–160°F), to about 15 cm (6") depth. Roots, thin slices of lean meat, whole fruits, vegetables, etc. can be dried by wrapping and burying them in sand, and allowing the sun to heat the sand. Again, sand size should be selected to be easily brushed or sieved from the material to be dried. The obvious advantages are even heating, and insect-proof storage while drying takes place. Much higher temperatures are achieved if the sand is boxed and covered with a sheet of glass. Salted fish are cured in sun-heated mud in Egypt.

Grain Storage Techniques

- Airless atmosphere—if tight-fitting lids are available, almost fill a press-top drum or can, light a small candle in the seed box, and press down the lid, having first painted a thin varnish on the opening. The candle burns off all the oxygen and goes out, and the varnish seals the lid. No pests can bother the grain.
- Carbon dioxide—store cobs, grains, and seed in a sound leak-proof silo, a bin, or large plastic bag. Fill the containers with carbon dioxide (from a beer ferment tube) or dry ice. Seal the door well; or close the mouth of the bin with mud; or close the bag with a tight ring. Pests perish in a carbon dioxide atmosphere, lacking oxygen.
- Heat treatment—grain or seed can be heated to 65°C (150°F). This kills all insects but leaves the grain germination unaffected. Dry grains can then be stored in silos, bins or bags.
- Cold treatment—grain stored at temperatures below 15°C (60°F) is pest-free. Many countries (Canada, Russia, USA) have naturally cold areas suited to long-term grain storage in icy conditions. Freezing kills the insect pests of seed. A day or two in a freezer protects your garden seeds. For small amounts of seed grain, first freezing for ten days, then sealing well with carbon dioxide or dry ice in dark containers at less than 6% moisture is a good general rule.
- Natural insecticides—storage in grain stores, where ash, dried neem tree leaf, magnesite, or diatomaceous earth is mixed with the seed also minimises grain loss and pest attack; ash and the earth dehydrate parts of grain, but do not affect germination or food value. The larger grains (wheat, rice, sorghum, maize, barley, rye) are the more difficult of the grains to store for seed (unprocessed), as they have many storage pests. Such grains also need to be freshly milled, as hulled flours and meals deteriorate over relatively short periods (three to four weeks). Wholemeals quickly become rancid, so it is better to store whole grains rather than milled. In Indian villages, termites are deterred by raising grain bins on short legs, so that chickens can find and destroy the termite earth tunnels. Rats are excluded by mushroom-like tops on the storage bin legs, see Figure 1.1, now used "decoratively" in the UK. Mice which attack hay stores in Australia, reducing protein by eating seed and soiling the hay, are excluded by sheet iron, turned down at the top. These same fences are used around gardens for mice and snails in Australia and Germany. In the long run, only grains regrown every one to three years have been kept for thousands of years. Just save some seed and sow the rest to keep your grains going, and join a seed-savers group or seed exchange.

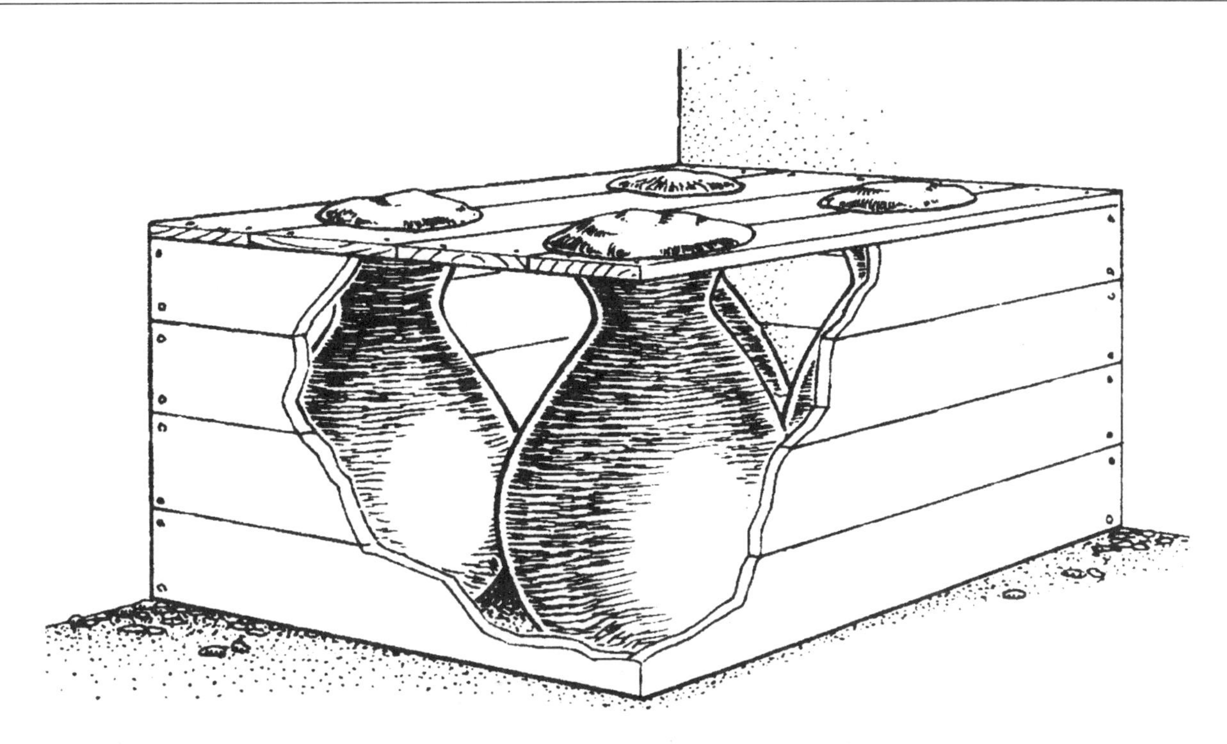

Figure 1.2—In India millet is often stored in large pots built into low walls and closed with mud.

Storage with Insect-Repelling Plants

Plants used to repel, deter, or kill insects in storage containers are used all over the world. They are used dried, as a loose powder, plastered on as a lining with cow dung and mud, layered in sequence with grains, or as a final layer in deep pots or earth stores.

Some examples of insect-repelling plants used in various parts of the world are:

- The pods of the snake bean tree (*Swartzia madagascariensis)*, powdered and used to line bins. Effective as a repellent and insecticide.
- The leaves, berries, kernels, and kernel oils of white cedar (*Melia azedarachta*) and the neem tree (*Azedarachta indica*). Both are effective on crops, in water, in grain bins, and as an aqueous extract against body parasites. Both have the same spectrum of effect as malathion, but are harmless to people.
- Root powder of sweet flag (*Acorus calamus*) is reported as being more effective than malathion if dusted into stored seed or grain (India).
- Ash from the burnt leaves of Cat's Tail Aloe (*Aloe castanea*), South Africa, is used for weevil control; also ash from *Aloe ferox*.

Storage of Small Grains

Small grains (millets, minor millets, amaranth, quinoa, teff) have two good characteristics—they store much better than large grains, and are therefore the common grains of village famine-reserve stores; and they are rarely much trouble to husk, being simply beaten into a sieve over a large drum after drying, and then winnowed, dried, and stored.

Ragi millet (*Eleusine*) in India is often stored in large clay pots built into a low internal wall of the house, with the pot lid mudded over in the bench, see Figure 1.2. Millet is said to store this way for four to five years. Similar storages are made for parboiled rice and roasted grains.

Well-dried small grains are suitably stored in screw-top jars, or in press-top tins in which a candle is lit just before the lid is pressed on and sealed with a gum, paste, or varnish.

In Africa, large pot-shaped holes or bell pits are dug in well-drained soils, and well plastered with cow-dung, mud, and powdered neem leaves. See Figure 1.3. These are filled with grain and the relatively small lid (to admit a slender child) sealed with mud. This can be earth-covered for security. Small grains keep this way for years. Ash can be added at last filling to deter insects.

Storing Potatoes

Lay straw in a dark dry shed, dust with burnt lime and lay on potatoes. In this way, build up a "clamp" layer by layer. The lime deters pests; ash can be used instead of lime. This method also for yams, sweet potatoes, and starchy roots.

In the field, alternate layers of straw and potatoes are built up, and covered for the winter with a thick layer of topsoil, see Figure 1.4. Heaps are 1–1.5 m (3¼'–5') high, up to 20 m (22 yd) long, and are uncovered and carted as needed. Root crops keep warm in such winter "clamps".

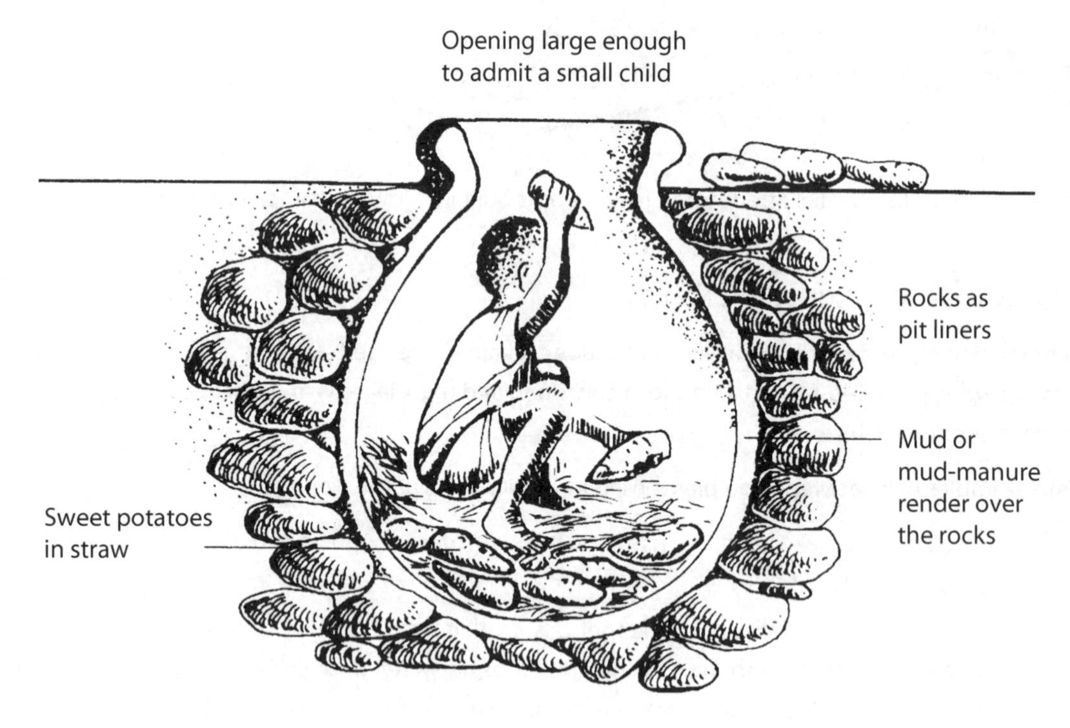

Figure 1.3—Sweet potato store (Zimbabwe) and for small grains.

Sealing Jars, Crocks, Tins

When preserving and fermenting food, it is important to "stop the action" at its peak. Otherwise, a product will just keep on fermenting and changing until acid, ammonia, or soft rot sets in. Cheeses not covered with wax and muslin will soon instead be covered with moulds. Beers go flat in their jugs. Wine turns to vinegar. Pickles fall apart. Cured meats rot. Bread dough collapses. The product **must** be sealed before this happens, usually by heating, canning, corking, waxing, drying, etc. Below are some methods of sealing jars and tins:

- Paraffin, vegetable wax or beeswax, melted, is poured on solids or semi-solids, like jam, packed pickles, and cheese; this wax is recovered for re-use by washing, boiling, cooling, skimming. Corks can be dipped in hot wax. When burying earthenware, seal lids with either mud (by itself), mud plus fresh cow-dung, or salty flour paste.
- Pour a layer of olive oil on liquids; mustard oils are often poured on hot chutneys or achars in India.
- Dip a fitted cloth or paper in oil, place over jar mouth, and screw down cap on jar.
- Use bottles or jars with rubber rings or seals in good condition.
- Cover large masses tightly with plastic, then bags, and then weight with a board or stone (e.g. for miso).

- Seal tops of storage jars (wood or pottery) with a thick flour-water-salt paste. In hot countries these are kept buried in the earth in the shade. Even large barrel tops can be sealed with melted paraffin or hard fats floating on brine.
- Make a medium thick paste of flour and water, and this can seal lids or even suspect (boiled) rubber rings.

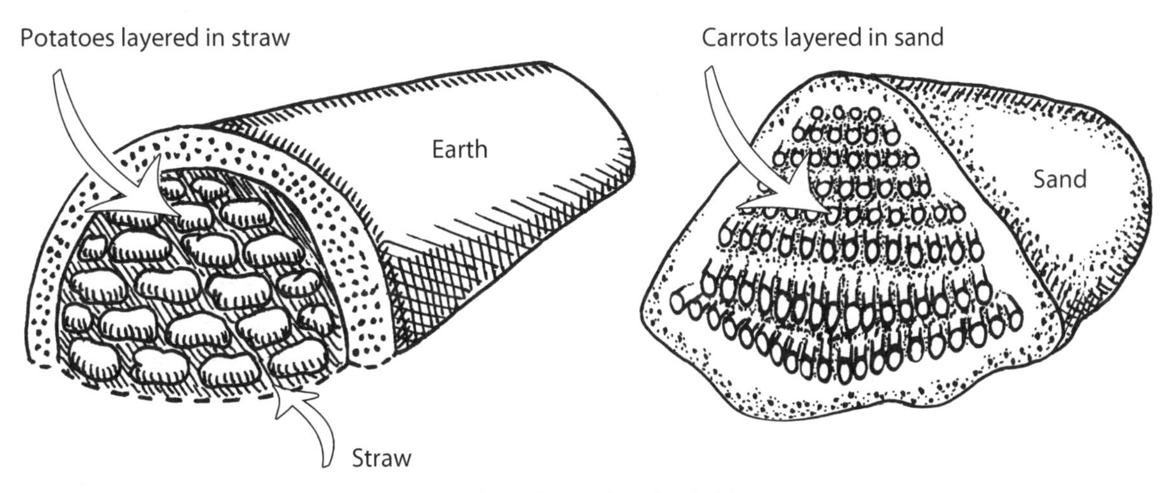

Figure 1.4—Potatoes and carrots are stored in "clamps" in the field over winter

1.3 DRYING

Sun drying is the cheapest and easiest way to preserve many foods. In the process of drying, the moisture content is lowered and natural sugars (in the case of fruits) are raised, so that no harmful bacteria can exist in the product. Meats and fish are rubbed with salt, so that salt rather than sugar provides the anti-bacterial effect. Vegetables, which have very little sugar to preserve them, must be dried to the brittle stage before storage. Once it is dried, the product can be brought back to its original condition (with little flavour or texture change) by soaking in a small amount of water.

The only real problem with drying is determining when the product is sufficiently dry. If too much moisture is left in, it may rot after storage. If the product gets too dry, it often cannot be brought back to its original condition. A sample of all products should be packed away and then examined after a two-day period (with a periodic check after several weeks). If there is any sign of moisture, dry the product longer.

Drying Methods and Containers

In warm climates or in summer, sun-drying is easiest. The usual containers are flat trays with mesh on the bottom. The mesh allows the excess moisture to escape. The trays should be made so that they can be set out in the sun individually or stacked on top of each other for shade and air-drying.

The trays can be kept covered with cheesecloth to keep flies and moths from depositing eggs in the drying food. Alternatively, freezing the dried product for a few days, then thawing and storing, will effectively kill off most insect eggs. Metal, plastic or snug wooden covers can be made for the stacked trays for fitting at night or in rain showers (e.g. for sweet potato drying in the tropics).

Andrassy (1978) records a good method of enriching dried food by first drying to 50% moisture, then plumping up by a soak in citrus, pineapple, or other fruit juices. This resembles the addition of passionfruit pulp, strawberries or acid fruits to bland pulps such as papaya and jak-fruit when making fruit "leathers". Vitamin C, if available, or citrus juice both preserve and fortify dried fruits. Similar methods are used in Africa for dry seaweed.

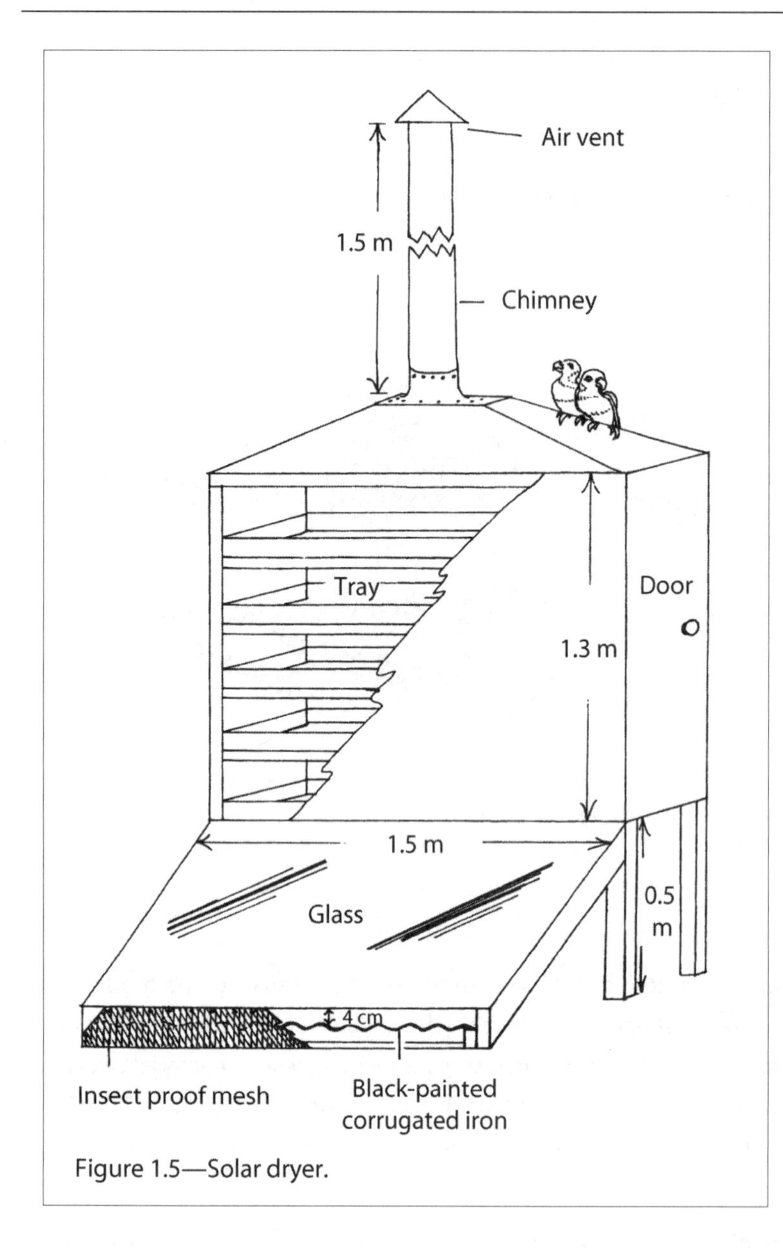

Figure 1.5—Solar dryer.

In cool climates or in areas where downpours are common all year round, a solar food dryer can be constructed. Figure 1.5 shows such a dryer (which can accommodate up to eight trays made from mesh in a wooden frame). The flat-plate collector is constructed with metal and glass (or tough plastic) and is located just below the drying chamber. The essential is a black metal solar chimney to suck up hot air over the metal collector and through the drying unit. The solar chimney also exhausts water vapour from the drying product.

Oven drying temperature for vegetables, fruits, meat, fish etc. should not exceed 50–66°C (122–150°F), even if oven doors are left open. Always check this with a thermometer. Frequent turning helps in drying.

To store dried foods, keep in a dark and dry place, seal well in plastic bags, press-top cans, and plastic or glass jars. Check periodically for moulds, and re-dry if necessary, in an oven at 66–70°C (150–158°F) if the sun is obscured, for an hour.

1.4 SMOKING

Smoking should not be used as a way to preserve any staple foods; stomach cancers in Japan and elsewhere are strongly suspected as arising from a too-frequent consumption of smoked products. Fish that are salted, sugared, and fried do not need smoking for preservation. Although hams, sausages, and meats have been smoke-cured for centuries, very little tar is taken from skinned sausages and hams. If one cares to smoke products there are three basic methods (using fish as the example):

- Smoke-Grilling—(South East Asia, China) whole, grilled carp and scale-fish are smoke-flavoured by scattering brown sugar on the coals. Skin the fish once they are broiled; only a slight flavour of smoke remains.
- Cold Smoking—immerse split whole fish for as little as ten minutes in 20% brine, or dry-rub with rock salt and brown sugar. First air-dry to just tacky to the hands, and then hang for twenty-four to thirty hours in cool hardwood smoke in the smokehouse. These keep for three to four weeks in a cool place but can be further wind and sun-dried in cool areas. Cold smoking temperature is held at about 28°C (82°F). Fire starts at 38°C (100°F). This gives long smoking preservation.
- Hot Smoking—smoke and cook eels, fresh and unsalted, in hot smoke from a hardwood fire. Thread eel through the jaws on a wire; the batch is smoked and cooked when a few fall off the wire (about thirty to forty minutes). These keep a week or so chilled, or (if brined and dry-smoked) a few months. As eels are smoked unskinned, removing the skin also removes most of the smoke. Hot smoking temperature is held at about 80°C (176°F). This gives a relatively short smoking for cooking.

Figure 1.6 displays elements for building a smokehouse, including an underground smokepipe for cold smoking and spirit burner for hot smoking.

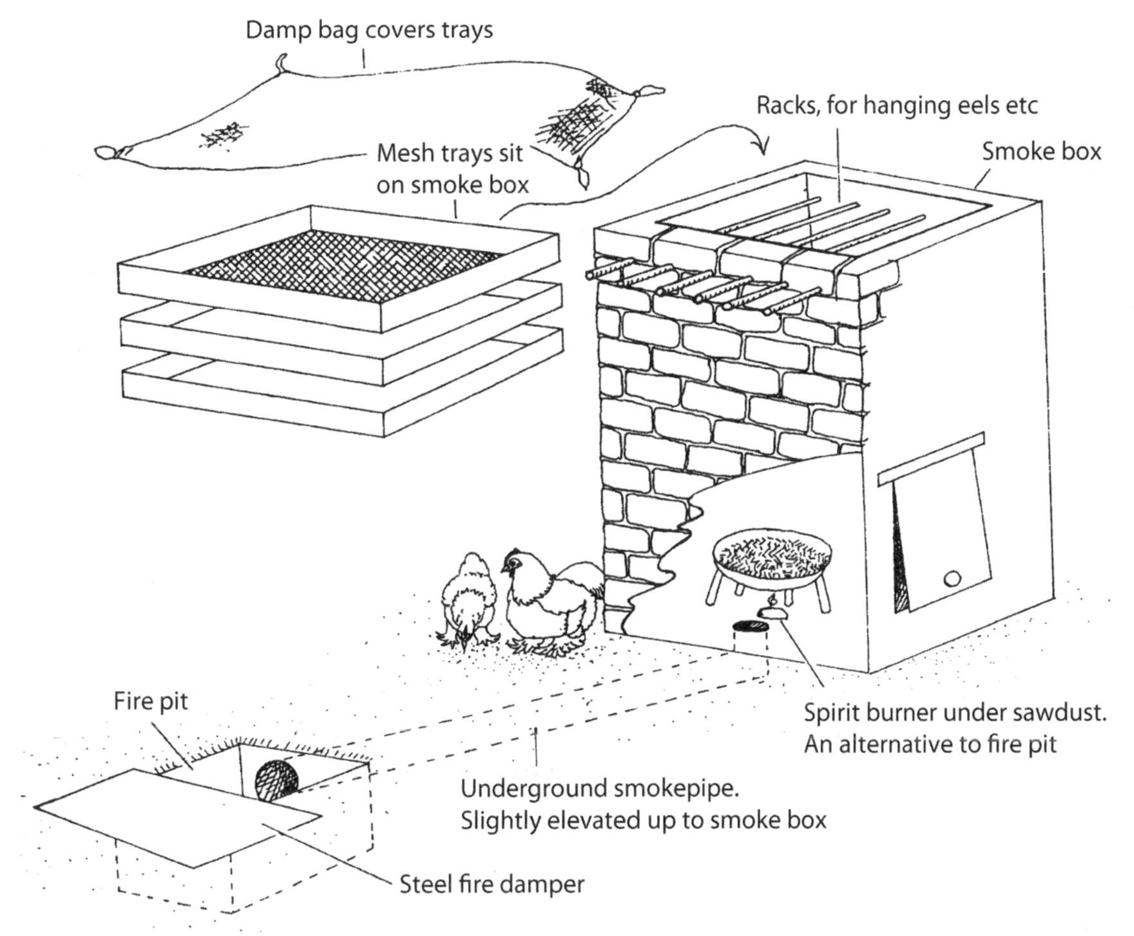

Figure 1.6—Elements to building a smokehouse using a fire pit for cold smoking or adding a spirit burner for hot smoking.

1.5 SALTING

Although salting meat and vegetables to preserve them has largely been discontinued in the west due to other storage methods, the east continues to use salt extensively to preserve such vegetables as kim chee, radish, and other greens. See Figure 1.7.

Brine is a solution of salt and water used to draw moisture and sugars from food. The solution forms lactic acids which protect food from spoilage organisms. When using moist food, brine may not be necessary. Salt added to fish, for example, will produce its own brine. During brining, liquid continues to be drawn from the food, and the brine may become weakened. More salt can be added.

Salt not only helps to dehydrate vegetables and meats, it also enters the tissues, destroying the autolytic (self-digesting) enzymes and halting the action of bacteria. Only some salt-loving bacteria (halophiles) survive, giving rise to the characteristic flavour of salted or pickled foods.

Srivastava (1985) asserts that pure, sterilised common salt is a better fish preservative than evaporated sea salt. Sea salt may contain low levels of calcium and magnesium which cause decay, hardness of flesh, and off-flavours. Special granular pickling salt can be purchased, or coarse salt from butchers or bakers. Refined table salt contains aluminium and is to be avoided.

All pickling solutions should be in glass, earthenware, wood, plastic, or concrete containers. Avoid aluminium and copper utensils, even iron for long periods; stainless steel utensils are appropriate for boiling.

Salt Brines

Saturation levels:

5%—1 cup salt to 5.7 litres (12 pints) water

10%—1½ cups salt to 5.7 litres (12 pints) water

18%—3½ cups salt to 5.7 litres (12 pints) water

25%—5 cups salt to 5.7 litres (12 pints) water

In the pickling recipes, I have left out all references to alum and often to sodium nitrate in pickles, although the latter may be of little harm and the former is rejected only because of its aluminium content. I have included lime, limewater, and lye; the latter may be leached from ashes or bought as a strong crystalline alkali. Treat strong lye as carefully as you would strong acid, with extreme care until diluted (as in olive pre-soaks). I do not use or recommend artificial colourings.

You may need to use sodium sulphite to clean or sterilise contaminated wooden vessels (casks or vats) that have picked up a vinegar or other unwanted micro-organism, before risking a large quantity of ferment. It is harmless if well rinsed out. I don't recommend any long-term preservatives; cool storage can be substituted.

Nigari (bitterns)

Bitterns is the oily liquid remaining in salt ponds after crystallisation of the salt. Flakes of salt are prepared from the evaporation of clean seawater. Typical percentage (%) of salts in seawater are sodium chloride (2.72); magnesium chloride (0.38); magnesium sulphate (0.17); calcium sulphate (0.13); potassium chloride (0.09); magnesium bromide (0.01). Over 60 trace elements are found in seawater. Seawater is used to coagulate tofu; also used in calcium sulphate (gypsum) or magnesium sulphate (Epsom salts). The problem today is to find clean seawater, free of pollutants, hence the increasing use of gypsum to set tofu.

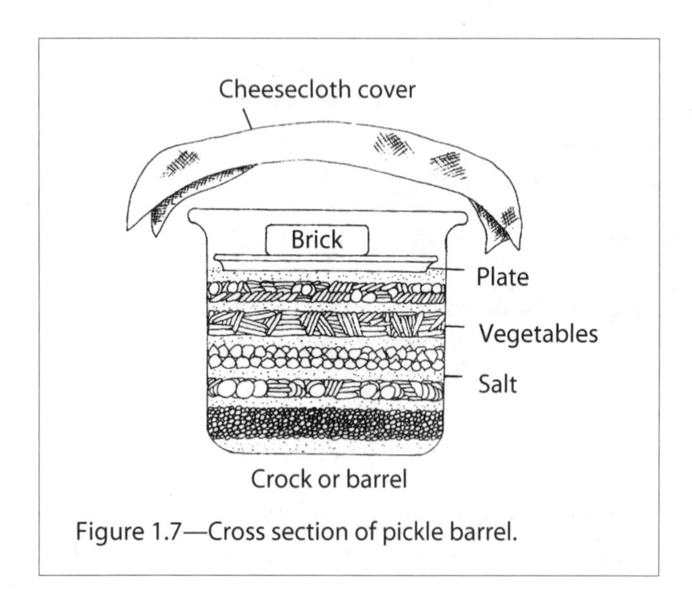

Figure 1.7—Cross section of pickle barrel.

Seawater boiled down gives moist solids which may be kept in glass jars. To prepare nigari, place this moist salt or moistened natural sea salt in a fine bamboo or nylon sieve over a non-corrosive jar and leave in a cool, dry place, or place in a large plastic bag together with a bowl of water. The reddish nigari will drip slowly into the jar. Larger quantities are made from strong cotton sacks filled with sea salt and supported on slats over a large wooden trough. Moistened, it drips into the trough or barrel.

For agricultural uses, bitterns are sold in plastic containers (the bromine content attacks glass) to be diluted 1,000:1 (water to bitterns). This provides trace elements on broadscale agricultural properties. They are an excellent source of minor elements if leached, old, or tropical soils are to be returned to permanent forest or pasture.

1.6 SYRUPS AND SUGARS

Sugar (or 50:50 sugar and honey) is used to make jams, preserve meats, fruits, vegetables and to provide the raw materials for yeast ferment to alcohols; and then further ferment to vinegars.

Sugar cane contains 11–15% of sugar by weight, is produced at from fifty to one hundred tons per hectare, and from 9–11% of this is recoverable sugar. In India, 60% of all sugar is diverted to jaggery production, and 10% of canes are reserved for "seed" or setts for the next crop.

Jaggery is an Indian word used to describe a coarse brown sugar made from palm sap or sugar cane. It is boiled at 116–118°C (240–244°F) and is about 65–75% sucrose, 10–15% invert sugar, 2–3% ash and 3–5% moisture. It crystallises in moulds after cooling.

Syrups

Light	(40% sugar)	1.8 kg (4 lb) sugar dissolved in 2.8 litres (6 pints) water
Medium	(50% sugar)	2.3 kg (5 lb) sugar dissolved in 2.3 litres (5 pints) water
Heavy	(60% sugar)	2.7 kg (6 lb) sugar dissolved in 2 litres (4¼ pints) water

A heavy syrup (1 cup water, 1 cup sugar, some lemon juice) can be made and kept, later thinned to preserve fruit; about 1 tbsp of lemon juice in 1 cup of water prevents the thick syrup from re-crystallising.

Vitamin C in cooled syrups prevents fresh fruit browning if sealed with wax in cartons, and frozen.

Note that fruits can be frozen on oiled trays and later bagged; small fruits or berries and mangoes are mostly kept this way, without syrup, and used just thawed. All frozen fruits should be consumed as soon as thawed.

1.7 VINEGARS

Vinegar is a basic acetic or tartaric acid solution derived from open-fermented sugars or fruits, and is produced in the fermentation of beer, cider, or wine by the fungus *Mycoderma aceti* together with bacilli. To make vinegars use glass, stoneware (no lead glazes), or wooden containers.

Most people who appreciate complex flavours appreciate brewed vinegars, brewed ginger beers, brewed soy sauce, and non-petrochemical foods. Pickling foods in vinegar is an alternative to the natural lactic acid preservation of vegetables such as sauerkraut and kosher dill pickles. Pickling with vinegar does not, in itself, involve fermentation, although fermentation occurs when making vinegar.

Wine, malt, and cider vinegars are the most common. Wine vinegar is used in salad dressings and as a base for various herb vinegars; it is usually more acid than malt vinegars. Rice vinegars are preferred in Asia. Cider vinegar is used by many country people as a home medicine. Malt vinegar is best liked by cold-country peoples, and for myself it is the essential vinegar for pickled onions or with salted meats and salads (although onions from salt-lactic fermentations are excellent when later stored in malt vinegar).

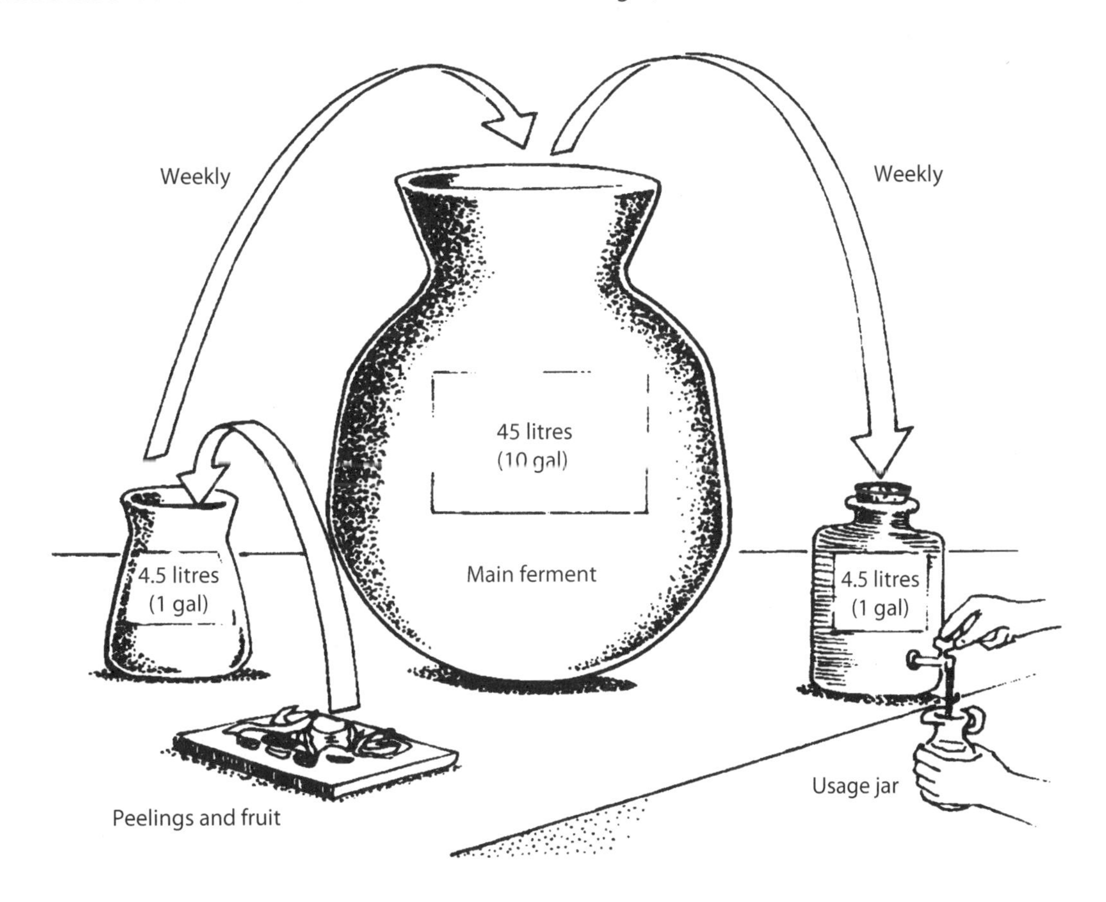

Figure 1.8—Making vinegar.

"Mother of vinegar" is a tough, fibrous, potentially perennial bacterial culture, appearing as cloudy stems or a fungal form in weak alcoholic solutions, or imperfectly-sterilised cider, wine, and beer containers. Some special cultures of yeast-bacteria form sheets and are used to make vinegars.

The cloudy (often bacterial) slime in natural vinegars can be removed with three drops of sulphuric acid per gallon. To clean vinegar, pour it through bone charcoal, add isinglass (dried fish swim-bladder) or white of egg before straining. Birchwood chips placed in barrels, or two to three drops of concentrated sulphuric acid per 4.5 litres (9½ pints) stops slime moulds or bacterial slimes forming.

Making Vinegar

To make vinegar from fruit peels or cores, place fruit wastes in a large jar and cover with lukewarm water to which has been added a little vinegar, beer, sour wine, or beet juice. Stir all together and tie a muslin top over the top of the jar to exclude vinegar flies. Leave in a warm place for two to three weeks. Strain and bottle off; sterilise to store.

Once vinegar has been made, a continuous process can be started if needed. This needs three containers, one large and two smaller, the small containers are just one tenth the volume of the large container. See Figure 1.8.

The large container is filled with home-made or bought vinegar; now one small container is filled from this, which leaves the large container one tenth empty (e.g. 4.5 litres [9½] pints is taken from the 45 litres [12 gal] container.) When the small container is almost used, the following fresh mixture is made up in a second small container.

In this, we can make up new mixes of beer, wine, fresh fruit juice, fruit peelings, tea, beet juice, sugar cane juice (good sugary juices can be expressed from apples, pears, plums, beets, sugar cane, honey, sugar, carob pods, molasses, or malt solutions used in season), or one part ethyl alcohol to ten parts water plus a little vinegar. This is poured into the large container. Each and every time a small container is emptied, the new mix is added to the large ferment cask. The usage jar should be the right size to allow one week of use. Never let the vinegar in the large container become weak. This is the best process to follow for regular use, and can be stepped up or down by altering container sizes, to suit demand.

An easy way to make vinegar is to let sweet cider stand open to the air for four to six weeks and it will turn to vinegar (or float some "mother of vinegar" and let it stand).

To make "mother of vinegar", the peelings and cores of mangoes, apples, pears, grapes, peaches etc. are washed and put in a wide-mouthed crock; covered with cold (once boiled) water and kept in a warm place (on a hot water system or on a rack above the back of the refrigerator). A scum will form and gradually thicken. Taste after two, three, four and five weeks. When the vinegar is strong enough, strain through a colander and save the scum (the "mother of vinegar") to transfer to a new brew. Bottle and seal the vinegar, having first briefly boiled it to sterilise, or having added a few drops of sulphuric acid.

Health Vinegar

Boil 4.5 litres (9½ pints) of water with 454 gm (1 lb) of dark sugar and let cool. Using 2 cakes of compressed yeast (or 1 packet dried yeast) and 2 slices of wheat toast, spread the yeast on toast and float on top of water mixture. When toast falls to the bottom in about five weeks, taste, strain, boil, and bottle.

Herb Vinegars

Fill jars with sprigs of fresh selected herbs or four to six leaves of garlic per quart, and cover with cider vinegar. Let stand for twenty days. Strain and bottle; a fresh sprig may now be included to "type" the vinegar.

Rosel (beet vinegar)

Pack beets without tops in a crock. Cover with cold water and stand in a warm place for three to four weeks or until sour. Used for beet soup *á la Russe*.

Oil with Lemon or Vinegar—Salad Dressing

Melt 1 tsp of sugar to 1 tbsp of vinegar or lemon juice in a pot, adding a pinch of salt per pint, stir in paprika. Now combine 1 part cooled vinegar with 3 parts olive oil, bottle and shake well; a crushed clove of garlic can be added before shaking and serving over leaf salads. Mustard, grated horseradish, chilli powder or crushed chilli, and chopped parsley are optional extras for a small bottle of the main dressing.

Pickles

Basic vinegar preservative is 4.5 litres (9½ pints) vinegar to ½ cup salt, ¼ cup sugar, ginger, 1 tbsp mace, cloves, cinnamon and 1–2 chilli peppers and mustard seed.

To pickle most "hard" vegetables, cucumbers, onion, walnuts, and butternuts, pick unripe nuts in first days of midsummer or the week before, and pierce them three to five times with a skewer, salt in brine for eight to ten days, changing them each day to fresh brine, and boil a few minutes in solution as above before bottling. For onions and cucumbers, soak in brine overnight only and pour on boiling vinegar and spices in jars, and fit with tight lids.

1.8 CHEMICAL PRESERVATIVES AND ANTIOXIDANTS

Tannin

Tannin is removed from, for example, pressed whole pomegranate juice (where it occurs because of the tannin in the rinds) by precipitation using a gelatine solution, then filtering, pasteurising, and bottling.

In acorns, tannins are removed by part germinating in soils or sawdusts, by the use of iron-rich earths mixed with the meal of pounded or minced acorns, by limewater, and by potassium permanganate. Most tannin is removed from acorn meals by repeated extraction with warm water, and only after this are earths or chemicals used to immobilise the residual tannins as insoluble ferric tannates.

Baking Powders

Baking powders are composed of sodium bicarbonate and acid (e.g. some tartaric or citric acid) and a little pure starch, such as cornstarch. When wetted, this mixture releases bubbles of carbon dioxide, like yeast, which aerates the dough.

When using cream of tartar or sodium baking powders for breads: 2 tsp to 1 cup of flour or 1 tsp of soda and 2⅓ tsp of cream of tartar to 4 cups of flour; 1 tsp of soda to 1 pint of thick sour milk (lactic acid); ⅜ tsp of soda to 1 cup of molasses in stiff doughs.

Antioxidants

Kamala, a golden dye from the red woolly coat of the ripe seed capsule of *Mallotus philippensis* is used as an antioxidant in India for butter, ghee, vegetable oils, and fats. Stable, harmless, tasteless, and odourless it colours food a light yellow, prevents rancidity, and retards the loss of vitamin A. If *Mallotus* fruits are stirred vigorously in water, the dye falls out as a sediment and can be sieved and dried, sifted again when powdered, and in an alkaline solution is used as a cloth dye (*Wealth of India VI*). Dried ginger is also added to fats and oils as an antioxidant in India.

Limewater

Pour 1.1 litres (2⅓ pints) boiling water over about 2.5 cm (1") crushed or unslaked lime. Stir well. Next day, pour off and bottle the clear liquid. Use 1 tbsp per 1.1 litres (2⅓ pints) water.

1.9 COOKING

General Procedures for Retaining Good Nutrition

"Invisible chemical losses of nutrients that occur during food processing and cooking are the special responsibility of, and common ground between, nutritionists and food scientists." (Davidson, 1979). Not, that is, to exclude gardeners, cooks, soil scientists, and microbiologists, one hopes!

Ferment and many like preservation techniques may not only avoid the need to cook altogether, but ferment usually adds nutrients to food staples as vitamins, amino acids, and enhances digestibility via enzymes.

As for cooking, Davidson gives this very good advice:

- Use minimal heat for the shortest possible time.
- Moderate heat at all times.
- Add fruit and vegetables to already boiling water or heated oil.
- Don't add food to cool water and then heat.
- Use a lid to exclude air while cooking (hence shorten cooking time).
- Do not use copper pots (they destroy vitamin C) or baking soda with vegetables.
- Fry in deep **hot** oil.
- Cook as soon as possible after harvest.
- Serve (do not keep warm or hot) as soon as possible after cooking.
- Use minimal cooking water, do not discard if gravies or soups can be made from this water.
- Pressure cookers can be used with confidence, for like microwave ovens, the time exposure to heat is short.

Do not rinse or wash clean grains, fresh vegetables, meats, or clean fish before cooking, and cook as soon as frozen foods are part-thawed (or use that microwave oven). However, salty foods (especially salt meats) are not well cooked by microwaves and may contain harmful bacteria, so cook such foods normally. Never peel vegetables, just scrub them clean.

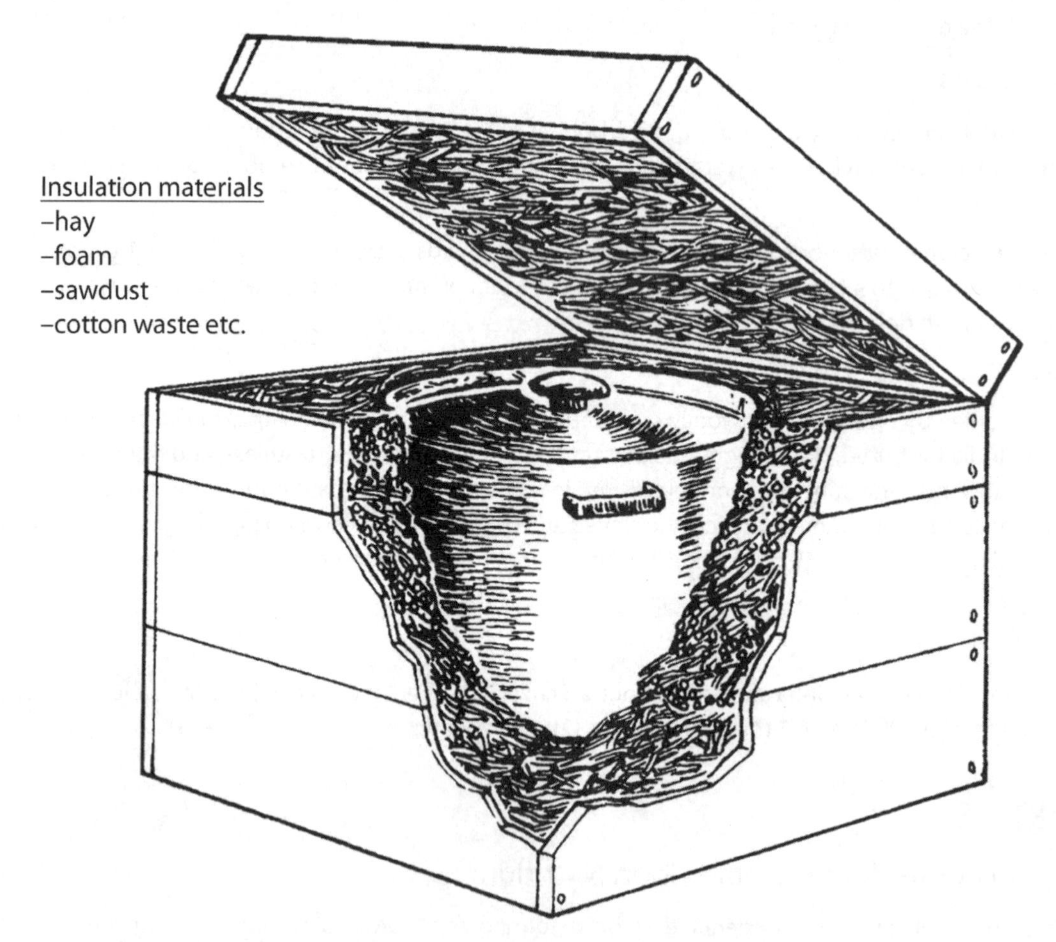

Figure 1.9—Haybox cookers are used to conserve energy. These can be made from a wide variety of materials, according to what is available.

The most economical cooking is either not to cook (as in sushi fish) or to quickly boil, oil-fry, or stir-fry over an open flame in thin steel. If long cooking is essential, do it overnight in a closed oven or hot-box or a large earth pit serving a large number of people at one cooking. Make a party of it, if you like, but remember that fuel is a central consideration in energy planning, and cooking methods are also capable of great energy savings.

The largest loss of vitamins and minerals occurs in foods where boiling plus discarding the cooking water is used as a cooking method. This is exactly the method used by my mother, and still widespread in Australia and the UK! Baking loses less minerals, and increases the energy value and protein content by reduction of water, a factor important for children, who eat less of bulky food.

Fireless Cooking

In places where wood, charcoal or other fuel is scarce, you can save energy by using a haybox cooker or solar cooker. Haybox cookers are easy to construct. See Figure 1.9. The idea is to first boil the contents of the pot (grains, legumes, stews) and then put it, pot and all, into an insulated box to finish cooking.

Stir-Fry (wok cooking)

This is the most economical of all cooking systems (Figure 1.10), hence, fuel saving, and also the least likely to destroy vitamins. A wok lid helps steam the food after a brief sizzle in oil (about 1 tsp of oil to 1 cup of food to be fried).

First, sort the food into plates according to cooking time, and chop into about 2 cm (¾") pieces. Add, in sequence, onion, garlic, ginger, meats, hard vegetables, and lastly spinaches or leaf; stir as you do this, or rather turn constantly using a bamboo paddle. After four to six minutes, add ½ cup soup stock and cover, let steam about five to six minutes, try for tenderness, serve immediately. The vegetables should be crisp, not limp.

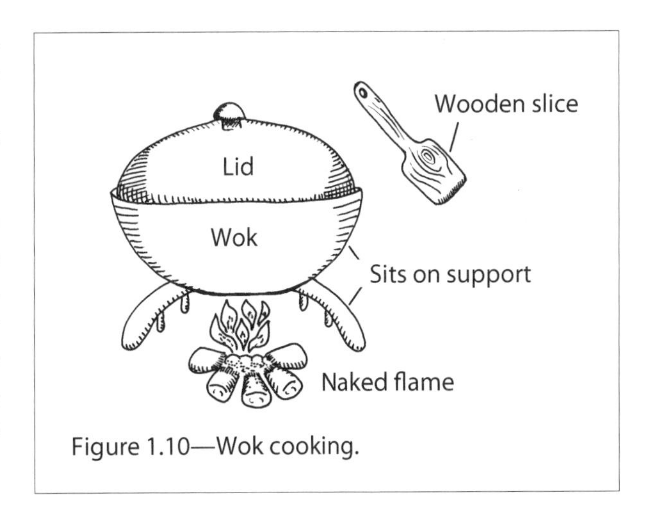

Figure 1.10—Wok cooking.

Cooking in Bamboo

Meats, roots, fruits or combinations of food can be packed into bamboo stems three to five feet long. These are plugged with a ball of leaf (banana is often used) and a leaf is also tied over the plugged end; the other end has an entire septum of bamboo left as a plug. These are roasted over a coal fire until steam escapes, and can be stored three to four days without food spoiling. Meat for longer storage is first well-cooked then sealed in bamboo with hot pork fat (keeps for weeks).

Pit-Cooking (in an imu or earth oven) — Polynesia, Pima, Salish Amerindians

Pit-cooking over four to twenty-four hours is an economical way to steam-cook very large quantities of food for a large group of people, or for later drying and storage. A pit is dug 1–1.5 m (3'3"-4'11") deep, 1–3 m (3'3"-9'10") long, and about 1.5 m (4'11") wide according to the bulk of the food to be cooked. It is about half-filled with wood, and on this wood are placed about a double layer of volcanic stones. The fire is lit, and burns four to six hours to ashes. More wood can be added in the last two hours. Then the whole is sprinkled with water until all the hot fire is gone, and a layer or layers of banana leaf or green aromatic leaves (lemon grass, papaya) is laid on. The meats (leaf-wrapped), root and leaf vegetables, bulbs, fruits and cucurbits (pumpkin) are placed on this layer and the pit so filled to near the top.

Now, clean bags, banana leaves, and old sheets are laid over the food and sprinkled with water, and if an old canvas or even carpet is available, it is laid over the whole pit. The earth is carefully returned to cover the canvas (no steam holes should be seen) and the whole is patted down. Whole stuffed pigs cook in these in five to six hours, or some pit-cooked foods (camass, agave) can be left overnight. When filling the pit, speedy packing preserves the heat.

For feasts, the pit is uncovered, the packaged foods removed in their leaves, with the meats and fruits. Traditionally, women are given first choice of the food in Polynesia. It is wise to use wire mesh baskets for large meats, for easy removal by hooks or poles, and to leaf-wrap most vegetable or fruits. *Cordyline* (ti) or banana leaves are favourite wrappings, lemon grass or ti a favourite flavour.

Wrapping Leaf Bundles for Ovens

In leaves, meats, starchy food, and root vegetables can be wrapped alone. Also, a thin layer of grated starchy root can be spread on leaves, and fillings of banana, chopped meat and onion, fish pieces and lemon grass, lengths of bamboo stuffed with savoury mixes, and in fact a variety of foods can be packaged for bake-steaming. The wrapping leaves give colour, aroma, flavour, and may themselves be edible.

Pork pieces can be wrapped with apple slices, beef with onion, rabbit with onion and prunes, lamb or goat with prunes and dried apricots; all these additions can be handily placed near the wrapping tables. Flavourings are:

- Leaves of lemon grass, ginger, mint, beefsteak (shiso), turmeric.
- Dried prunes, pawpaw, apricot.
- Fresh apple, onion, ginger root, garlic.
- Coconut cream, soy sauce, vinegar.
- Herbs such as thyme, oregano, coriander.
- Chilli, capsicum, black pepper, mustard, paprika.

Main wrap leaf—ti, banana, breadfruit, taro, giant taro, coconut frond.

Ties—fibres of Manila hemp, ramie, pandanus, New Zealand hemp, cabbage tree leaf, bark fibres, *Sanseveria*, all can be used to tie bundles.

Edible wraps—grape, cabbage, and taro leaves.

Colouration in Food

Red

Red beefsteak leaf (*Perilla*)—used in pickled sliced ginger, and fish ferment.

Annatto seed (*Bixa*)—used in cooking rice, as an aqueous extract of seeds.

Adzuki pulp—used in red paste ferments, or as an aqueous extract of whole beans.

Cochineal insects—alcoholic extract, used in sweets and cakes. Red mould (*Mucor*) in Indo-Malaysia is used to prepare ang-kak (red rice). Ang-kak is used in fish and paste ferments to give a red or orange colour. It is a form of *Monascus* or *Neurospora* which gives an orange colour, or a red colour, to ferments. The rice is separately inoculated, and added late in the main ferment.

Yellow

Yellow colouration comes from *Crocus* anthers (saffron) and from turmeric (grated or sliced root or powder). Grated cassava steamed in ti leaves is yellow. Flowers of the lipstick plant (*Bixa*) add yellow to rice.

Black

Black-brown comes from cuttlefish or squid ink sacs, and is used in shellfish ferments. Black sauces are made with black soybeans as a base, although other black legumes (some four-winged beans, some runner beans) could be tried. "Black" breads are made from rye, and blue-black breads include iron-treated acorn meals.

Green

Green dye is extracted from spinaches, citrus leaves, and can be settled and used in alcohol extract for keeping. It is used in sweets and starches. Colours are also added to salads as edible flowers of violets, marigolds, nasturtiums, hibiscus, etc.

Texture

Texture is provided by water chestnuts, bamboo shoots, fresh celery slices, fresh onion, pickled onion, pickled radish, fresh small radish, coconut meat flakes, roasted peanuts, almonds, or walnuts. All are used in specific stir-fries or stews for this reason, and in salads and soups.

Pork cracklings and crisp bacon slivers can likewise be added to chowders, sashimi, sushi, or other soft foods as they are served. Kim chee, with seaweed, radish, cabbage stalk should always be a little crisp.

1.10 FERMENTATION

As an industry, fermentation and its products may be one of the largest in the world next to motor vehicles (*New Scientist*, 4 September 1986). Products range from beer, wine, cheese, soured milk products, bread, and other yeast products to antibiotics. Not so well-known are the fermented foods of Asian countries, notably soy sauce, miso, and sake. Almost unknown in the west is the fermentation of vegetables, including soybeans and cabbage (tempeh and kim chee respectively), and the southern African fermentation of grains for porridges.

Although soy sauce, miso, and sake are produced on a large scale, Asian and African fermented foods are still manufactured locally. Because traditional methods are relatively simple, these processes are usually neither expensive nor do they require complex machinery. These village-scale or home-scale fermentation recipes are what interest me, and what this book is mostly about.

Fermentation techniques use microbes to transform raw materials (the substrate) into a useful product. In this use it includes:

- Producing food items that prolong the life of the material.
- Freeing seed from fruit pulp.
- Breaking dormancy in seed.
- Better digestion of food using caecal or rumen inoculants in animals.
- Fixing nitrogen or minerals by plant root associates.
- Blocking harmful micro-organisms by beneficial digestive tract organisms.
- Detoxification of food.

Applied to foodstuffs, fermentation and associated enzymes can:

- Make vitamin-enriched and very stable foodstuffs from low-value carbohydrate bases.
- Preserve perishable resources such as fish protein as sauces which can be used for years, thus extending the time of use and improving or salvaging nutrition in foodstuffs.

In the beginning, microbes would affect foods in various ways, with humans noting which methods made food unacceptable, and which ones preserved and actually enhanced flavours. The latter were encouraged and exploited to yield the various fermented foods we eat today. Stanton in the UNU *Food and Nutrition Supplement* No. 2, November 1979, notes that tribal peoples developed locally ingenious and sophisticated physical, chemical, and microbiological methods to make plants more digestible, to remove toxicity, and to enhance nutrition in plant foods.

It makes a lot of sense for one family or a small cooperative to make the basic koji (mould) based on barley, rye, wheat, rice, potato, or sweet potatoes, as this needs a small heated room, several shallow trays, and a large steamer. All the rest of the ingredients can be prepared at home, and miso and soy sauce stored there. An analysis based on a survey of local consumption and preferences would enable a village to develop and co-fund a small ferment business to supply local needs. When and where milk is available, yoghurt, koumiss, and a few cheeses can also be manufactured.

In any regular production facility at village level, especially those involving bread and cheese, or complex ferments like miso, care must be taken in obtaining and preserving species cultures, frequently by re-inoculating new materials from a previous batch. To begin with, reliable inoculants should be bought from another producer who specialises in that activity; this person may sometimes have access to a more centralised institution which holds large stocks of mirco-organisms intended for a specific industry. Some of these key institutions are listed at the end of this book.

1.11 USES OF MICRO-ORGANISMS

How Can Ferment Improve Nutrition?

- By preserving foods for long periods (lactic acid ferment) and at the same time conserving vitamin C and synthesising B & K vitamins so much needed by vegetarians (B12). Vitamin C is preserved in fresh vegetables that are quickly fermented (e.g. sauerkraut).
- By creating portable and storable substances, such as fish sauces, which are high in protein, and which contain iodine, e.g. seaweed.
- In the creation of "complete" protein from low-protein or unavailable protein foods.
- By creating flavourful sauces, condiments, and pickles as an accompaniment to bland foods.
- In making indigestible starches far more digestible through enzymatic action (which could be said to be a form of pre-digestion).
- By lowering poisons such as hydrocyanic acid, nitrites, prussic acid, and oxalic acids, or the outright destruction of nitrites and nitrosamines or glucosides.
- By reducing or decomposing mycotoxins (e.g. aflatoxin and other toxins secreted by moulds).
- Some ferment organisms (yeast, bacteria) stimulate or augment the human digestive flora, and are used in cases of enteritis, dysentery, and alimentary upset or fevers.
- Lactic acid ferment reduces the incidence of pathogenic (disease) bacteria after three days and is often prescribed after the use of antibiotics.

In industrial ferment processes, products are as varied as:

- Antibiotics such as streptomycin, penicillin, tetracyclins.
- Organic solvents such as acetone, butanol, ethanol.
- Gases such as carbon dioxide, hydrogen, methane.
- Flavours such as monosodium glutamate, nucleotides.
- Organic acids such as lactic, acetic, citric.
- Amino acids such as lysine, glutamic acid.
- Vitamins such as vitamin A, C, riboflavin, B12.
- Hormones such as steroids, insulin.
- Enzymes such as amylases, proteases, invertase.

Also:

- Agricultural aids such as gibberellins (plant-growth regulators produced by a genus of fungi), legume inoculants, bacterial and fungal insecticides.
- Miscellaneous, such as fats, glycerols.
- Beverages such as beers and soured milks.
- Foods such as cheese, bread, pickles, miso and so on.
- Minor uses, as with seeds, inoculation of plants, animals and people.

Uses of Micro-organisms (mainly yeast and bacteria)

The uses of micro-organisms are numerous; apart from enhancing food value, they are used to:

- Produce enzymes for digesting oil, stains in clothes, clearing drains, and as rennin in milk coagulation.
- Grow amino acids as nutritional additives to foods.
- Deactivate dangerous chemicals such as PCBs, phenols.
- Produce yeast cells from oil for animal food or possible human food.
- Help mine oil by thinning out crude oil in rocks, and to leach out uranium, arsenic, cobalt, nickel, zinc, and lead from ores, mainly as sulphates.
- Produce medicinals. Altered bacteria with human or other animal genes can produce insulin, interferon, vaccines (viral protein).
- Industrial feedstock. Fermentation produces hydrogen, methane, carbon dioxide, ethanol, vinegars, lactic acids.

1.12 THE ORGANISMS

Although fermentation has been occurring for many thousands of years, it is only recently that we have been able to isolate the various micro-organisms responsible and to describe the processes used to achieve the desired product.

Micro-organisms used in fermenting are basically fungi (yeasts and moulds) and bacteria.

Yeasts, fungi, algae and bacteria are useful for ferment, enzymatic action, or enhancement of protein and vitamins in ferment. They may be gathered from soils, flowers, stomachs of animals, livers, saliva, plants, body exudates, various body organs, or previous ferments; all may be used to curdle, hydrolyse, ferment or add vitamins to low quality foods or waste products to enhance nutrition.

However, most microbial cultures (once obtained) are kept as starter cultures for home and village ferments, and almost every common yeast or milk bacteria is a truly domesticated strain, rarely found in nature and only common in human-created kitchens or factories. We have, in fact, domesticated them.

Yeasts

Yeasts and moulds are fungi. They cannot make their own food as they contain no chlorophyll, so they live on other living, decaying, or dead substances such as skin, wood, earth, leaves, fruit, and leftover food. The way they take in food is by secreting powerful enzymes which break down organic materials into smaller molecules.

Yeast converts carbohydrates such as grain and legumes to many products, the most well-known being bread and beer. The action of yeast feeding on the sugars in the carbohydrates produces two by-products: carbon dioxide and alcohol.

In bread, the carbon dioxide released causes dough and crushed grains to stretch, raising the bread. The alcohol is evaporated off and the heating process kills the yeast. For beverages, oxygen is controlled, so that alcohol results from fruits or grains with high sugar contents. If wines or ales are then exposed to air, yeast turns the ethyl alcohol to vinegar (acetic acid). For dairy products, a combination of moulds, bacteria and yeast produces such cheeses as Roquefort and Camembert. The bacteria starts the conversion process, while the yeast and moulds provide the distinctive flavour.

Wild or cultivated yeasts may exist as hundreds of strains of one species. This is true of brewer's yeast (*Saccharomyces cerevisiae*) in the UK, where some three hundred industrial strains are held. Brewers get different results not only by using different recipes and timing, but by the subtle flavours added by their favourite yeast strain. Similarly, one can buy thirty to sixty yeast strains to make soy sauce, miso, bread, and so on. Some yeast species are consistently gathered from specific fruits or vegetables, others from flowers or the stems of grasses, or from the air as spores (usually in a stillroom, on a farm, in a brewery or a bakehouse).

Yeasts dried or as extracts are rich in soluble vitamins, e.g. thiamine, riboflavin, niacin, vitamins B6 and B12, pantothenic acid, folic acid and biotin. This group of B vitamins joins with enzymatic proteins to activate enzymes. B complex factors (choline, inositol) are not vitamins for people, but may act as co-enzymes for biological reactions. Brewer's yeast is also rich in chromium and selenium. High levels of purines mean gout sufferers should avoid this yeast.

Irradiation, antibiotics, salting, alcohols, sugars, drying, smoking, vacuum packing, and freezing all favour the growth of yeasts over bacteria, as do acid substrates and "antibacterial metabolites" produced by yeasts. Yeasts are favoured or encouraged by an alcohol content, pH of 6.0–7.0, available sugars, aeration or at least agitation and stirring, and precursor enzymes such as malts. Bitter oils, as found in hops, favour yeast over *lactobacilli*.

Disfavouring or inactivating yeasts are the antioxidants (benzoic acid, potassium sorbate, parahydroxy benzoic acid, propionic acid, sorbic acid, sulphur dioxide), or heating for twenty minutes at 62.5°C (144.5°F). The acetic, citric and lactic acids also inhibit many yeasts. The oils of garlic, oregano and thyme are all inhibitory on yeasts.

Bacteria

Bacteria and fungi both attack cellulose and lignin, but as bacteria have no method of entering hard cells, their effects are limited to surface erosion. Bacteria are most active in acid environments and anaerobic conditions (peat bogs, stomachs of ruminant animals). Thus herbivores harbour a rich microflora of bacteria which (using cellulase enzymes) make grasses and twigs digestible, and further down the alimentary tracts the bacteria themselves are dissolved by protease, and digested. Due to the fine grinding (chewing the cud) and body heat, bacteria are most effective and speedy at cellulose digestion inside ruminants, where many plant cells are broken up.

The most useful bacteria are the lactic acid-forming bacteria.

When fermenting foods like meat and vegetables, the idea is to add enough salt at the beginning of the fermentation process (20–30% by weight) so that harmful bacteria are suppressed long enough to encourage the lactic acid-producing bacteria that are necessary in fermentation. Such bacteria rapidly multiply in the juice formed by the salt and vegetable combination to produce carbon dioxide and organic acids, which lowers the pH. A lowered pH (higher acid content) hinders development of undesirable bacteria, as does the formation of carbon dioxide, which provides an anaerobic environment (suppresses aerobic micro-organisms). Lactic acid bacteria thrive in anaerobic conditions.

The combination of low pH, anaerobic conditions, and the interaction of salt and acid have a higher success rate in inhibiting harmful organisms than salt alone, and also impart characteristic flavours to the food (e.g. sauerkraut).

In short, if the substrate is good, ferment may be successful, but above all, success is a question of governing the temperature to near optimum for that product, in observing good hygiene, and in selecting and using good strains of inoculants for that purpose.

Bacillus subtilis, occurring naturally on the skins of citrus fruit, inhibits the fungi causing blue and green moulds, and sour rot. Thus, unwashed fruits develop less moulds than washed fruits.

Bacterial Cultures

In Russia, *Streptococcus faecium* is cultured for beet juice and lactic acid ferments and cultivated in a molasses medium of: 1.5% to 1.8% molasses, 40–60 mg of amine nitrogen (yeast enzymolysate), and 0.75–0.9% maize extract. The pH is kept at 6.4 to 7.2, and temperature at 37–43°C (98–109°F) for seven hours.

For *Lactobacillus salivarius*, the proportions above are: 2–2.4% sugars of molasses, and 66–88 mg of amine nitrogen.

Cultures of the above forms can be isolated from the epithelia of the intestine of day old chickens (*S. faecalis*) and later (chicks a few days old) *S. faecium*. After the fourth and before the eighth day, *L. acidophilus*, *L. salivarius*, and *L. fermentum* appear, as does *L. buchneri*. All will grow at 45–50°C (112–122°F). Except for vitamin B12, and biotin (trace), the biotamines are synthesised by these bacteria in the intestines of chickens.

Very dense counts of lactic acid bacteria are also found on the integuments of bees collecting nectar. (Data from papers by N.K. Kovalenko, and others from the Russian Academy of Sciences, Kiev.)

Yeast and Bacteria Mixes

Just as many very stable life forms, from lichen to oaks, demonstrate a constant association of, for instance, a fungus and algae, or a tree and fungi, so many ferments (kefir, vinegar) have inoculants of yeast-bacteria associates bound more or less inextricably together. The "mother of vinegar" is a tough, even leathery membrane of polysaccharides secreted by the bacteria, in which lie yeast cells, close to but separated from the bacteria by polysaccharide sheets or nets of fibre.

This tough inoculum is also found in sourdoughs, where lactic acid bacteria and yeasts co-mingle, and is thought to give some strength to an otherwise weakly-bonded rye dough. (*New Scientist* 22/29 December 1983, p22).

The great improvements to efficient ferment are:

- Temperature control, so that optimum temperatures are available over various stages of the fermentation process.

- Cooking, mashing, sprouting or denaturing the basic grain or grain legumes. To some extent, an even turning of grains while fermenting and cooking is needed. Any liquid or even solid phase is more efficiently aerated, or dried, via perforated floors or by bubbling clean warm air through the ferment.

We still need superior ferment inocula.

One area not often mentioned in published papers is that of the co-enzymes, or metallic ions (trace elements) needed by micro-organisms either as catalytic systems working with enzymes, or as basic nutrition for the rapid increase of micro-organisms. Such trace elements are readily available in seawater or "bitterns". Some peoples use dry-roasted pigeon manure to fertilise ferment organisms or to activate enzymes (Tibet).

Microbial Nutrition and Preparation of Substrates

The basic needs of micro-organisms must be provided if an active ferment is to be successful. To be very basic, all micro-organisms need water, or at very least a moist substrate; to this we would add whatever vitamins, trace minerals, major nutrients and sources of growth energy (carbohydrates) such as sugars and starches that are required for the specific product that we are attempting to create.

As for pH, optimums at 6.5 to 7.5 and ranges from 4.0 to 9.0 are found in bacteria. As micro-organisms themselves shift the pH towards (commonly) acid or (less commonly) alkaline, basic buffering by lime, proteins and phosphates are common in ferment mixtures.

Major nutrients (CHONSP) are those of carbon (C) and hydrogen (H) (from carbohydrates); oxygen (O) and nitrogen (N) from the atmosphere; sulphur (S) from sulphates and phosphorus (P) from phosphates and are supplied if deficient in the basic materials. Anaerobic bacteria work under CO_2 atmosphere, and can be supplied with an airlock to exclude or reduce oxygen.

Trace or minor elements (sodium, potassium, calcium, magnesium, manganese, iron, zinc, copper, tin, cobalt, vanadium, molybdenum, chromium, etc.) must be present in minute amounts, so that enzymes can function. Some (zinc) are critical to very many enzymatic functions, others (molybdenum) to a relatively few catalytic functions with enzymes, especially in nitrogen fixation in legumes.

Warmth has a profound effect on enzymes, while as with pH, normal or medium temperatures are from 21–41°C (70–106°F), low temperatures at from 15–21°C (59–70°F), and high temperature processes at 42–60°C (108–140°F).

Gases (oxygen, carbon dioxide, methane, nitrogen, air, hydrogen) can be excluded, part-excluded, or bubbled through the medium, to control the gas supply, to stir the medium, or to supply plentiful elements to the microbes. Microbes can be stored alive as working stocks, broths or gelatins ready to add to media and these are often stored in a refrigerator. Primary stocks or varietal collections are stored as freeze-dried cultures or spores held at very low temperatures. Many peoples catch cultures from reliable sources such as the surface of fruits or vegetation, from foliage or flowers, more rarely from insect digestive tracts or more commonly from purchased ferments or cultures.

Microbial Interaction

Apart from the constant association of yeasts and bacteria in kefir grains, and vinegar cultures, there are several beneficial associations in other foods. For example.

- In Indonesian tempeh, the mould *Rhizoporus oligosporus* is often contaminated with bacteria (*B. subtilis*), which produces vitamin B12 (Steinkraus, 1989).
- Cassava and rice ferment in Indonesia use the mould *Amylomyces rouxii* with at least one yeast such as *Endomycopsis burtonii*. Together they increase protein, triple the amount of thiamine, and synthesise lysine (a missing amino acid in starches). (Steinkraus, 1989).
- In mushroom cultures, fruiting may be aided by bacteria which act as a fruiting stimuli. For example, totally "clean" substrates and casings (substrate soil or peat covers) of *Agaricus brunnescens* fail to fruit in the absence of bacteria. A key bacterium is *Pseudomonas putida* (also assisting *P. silocybe*). Casings can be inoculated with 5% spawned compost in order to increase bacteria. As some strains of bacteria do not increase or stimulate fruiting, the inoculation must be from successful cultures.

Nitrogen-fixing bacteria (*Bacillus megaterium, Anthrobacter terrigens, and Rhizobium melilot*) all stimulate abundant mushroom fruiting in sterilised soils. The blue-green algae *Scenedesmus quadricauda* also increased fruiting 60%.

Activated charcoal in casings or substrates also enhances bacterial efficiency. Note that any cultures of bacteria are not allowed in the substrate mycelia in the laboratory, but are later added to the casings, as they may well compete early in the mushroom growth. As in milks or cheeses, selected strains for selected mushroom species are indicated.

Enzymes from young button mushrooms, extracted in water or methanol, have also increased fruiting yields in the same or other species (Stamets and Chilton, 1983).

It is becoming obvious that algal-fungi-bacterial associations will be increasingly specified and used in future cultures, agricultural or food-based.

The Economical Production of Heat for Ideal Ferment Temperatures

Fast ferment at ideal temperatures, with a known microbial wort, is ideal, in order to prevent other and undesirable (spoilage) organisms invading the ferment. Where salt prevents spoilage, less salt plus acidification to pH 4.5 would make a more healthy product which could be consumed in bulk. The low heat levels needed can be economically produced by simple solar collectors or by compost heaps, or where fires and hot water tanks lose heat.

To achieve "clean" ferments, heat plus airlocks are needed; heat sources available in villages can be produced cheaply by solar panels, in small glasshouses or plastic tents, over fires or in open ovens, or in containers in compost heaps. Even a sunny windowsill helps in making achars, sauerkraut, or kim chee.

Very large heaps of foods such as cassava, or pits of fermenting material will produce their own heat by bacterial action. Of all of these heat sources, the most reliable, long-term and least energy-consuming may be the solar panel (or a sand-bed under glass), or compost heat.

Fermenting New Foods or Wastes

If we study a whole series of ferment recipes, it becomes obvious that a yeast such as brewer's yeast (*Saccharomyces cerevisiae*) can be involved in a great variety of liquid and semi-solid ferments such as those for breads, beers, wines, and even koumiss.

Thus when we want a starter for any such endeavour, brewer's yeast will do the job, and over a relatively short period, will adapt to the substrate. If we then keep a yeast as a dough or dried flour-based biscuit, we can use our own strain. Similarly, a little yoghurt or yoghurt whey helps any lactic acid ferment of starches, milk, vegetables and so on, and can be used as marinades for meats or fish before cooking. Often, whey from milk cultures is included in other ferments, starters or gravies.

The same thinking applies to vinegars, achars, and the mixed cultures of koumiss; given that a ferment works for one vegetable, the method can be tried across others. One fruit recipe can be applied to many types of fruit, and one meat preservative technique to any other meats. The factor that has, in fact, limited the ferment systems is that they have been secluded in traditional societies, and thus limited to traditional products. However, a recipe book such as this one broadens the scope enormously.

CHAPTER TWO

THE FUNGI • YEASTS • MUSHROOMS • LICHENS

2.1 INTRODUCTION

"... few people realise how intimately our lives are linked with those of the fungi...scarcely a day passes during which (we) are not benefited or harmed directly or indirectly by these inhabitants of the microcosm." (Alexopoulos, 1962).

Mycology is the study of mushrooms and fungi. Fungi includes the rusts, smuts, moulds, slime moulds, mildews, yeasts, thrush, jelly fungi, mushrooms, puffballs, and stinkhorns. Doubtful inclusions are the cellular and netted slime moulds, but these too are often studied by mycologists, and occur in ferments.

Fungi like aerobic and less acidic media, and have a fast and efficient way to penetrate cell walls, combining mechanical force with wood-dissolving enzymes. Not all fungi can use cellulose, and even less species effectively digest lignin. Fortunately for us, many of those that do are those we can eat as mushrooms.

Unlike plants, fungi lack chlorophyll, nor have they developed vascular systems. Thus they live by degrading or invading other living or dead organisms. Some are parasites on or within other living things, many are saprophages (eaters of decayed material), and the majority can be cultured on one or other forms of dead, organic material.

The very common fungi-plant-root associates in soils, called mycorrhizal associations, benefit both the plant and the fungus. The fungus mobilises essential mineral elements for nutrition of the plant, while the plant feeds the fungus with sugars made by photosynthesis. Unlike plants, fungi need complex foods, not just simple elements, but given sugars they can synthesise their own vitamins, proteins, enzymes, and cell structures.

Fungi like an acidic growth medium (pH 6.0 or less) and flourish best between 20–30°C (68–86°F). As a communal cellular body, it is probable that fungi can exist for thousands of years, and under the right conditions fungal growth is indefinite.

Recently, forest fungal masses spreading across 35–100 ha have been found. These (not whales!) are the largest living organisms, and may be from one to five million years old.

2.2 THE CAPTURE, DOMESTICATION, AND STORAGE OF YEASTS

The yeasts we buy commercially are raised in a sugar or molasses mix, with ammonium, phosphate salts, and a low quantity of mineral salts or seawater. Ferment is at 30°C (86°F) or less, pH 4.0 to 5.0. The mass is aerated and the yeast cells centrifuged, washed and filtered. Then the yeast mass is mixed with small amounts of glycerine or vegetable oil and extruded as blocks for sale. These are kept refrigerated for up to three weeks, whereas dried pelleted yeasts can be kept for months in sealed packets or jars.

Other home storage methods include painting the inside of a wooden tub with coat after coat of yeast solution, and storing it upside down, sometimes for months, or more handily dipping a twiggy branch or branches into the yeast, and hanging these inside paper bags, freshly dipping them each time a brew is started up.

Some yeasts can be reliably captured on grain flours, the surface of sugary fruits, or in the nectareous washings from honey-providing flowers. However, yeasts like *Saccharomyces cerevisiae* (brewer's yeast) may be rare or absent in nature, and may only flourish on the substrate of grape must, or crushed grapes, which has a low pH (3.5) and high sugar content (16–20%). Such substances are only found in wineries. Low pH and high sugar content inhibit many other yeasts and bacteria, and favour the ferment yeasts such as *Kloeckera ariculata* (which occurs early in the ferment), *Saccharomyces cerevisiae* or more rarely *Torulaspora tosei*. In Italy and Spain, *Metschnikowia pulcherrima* occurs frequently and in Bordeaux (France) *Torulopsis bacillaris*.

"Spontaneous" or wild yeast spores were first caught from the air. Once a satisfactory strain is available it is used for "worts" for beer, bread making, or paste ferments.

Making Yeasts (Worts)

The nutrient solution for home capture or growth of beers or bread yeasts is made up of a litre (2 pints) of boiled water to which a handful of hops has been added. This is strained and about a quarter of the volume added as grain flour (wheat, rye, barley) with salt, sugar and some ash. When just warm, baker's yeast can be added, or the solution is left open in a jar in a warm place until it ferments naturally. This is called the wort. To keep the yeast, make the wort into a dough with barley flour, roll out thin and cut into discs, and sun dry as fast as possible. These disks, stored dry, can be broken up and used like baker's yeast. For short-term use, rolls of this dough about 3 cm (1") in diameter may be kept in cold water in a refrigerator.

Otherwise, make up a solution every week and transfer a cupful of last week's solution to the fresh base, making bread each week to which the yeast solution is added. Bread made this way needs six to eight hours to rise, or four hours in warm weather. The solution can be kept under an airlock or covered with muslin to keep out "wild" yeasts.

A Wort

Boil together 4 litres (8½ pints) of water, 250 gm (9 oz) sugar, and ¼ cup of salt, and let cool to blood heat. Add 250 gm (9 oz) flour (mixed with some of the water) and float 1 piece of toast on top. Cover bowl with muslin. Stir occasionally over three days.

Optional—Boil 1 oz hops and 1 root of grated ginger in a little water and add malt. Sprout and boil barley if no malt is available.

After three days, add 1 kg (2¼ lb) boiled and mashed potatoes. Ferment should now start and brown yeasts rise to the top. If so, strain and bottle by nightfall or the next day. Cover bottles with a few layers of muslin or cork lightly (bottle two thirds full). Keep bottles in a cool place. Viable for six to eight weeks.

For bread, add 1 cup yeast wort to 2 kg (4½ lb) flour. Ferment in a warm place until risen (two hours), knead and place in bread tins until again risen. This first batch will give yeast for a second batch (see also yeast storage).

Note—For rolls, a quarter of the bulk of the flour can be mashed potato or minced acorn or chestnut, and if ½ cup butter or 200 gm (7 oz) olive oil is added, this bread remains moist for seven to ten days. (David, 1979).

Bulk Processing to Increase Yeasts

Boil together 30 gm (1 oz) hops, 60 gm (2 oz) ground ginger, and 1.5 kg (3.3 lb) malt for thirty minutes and strain. Let stand for six hours. Add flour to bring this to a creamy consistency, then add 1 cup brewer's yeast. Let stand all night and bottle next day. Store bottles loosely corked.

Increasing Yeasts

Boil 12 large mealy potatoes, just covered in one pot; strain and mash. In another pot boil 1 handful hops, ½ cup sugar, ¼ cup salt, ¼ cup ash. Skim and strain. Combine potatoes, potato water, hop solution (discarding the hops), and add 1 cup of flour, and a knob of yeast. Store in a large earthenware jar at first at 23.8°C (74.8°F), then in a cool place for ten days of use.

Add ½ cup of this to new worts, or ½ cup to 9 kg (20 lb) flour for bread.

Note—Peach leaves can replace hops. If no yeast is available, take 2 cups milk, add 1 tsp salt, ¼ cup flour and mix well. Keep warm by the fire. If it foams, yeast is present; if not, try again.

Dried Yeast

Yeast solution or dough kept cold can be preserved for weeks (the dough is kept in cold water). Otherwise, a very large quantity of yeast is made, fine filtered, pressed, and thoroughly dried. It is then rolled thin and placed on an absorbent surface, like blotting paper and dried out of the sun. These pieces are grated and the fine gratings dried and kept in dry, dark glass bottles; use as per fresh yeast.

Cake Yeast

Yeast as pressed cake can be kept under glycerine, or a liquid yeast kept if one eighth the volume of glycerine is added and the solution is kept cold.

Bone or willow charcoal (fine-ground) added to pressed yeast can be made into a dough and dried in sunlight. To use, make a wort solution as before and add charcoal yeast mixture; the charcoal falls out as yeast is released into the solution.

Testing Yeasts

Good yeast added to boiling water floats. If it sinks, it is no longer viable. Add flour and a little salt to a warm solution of sugar. Add some yeast. It should bubble in about one hour. A little malted grain or sprouted grain, ground up, may help.

Adding Yeast to Flour

For dried yeast, 8 gm (¼ oz) dried yeast is added to 454 gm (1 lb) flour (for brioche or bread)—about one third that of baker's compressed yeast.

For baker's compressed yeast:

25 gm (1 oz) yeast to 1 kg (2¼ lb)	flour for 3–4 people
35 gm (1¼ oz) yeast to 7 kg (15.5 lb)	flour for an 8 hour dough
210 gm (7½ oz) yeast to 120 kg (265 lb)	flour for a 14 hour dough

Yeast Drink

Add a small amount of yeast to a glass of lukewarm water together with 2 tsp of flour and 1 of sugar. Leave in a warm place for four to six hours, then boil, cool, and flavour with lemon, ginger, or fruit juice. This is a nutritious drink supplying thiamine and niacin.

2.3 MOULDS AND MIXED INOCULANTS

Koji

Koji is a starter culture for miso, tempeh, and a score of other fermented grain and legume recipes. It is also called marcha (Nepal) or ragi (India). Koji is a diastatic mould, with powerful enzymes reducing starches to sugars, maltose to sucrose.

Koji mould is not effective on grain starches, but is popularly used to break down sweet potato, sago palm, breadfruit, or other starchy foods (such as cooked and mashed starchy fruits and vegetables).

The moulds are of wide natural occurrence and can be caught by wrapping 8–12 cm (3"–4¾") balls of steamed and moulded substrate (rice, barley, millet, soybeans, and so on) in leaves or straw. These are bound fairly tightly with cotton or string, and hung in the air below eaves, where over one to three months the breezes bring spores. They are ready for use when covered with a white mycelium or darker green spores.

Apart from straw, banana leaves, the leaves of rushes and sedges (e.g. *Phragmites*), *Hibiscus*, *Xanthium* leaves, sorghum, corn, and millet leaves, etc., can be used as wrapping or strewed on cakes of cooked substrate, as they all have mould spores on them.

Traditional koji (Korea) was made from steamed and mashed beans or steamed rice. These were formed into balls and hung in rice straw slings under eaves or in kitchens for thirty to sixty days, until covered with green mould spores, then soaked, washed, and (the soak water rejected), pounded down to join the other ingredients of soy and miso.

In Nepal, boiled rice-flour cakes are spread across tiled floors over an 8 cm (3") layer of clean rice straw, and then covered with slightly dampened rice straw to mould as marcha.

Nutrients added to the substrate to encourage the moulds in making spores are:

- Ash added to cooking water.
- Nectarous blossoms (cooked pulps of the Mahua flowers).
- Dried leaf proteins (e.g. fig leaf).

Once yeasts and moulds have fruited to spores, the spores can be tapped off onto clean paper, and stored in jars. These spores are used to inoculate new batches of rice, or tipped into sterile (boiled and cooled) water for refrigerated storage. The spores and substrate can be mixed with a paste of flours, formed into small cakes, and dried.

The only dangerous mould (*Aspergillus flavus*) occurs in rancid oily substrates such as old or damp peanut press cake, pressed bran, and coconut meats. Use only fresh milled grains and soybeans as a substrate. No such dangerous moulds have been located in miso, tempeh, soy sauce, or chiang, which all use grains and beans as substrates.

Moulds as commercial cultures can be purchased from government or private agencies. Some of these are selected to include lactic acid bacteria or yeasts for a specific ferment for sauces, misos, and chiangs (see Appendix D—Resources for these suppliers).

Culture of Tempeh Mould

If a small piece of tempeh is spread over leaf surfaces (teak, banana, *Hibiscus*) or wrapped in leaf, spores of the *Rhizoporus* inoculant develop, and a leaf can be used to spread spores over 10 kg (22 lb) of beans, by rubbing these spores on the substrate of beans or a piece of tempeh may develop spores in the open air. Leaf wrapping produces spores in thirty hours. (Gandjar in Hesseltine, 1986).

Steamed tofu press-cake can also be inoculated for tempeh in this way, or the whole mass kept twenty-four to forty-eight hours in a banana-leaf wrap.

Preparing Grains for Koji

Grains or root starches for koji are always steamed to cook. This can be done in bamboo steamers, or in a bowl (covered with coarse muslin) in big pots. Try 6 cups of grains or 8 cups of peeled and cut roots. Use polished rice or barley, whole wheat, rye, and millet. If using coarse corn meal, first soak for one hour, then steam for forty-five minutes. Soak all grains before steaming, then wash and drain. Steam all materials for forty-five to fifty minutes, barley ninety minutes.

If using a starter, use a level tsp of koji spores or 1½ tsp of whole grain starter. Cool and spread out cooked grain on a clean sheet and cool to 43°C (110°F). Mix starter with about a quarter cup of lightly roasted flour, and sprinkle half of this over the spread grain. Turn and stir, add the rest, mix well, and mound into a hemisphere. This "pudding" can be thickly wrapped in blankets or otherwise insulated in a haybox (see Figure 1.9), with a hot water bottle below.

The rice should stay at 22–25°C (72–77°F); this can be adjusted every two hours by replenishing the hot water bottle or adding cooler water if necessary.

At evening, refill the hot water bottle and go to bed. Next day, check the temperature and bring to 59°C (138°F), removing any bluish, black, or pink moulds carefully. Mix, rebundle, keep at 25–59°C (77–138°F) (never below, and rarely above 40°C [104°F]) for forty-five hours, then cool mass to room temperature. All grains should be bound together by white mycelia of *Aspergillus oryzae*. Store excess in a press-top plastic container in the refrigerator.

To dry koji, spread it out or form into small cakes on newspaper to dry for ten to twenty hours in a clean dry place. Seal and store dry for one to two months (up to six months), or spread on trays in a 45°C (113°F) oven until thoroughly dry, then store in a sealed container for a year.

"Wild strains" are caught on the steamed grain, left partly uncovered overnight with banana leaves or rice straw, then transferred to the haybox for forty-five hours incubation. Try to choose a cool, clean-air period. The time to make koji in Japan is when the peaches come into blossom. Koji is purchased dried or fresh; dried it is as whole grains of rice or barley, or as soft "mats".

As fresh koji is 110% or so heavier than dried koji, it requires only 60% of the liquids used for dried koji in recipes.

Fresh koji is crumbled down and mixed with all the salt in a particular miso recipe, then stored cool and dry (not refrigerated). Never freeze the spores, nor the koji. Sporulated (olive-green) koji is best used as a koji starter; crumble this in trays of boiled rice.

Ragi

Ragi, the equivalent of koji in Japan or marcha in Nepal, is prepared as starter balls in much the same way. All have "secret ingredients" or a unique balance of them. A thick paste is made with clean rice flour and formed into balls. These are then sprinkled with a powdered ball from a previous batch, and cultured on bamboo trays covered with a cloth (or straw in Nepal). As the culture grows, the rice balls dry over three to five days, and are stored thoroughly dried.

In Thailand, the rice flour is mixed with the powdered yeast ball and sugar cane juice or honey water (flower water in Nepal). No sterilisation is used, so the culture is a mixed one. The dried cakes keep for months in dry press-top tins or sealed dry pots, and are sold in markets. The cakes are kept two to twelve months for use or market, and are used for tape (a rice ferment).

Ragi always contains *Aspergillus* or *Amylomyces*, plus a great variety of bacteria, yeasts, moulds, and fungi. *Aspergillus rouxii, Endomycopsis fibuliger,* and *Hansula anomala* are essential for a good ragi in Malaysia.

With a commercial starter, 1 gm (a pinch) of spores is used with 100 gm (3½ oz) of sterile rice flour and 30 mL (1 fl oz) of sterile water, incubated at room temperature for four days. This can be refrigerated for two weeks, or dried as ragi and used at 0.2% to the dried weight of rice or roots.

There are various proportions of rice flour and spices in ragi, with spices ranging from less than 1% to 50%. Ground spices used include: garlic, galangal root, white pepper, red chillies, fennel, sugar cane, lemon juice, coconut juice.

Marcha — Nepal, India, Bhutan

Rice or wheat (or a mix), and "wild plants" are used. Ground rice or wheat is mixed with plant powders or flower pastes and made into a stiff paste. This is fire-heated to 35–37.7°C (95–100°F) for twenty minutes, then moulded into small lens-shaped balls and fermented under damp straw at 15–25°C (59–77°F) (room temperature) for three to five days. Then it is sun-dried for market. It keeps for a year and contains *Saccharomyces, Rhizoporus, Endomycopsis, Pediococcus*, and *Lactobacillus plantarum*.

Nuruk — Korea

This starter is used for brewing and has a "delightful flavour" (Saono, 1986). Crushed wheat is mixed with 30–40% water, steamed, moulded into balls, and left to ferment box-wrapped in leaves in straw. After ten to fifteen days the balls are sun-dried, aged three months, then marketed or used in the home. They have a shelf life of six months. Moisture is 10% when dried. They contain the genera *Aspergillus, Candida, Rhizoporus, Penicillium, Mucor, Hansenula*, and the bacteria *Leuconostoc* and *Bacillus subtilis*. (Saono, 1986).

Look-Pang — Thailand

In Thailand, look-pang uses 95% rice flour and 5% of a mix of galangal (*Alpinia*) root, *Albizia* leaf, and garlic.

The look-pang used in the preparation of alcohol mash also uses 95% rice flour, garlic, ginger, galangal, *Myriopterum*, black pepper and chives (together less than 5%).

Other plant leaf ingredients (given the 95% rice flour) are *Diospyrus, Amomum, Eugenia, Lawsonia,* and coriander. The dried leaves often provide protein, and probably micro-organisms, to the rice.

Rice flour and dried or ground spices are made into a thick dough with water, then moulded into small lens-shaped balls. These are left at room temperature for a few days, then sun-dried. These "yeast balls" keep for up to a year. They contain up to 10–12 micro-organisms of the genera *Rhizoporus, Mucor, Chlamydomucor, Penicillium, Aspergillus, Endomycopsis, Hansenula, Saccharomyces*, and a variety of bacteria. Obviously, many of these yeasts and moulds are either in the rice flour, the water, or the spices. (Saono, 1986).

Mejo Korea

Soybean ferment starter is made of cooked and crushed soybeans. These are moulded into cakes which are surface-dried for a week, hung indoors, then fermented one month in a straw rack (or bound with rice straw). The moulded balls are then again more fully sun-dried to yeast balls, appearing dark brown and solid, smelling mouldy. They keep for a year, and contain (minimally) *Aspergillus oryzae* and *Bacillus subtilis*. Uses are for starting chiangs or misos, soy sauces, etc. Finished moisture is 10.4%.

Moulds and bacterial inoculants, algae, and yeasts are so often "caught" or are inoculated under dried or fresh leaf and flower materials that the following short list will suggest other sources.

Slimy Bacteria (*Bacillus natto*)—Caught in China on soaked and steamed soybeans placed in cloth bags and covered with cereal straw. Inoculate one to two days at 25–30°C (77–86°F). Straw also provides *Mucor* moulds in cool weather on beans or bean press-cake. (15–18°C) (59–64°F).

Leaf Covers—Cooked and steamed grains or legume seeds are cooled and covered with leaves of:

- Water reeds and rushes
- Rice straw
- Banana leaves
- Ti (*Cordyline*) leaves
- Bamboo leaves
- Lemon grass
- *Artemisia* leaves
- *Hibiscus* leaves and flowers
- *Tectonia* (teak) leaves
- Sugar cane leaf

Mixed in with doughs or grain masses are:

- Mahua flowers
- *Hibiscus* flowers
- Ginger or *Alpinia* roots (galangal)
- Bean or kudzu juices
- Chopped medicinal herbs
- Fennel and coriander seeds

Plants used to reduce aflatoxins are:

Oxalis spp. (bruised leaves), giving oxalic acid.

Plant extracts used to aid fermentation are:

Sugars (palm and cane) and molasses (sugar cane) or malt (barley).

Chu China

Wang in Hesseltine (1986), gives a variety of cultures for Chinese ferments. In China, the cultures or yeasts or moulds go under the general name of chu. See the following:

- With a basis of wheat—in summer, crushed wheat is washed, soaked until sour, then steam-cooked. The steamed wheat is spread in beds 6 cm (2½") deep on clean mats, allowed to cool, then covered with leaves of *Miscanthus sacchariflorus* or *Xanthium sibirium*, and if, after seven days of incubation, a yellow mass of mould forms, the wheat is rolled into balls, sun-dried, and used as a source of moulds for chiangs or ferments.
- Where wheat flour is used—a dough is made, rolled into balls, then steamed, cooled, fermented and dried as above. These yeasts are also used for alcohol (beer), soybean pastes, and fermented or brewed sauces (see chapter four for further information). *Aspergillus oryzae, A. sojae* and other moulds are produced in chu.
- With a basis of rice—rice flour is mixed with grasses and herbs, and the juice of kudzu *(Pueria lobata)*, stem and leaf also added. Formed into oval balls, this base was fermented thirty days under a leaf mat of *Artemisia apiacae*. Grasses such as lemon grass, water rushes, and medicinal herbs were included in the mould base. *Polygonum* became a common ingredient. Such moulds (with *Miscanthus, Artemisia, Polygonum*) were used, with malt, for digestive upsets.

2.4 FUNGUS ON PLANTS

Mycotoxins

Several fungi can produce toxins able to survive normal cooking or food preparation, and the four groups of these toxins that are common in foodstuffs are:

- Aflatoxins
- Ochratoxins
- Trichotoxins
- Zearalenones

The occurrence of these poisons in stock feed may carry over into the mills, and also to meat and eggs. "Allowable" levels in stock feeds range from 20–1,000 ppb, a range due more to the inability in some areas to find toxin-free food than to health concerns, which always "allow" very low levels (<20 ppb, as in the USA).

Particular care should be taken to rapidly process oily seedcake, to rapidly dry fruit kernels cleansed from the pulp and left wet, or to use nuts while they are still fresh. Rancid oils or nuts, sunlight, and a humid atmosphere produce aflatoxins.

Corn Smut

Smut is a fungi (*Ustilago maydis*) which invades the ears and cobs of green maize. It is considered a valuable food of the Mexican farmers of Oaxaca, who gather it in the black-spore stage. Some of this is roasted on the cob, but traditionally the green ear with its smut has the cob removed, and is boiled with coriander, onion, and garlic. The corn seed and smut can be ground together with chilli as a sauce. It was an official remedy as a vasoconstrictor (to prevent excessive bleeding) in the USA, and used in cases of ulcers, sores, and general haemorrhages.

Ears of corn in smut can be dried, then the infected seed is ground with cornmeal, chilli, onion, and cilantro mixed in. The whole is made into a pancake and fried in oil as a dish called amarillo. Dried smut and cornmeal are stored together in earthenware or bags.

Do not confuse maize or corn smut with the ergot of rye (*Claviceps purpurea*), which can be fatal if consumed. (The active principle is LSD, once known as a hippy drug.)

Fungus on Bean Leaf Shoots

Sometimes the young terminal shoots of the four-winged bean (*Psophocarpus*) are attacked by the fungus *Woroninella psophocarpi*, which deforms them and turns them orange. They are collected and eaten as a vegetable in Indonesia.

2.5 MUSHROOM LICHENS

We have gathered large, fruity, fungal bodies from fields, forests, termite mounds and straw heaps or rotten logs since our earliest foraging days, no doubt learning from the many game animals (pigs, monkeys, wallabies, rodents) that also seek them. Pigs and dogs are still used to locate truffles, and marsupials like wallaby and potoroo dig for them. Even some invertebrates culture mushrooms, e.g. the foraging ants (*Atta* species), some wood beetles, and the termites. We eat dozens of fungal species, some local, many now cultivated. Amerindians, Vikings, and other groups have used or still use hallucinogenic mushrooms for religious visions, and amateur mushroom enthusiasts are sometimes poisoned by consuming poisonous varieties.

People have cultured field mushrooms since 1750 on composts, and in Japan and China, culture of straw mushrooms and shiitake (wood mushrooms) is ancient. But only in the last few decades have a wide variety of mushroom cultures been supported by the modern understanding of spore cultures and sterile substrates.

Many people in Japan and China produce very large quantities of ligno-cellulose wastes that mushrooms can convert to food. Many traditional cultures, such as the Hawaiians, use logs and woods of kukui *(Aleurites)* and hau *(Hibiscus)* stuck upright in the mud of taro terraces to grow ear fungi (hei-hei-au). This is probably an *Auricularia* species, similar to *Auricularia polytricha* grown on broad-leaf woods in China. Logs of hardwood are used to culture shiitake (*Lentinus edodes*); softwoods like poplars grow *Pleurotus* spp. in earth-covered pits in Canada. The spores of the desired species are introduced into holes bored into the wood, or a slice of a log with mushroom hyphae is nailed to fresh logs to inoculate the wood. Thus it is probable that many waste woods could be used in these ways to produce food, as part of normal farming practice, if inoculants were available.

Special groves (e.g. of nut pine, *Pinus pinea* and species) are easily inoculated with forest-floor fungi such as ceps (*Boletus* spp.). Wild-grown mushrooms are incomparably better than cultured strains for flavour. Acorns, sprouted in soils containing only the spore of truffles (all else having been sterilised) are planted out as oak forests, and truffles are produced in about ten years.

Never eat fungi collected in the wild unless you are certain of their edibility and for this reason a local guide is essential in strange areas, or buy excellent, well-illustrated books. Even normally edible fungi can bring on allergic reactions. Fungi can even take up biocides and radioactive substances in polluted areas. Local processing methods (some fungi are toxic raw, but not cooked) are as important as local identification. All that said, careful work will extend your range of food from fungi.

Wild mushrooms occur in such numbers at times that there are many well-tested ways to preserve them. Wherever possible, they are picked as clean as possible, and the stem ends cut off as pulled or gathered, the caps wiped off, (but only if soiled, washed) then they are best dried for high-protein food but can also be reduced to a ketchup, salted or fermented to lactic acid, even made into vinegar.

To grow mushrooms, obtain natural or commercial spore or spawn. Spawn is the mushroom organism in its mycelium form, before it begins producing its fruit (the mushroom).

For ground mushrooms, holes are dug or bags of spawn or mycelia placed on peat, then the hole filled with chopped straw, a little gypsum or lime, peat, and cowpats before covering with earth or mould of leaf material. Washings of adult mushroom heads can be used for spores, although once the mushroom head is open, most of the spores are gone. Bags are buried filled with mycelia and sawdust, leaf-covered, and some mushrooms allowed to ripen and spawn. Full lists of species and special techniques are given by Steinbeck (1981) who includes a lot of illustrative material showing successful naturalised species in gardens, at least for the cooler northern hemisphere.

All heavily mulched gardens grow "flushes" of fungi, but by suiting cultures to shrubs, trees, and locations, more edible species will be present. It is an excellent idea to naturalise many species of mushrooms, and as Steinbeck notes, gardens designed to grow some varieties of mushrooms also collect other useful species from wild spores, so that as you start the process you create conditions for it to continue and improve.

For serious growers, I strongly recommend books such as that by (Stamets and Chilton, 1983) which is suited to the small or large grower. One of the very useful tables in this book is that of the analysis of substrate materials used to prepare beds for mushroom culture. These range from sawdust, woodchips, dry straws, foliage hay of legumes, hulls of grains, grains themselves, corn husks and silks, spent brewers' wastes, chaffs, bagasse, molasses from sugar processing, cotton wastes, oil-seed cakes or wastes, acorn meals, fruit pulps, seaweed agars and wastes, fish and bone or meat meals, citrus skins and pulp, sugary saps from trees, potato and sweet potato flours, dried milks and wheys, brewer's yeasts and yeast powders, flour screenings, brans, and so on.

A point of interest in mushroom culture is that many other organisms can assist in or initiate fruit bodies if present in mushroom substrates. Here is a short list:

- Yeast (*Saccharomyces cerevisiae*) added to agars or composts.
- *Rhodotorula glutinus* for chanterelles, plus active charcoal to germinate spores.
- *Torula* spp. form food for mushrooms in cooled composts.

Mushroom Compost Materials

Note—One tonne = 1,000 kg = 2240 lb.

Per dry tonne of cereal straws, chaffs, hulls (this may be up to 10–20% horse manure mixes from stables):

- Legume hays can be added at 20% weight.
- Urea at a maximum of 11 kg (25 lb) per dry tonne.
- Meat and fish meal at about 14 kg (30 lb) per dry tonne (not often added because of cost).
- Brewers' dry wastes, oilseed meals, and chicken manures can be 30% or more of straw weight. These are often mixed with sawdusts and added at low rates (45 kg [100 lb] per tonne).
- Fruit pulps and molasses (often mixed with hulls) at 112 kg (250 lb) per tonne.
- Horse manures are excellent with hay or wood chips as part of the dry bulk.
- Gypsum is usually added at 22–45 kg (50–100 lb) per tonne, the higher rate if chicken manure is used.
- Limestone flours are always added to offset acidic fruit pulps, up to 45 kg (100 lb) per tonne.
- Moisture is added to 71% of content, and makes up a little under half the total weight.

This sort of mix is composted and cooled before inoculation in beds; beds are often "cased" with soil or peats as a cover to the compost bed.

Solid Substrate Ferments

Ferments of soybeans, grains, and the composts of mushroom culture, are grown on substrates at about 70% water, which although it inhibits bacteria, allows yeasts to operate effectively, as they do in many cultures of mushrooms and moulds.

As most solid compost ferments are aerobic, and waste products (heat and carbon dioxide) may build up, aeration is essential to reduce heat and to remove excessive carbon dioxide. The advantages are that many such ferments are more simple, eliminate liquid wastes, and often save energy or use solid waste products, hence are economical. Tray aeration of beds is also usual, and air supply regulated to control heat, or aeration in mushroom beds occurs with loose stacking of the substrate, which in any case is difficult to compress.

The Culture of Mushrooms

A considerable technical skill is needed for successful mushroom culture and efficient production can mean tight environmental and financial control.

The species frequently cultivated are:

- *Agaricus bisporus*—field mushrooms; in manure composts.
- *Lentinus edodes*—shiitake mushrooms; in hardwood (oak) logs or "logs" of compacted chips and sawdust.
- *Volvariella volvacae*—straw mushrooms. Ideal culture temperature is 32°C (90°F), stock culture kept at 20–25°C (68–77°F). Substrate is compacted straw or cotton waste with 2% wet weight of lime. Composted in clear plastic for two days, turned, then again and made up as 10 cm (4") deep beds, sterilised with steam at 60–62°C (140–144°F) (or inoculated at 2% wet weight), spawn is inoculated at 32–34°C (90–93°F) under plastic for four days. Remove plastic and spray water to get a mist at the fifth day after inoculation.
- *Pleurotus sajor-caju*—is also cultivated on cotton waste with 2% lime (wet weight) and 70% of water; these mushrooms grow well on wastes already used by *Volvariella*.
- *Pleurotus ostreatus*—oyster mushrooms; are grown on poplar logs or chip wastes.

Straw Mushrooms

R.V. Alicbosan (*Food and Nutrition Bulletin Supplement #2*, November 1979) discusses simple farm methods to grow straw mushrooms using a variety of wastes.

Newspaper is "fertilised" by soaking in 3 gm (a large pinch) of urea per 4.5 litres (9½ pints) water. The spawn of the straw mushroom (*Volvariella volvacae*) is pressed into palm-sized pieces of this newspaper, every 5 cm (2"), and 5–8 cm (2"–3") from the edges of the bed. Three 2-cup bottles of spawn will plant a 4 m bed, with the first layer of spawn (½ bottle) pushed down 4 cm (1½") into each layer. Straw can be mounded on top to 10 cm (4") deep at the centre. Each bed is roofed over, often with plastic, so that the condensation drips or rains fall outside the bed. In dry seasons, use only four layers of straw, and in the wet six layers are used as humidity is higher.

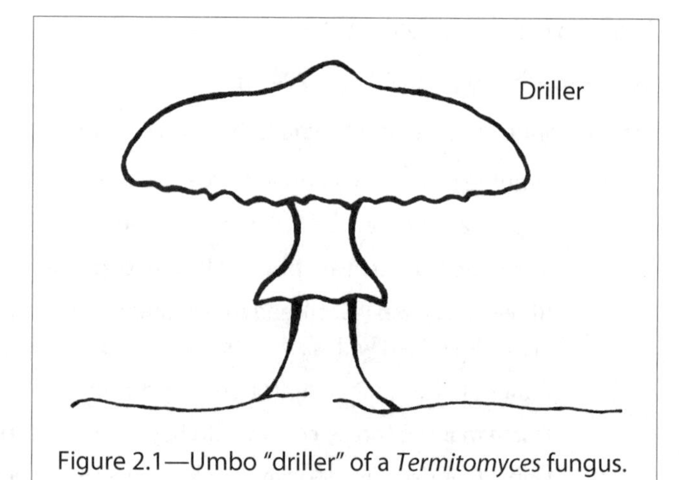

Figure 2.1—Umbo "driller" of a *Termitomyces* fungus.

Dry banana leaves, cut from the plant, can replace straw in bundles. Beds are not watered for five days, but in dry seasons generous water is given on the sixth and seventh days until mushroom pinheads develop. If water is applied in the wet season, it is on the sides of the bed. When mushrooms reach corn-seed size, the beds are watered again. The first harvest is taken at ten to fourteen days, for three days. Average should be 1–2 kg (2¼–4½ lb) per day.

The beds now rest for five days, and again three days of cutting (average now of 454 gm [1 lb] per day). Production may continue for a month or more. All mushrooms are pulled whole from the bed and later trimmed to prevent bacterial rot. A 4 m (12'), six-layer bed should produce 12.8 kg (28 lb) of mature, or 7 kg (15.5 lb) of button mushrooms. Mushrooms in such open beds convert 10–11% of the dry weight of the straw to food.

The rest of the straw can be composted, fed to stock, or mulched around trees and crops. This very basic farm system is a model of food production from waste at village level. The mushrooms are good sources of riboflavin, niacin, and phosphorous. Protein content is close to that of legumes (15–25% for dried mushrooms, 3–5% for fresh mushrooms).

For mushroom growing, substrates such as straw, cotton wastes, bran, paper, and mixes of these (with 5% lime, 5% bran) are steamed for thirty minutes, usually in a drum over a fire. Alternatively, they are composted ten to twelve days at 60°C (140°F), then pulled apart and cooled before inoculation.

Truffles

In autumn, the San of the Kalahari dig for the desert truffles *Terfezia pfeilli* in quantity, although if there are no rains they may not come. They are detected by surface cracks in the ground. Similar desert truffles form part of Australian aboriginal food.

Termite Fungi

An excellent edible parasol fungi (*Termitomyces letestui*) and a pink-spored fungus (*Podbrella*) thrust their way through the hard materials of termite mounds, where they are cultivated by termites in compost heaps of plant material. The fungi use a "driller" (a pointed knob called an umbo) to pierce a way to the surface. See Figure 2.1. They are gathered in quantity in Africa and eaten fresh, cooked, or dried. My African friends tell me that it is discourteous to eat the fungi (food of the termites) in sight of the termite mounds, and therefore they take them to huts or compounds for cooking. Many are of great size, some quite small.

Processing Mushrooms

Mushrooms may be processed in these ways (and probably others):

- Boiled in 5% salt water (parboiled may reduce slight poisons).
- Fermented in barrels (lactic acid ferment) after boiling.
- Dried in the sun or over stoves, threaded on strings.
- Fried in butter, peppered—bagged and frozen for long storage.
- Stewed in milk—bagged and frozen for long storage.
- Boiled in weak salt, stored under oil to preserve.
- Dry salted, processed, and juices reduced by boiling, as mushroom sauce and ketchup.
- Eaten raw in salads or alone.
- Preserved in vinegar sealed by oil (pickled).

Drying Mushrooms

Edible fungi are sliced into pieces about 1 cm (½") thick, and spread on clean boards either in the sun or in ovens at 48–60°C (118–140°F). They must not cook, or mildew, but be fairly rapidly dried to a crisp. Often they are threaded whole, not touching, on nylon or tough thread, and dried to leathery.

Crisp dried mushrooms are ground to powders and flours, and used for sauces and gravies. They are added to grain meals for protein and calories, trace elements, and vitamins.

Powders properly dried can be stored in sealed jars for up to a year. The volume of dried mushrooms is 10% that of fresh mushrooms. Soak in a small amount of water before cooking (one to two hours' soak).

Mushroom Ketchup—Variation One

Pick over and clean the mushrooms, but do not wash if unnecessary. Layer them up in a bowl, sprinkling with salt. Continue this way until all mushrooms are salted.

Leave for four hours then break the mushrooms up into pieces and leave for three to four days, turning, stirring and pressing occasionally. Try to extract all the juices.

Next, measure juice, plus mushrooms and to each 2 cups add a little cayenne, ginger, garlic, allspice. Simmer for about thirty minutes in a heavy pan, strain into a basin and fill bottles. This is a very salty mushroom sauce, and I use it in lieu of salt in meat dishes or in stir fries.

Mushroom Ketchup—Variation Two

Mushrooms are lightly salted and pressed in an ordinary fruit juice press, or they are lightly cooked over medium heat, and press-filtered through colanders to extract the juices.

Reduce the juices, with pepper, salt, cloves and bay leaves, until thickish or syrupy and bottle hot, well sealed. Mushroom ketchup can be blended into soy, fish, or Worcestershire sauces or used alone over rice, meats, or in casseroles.

Mushroom Ketchup—Variation Three

A glut of field mushrooms can be layered in a jar, each layer salt-sprinkled and the whole well compacted or pressed down. Leave in brine for four days, stew in brine for two hours, strain and press out lightly, and boil the liquor. Reduce the liquor to 1/3 before bottling; seal well. This ketchup can be used as a base for sauces. Seasoning can be simply peppercorns, or these and a little allspice. Store cool in the refrigerator.

Mushroom Ketchup—Variation Four

This very salty sauce is often added to Worcestershire or soybean sauces. Lay down a bulk of chopped mushrooms in a barrel or deep casserole. Sprinkle 5% by weight with salt, weigh down and leave for one to three days. Then drain and bottle in pre-warmed bottles. Cover with warm olive oil before tightly corking to keep. Use small bottles as the concentrate is salty and is usually added to other sauces and gravies.

Lactic Acid Preservation and Ferment

Boil mushrooms until they sink (medium heat at simmer) and add 3% salt **or** 2% salt and 2% sugar (by weight). Cool to blood heat and add a little yoghurt or yoghurt whey after adding boiled and cooled water equivalent to 60% of the mushroom juice.

The ferment takes fourteen to fifteen days, and should be via an airlock system, or in a close-weighted cask. When ferment ceases, full barrels or bottles can be sealed. If a little mould forms, a few drops of sulphuric or benzoic acid must be added.

Well sealed mushrooms keep a full year, but for households, about ½ gallon bottles are ideal. If too sour, a slight rinse before use suffices.

Vinegar Preservation

- Simmer mushrooms in a little salt and water with some chopped onion. Cool and cover with malt vinegar which has been separately heated with pickling spices (pepper, bay, cloves, cinnamon, mustard seed). If the juices are strained off before pickling, and more salt added, this can be boiled down to a ketchup.
- Sprinkle mushrooms with salt and leave overnight, and in the morning cover with a good vinegar and a bag of pickling spices, and gently simmer for five to ten minutes. Remove the spice bag and bottle hot in hot jars.

Pickled Mushrooms—Variation One — Russia

Russians and Finns collect, preserve, and ferment some forty to fifty species of forest mushroom, many of which are subjected to lactic acid ferment for preservation.

Mushrooms are boiled in very little water until they sink (eight to twenty minutes), often with grated carrot, bay leaves and coarse black pepper to flavour. Salt is added at 3–4.5% and vinegar at 18% of the weight. (Vinegar and mushrooms are separately boiled and cooled.)

Pickled Mushrooms—Variation Two

Lightly sprinkle salt on layers of chopped mushrooms and press gently with a stone or board overnight at room temperature. Next day add 50:50 wine vinegar and cider vinegar to cover. Add pepper, cloves, a little chilli, and boil for five to six minutes. Then seal (covered with heated olive oil) in small jars and refrigerate. Older recipes use pure vinegar over cooked mushrooms, spiced with pepper, mace, and cloves and dry-cooked until tender, shaking the pan to prevent sticking. The vinegar is added to the pan just before briefly boiling, bottling, and sealing.

Salted Fermented Mushrooms — Russia

- Mushrooms are cold-soaked, and the water changed two to three times, then drained. 100 kg (220 lb) of mushrooms are layered with 4.5–5 kg (10–11 lb) of salt, with bay leaves, pepper, garlic, dill, blackberry leaves to flavour. First layer with 50% of the salt, then after a day, boil and cool. Then add the rest of the salt (2.5% salt solution). Cover crock or barrel, weight a wooden disc with a stone, and now (boiled and cooled) mushrooms can be added as gathered. The pH should be kept above 4.0. They can be fermented up to two days, then sealed and stored cool.
- Delicate mushrooms can be blanched in boiling water for five to ten minutes, using 2–5% salt. Drain, wash in cold water, cool quickly. Salt as above. Let ferment in salt for seven to sixty days. 1% sugar may be added. Lactic acids form in the first fifteen days (temperature 0–2°C [32–36°F]). Some varieties are fermented up to two years. Check and add 4% salt solution if drying out.

Field Mushrooms (*Agaricus* spp.)

This is a traditional Tasmanian recipe for a glut of field mushrooms; I have used it often. Halve and cut the stalk off mushrooms (but also add the stalks as that's where much of the flavour is stored). Fry a large chopped onion in butter until soft, add mushrooms and one cup of milk and cook covered for five to eight minutes. Finally, mix a tablespoon of cornflour with another cup of cold milk, stir in, and add plenty of pepper, salt to taste. Cook until well thickened, serve cold on toast. When there are large quantities available, freeze in small containers of this stew and melt down in a pan with a little milk to serve alone, or with meats.

Cooking sherry can replace milk in this recipe with success.

Dry-Toasted

Sliced mushrooms are placed one layer thick on an oiled pan and toasted in an oven at 180°C (356°F) until crisp. They can be salted lightly and stored under oil in sealed jars, or dry-salted and stored in a pre-warmed jar as snacks, with crisp bacon.

Combinations

Mushrooms and onion base (peppered) cooked in olive oil or butter can be folded into dough rolls, or mashed into soft cheeses as spreads, or have beaten egg poured over as an omelette, topped with chopped parsley.

As a general guide, mushrooms combine well with onion, pepper, crisp bacon pieces, and dishes benefit from a sprinkle of chives and parsley before serving.

Polypore Fungi for Tinder and Styptics

The bracket fungi (only softish when young) have been used in all cultures for starting and carrying fire, and applied to open wounds as a styptic (to stop bleeding).

For wounds, the softer brackets are boiled or soaked, beaten with a pounder, or mallet, and worked until soft, ready to be bound on wounds.

Tinder is made from tougher parts, soaked in hot water and ashes (lye), beaten and dried, then soaked in saltpetre and ash solution so that (dried) they are easily lit from a flint spark or by firesticks. They glow for a long time, enabling fire to be carried if necessary to a new fireplace, where they are blown to red heat to ignite splints. Genera such as *Fomes* are used in these ways.

Lichens (fungal-algael associations)

Many lichens are edible, e.g.

- *Catraria islandica*—Iceland moss, alpine and arctic, contains iodine.
- *Evernaria prunasta.*
- *Lecanora esculenta*—The manna of Biblical times.
- *Gyrophora esculenta* and spp.—Tripe de roche, spp. in China and Japan.
- *Umbilicaria* spp.
- *Parmelia* spp.—Used in India as a food adjunct.
- *Peltiger* spp.
- *Ramalina* spp.
- *Usnea* spp.—Contains anti-bacterial acid for *Streptococcus*, tubercular bacteria etc. (beard lichens). These are used for food and fodder.
- *Roccella montagneiis*—A good fodder and food lichen.

Dye lichens (*Roccella, Lecanora*) yield—Orchil, cudbear and litmus dyes, the latter used in pH tests for acid-alkali balance; many are used as biological stains.

Scented lichens (Oak moss, *Evernaria prunasta*)—Used for scenting soaps and for scents in cosmetics.

Fermentable lichens—Iceland moss and reindeer lichen (*Cladonia rangifer*) have up to 60% sugars, readily hydrolised to glucose, thence via yeasts to beers or alcohols distilled to spirits (ethanol).

Bitter lichens—Soaked in dilute carbonate of soda solution to remove lichen acids, dried, powdered, and added to bread flours. Protein content of dried flours exceeds 9–15%.

Iceland Moss Jelly

In 50:50 milk and water, sweetened with honey, a tsp of powder (as above) will set to a jelly, and can be left in a mould to cool.

Iceland Moss

Soak 2 gm (medium pinch) of dried lichen in cold water for one to two days, then boil for one to two hours, strain, stand for ten minutes, pour off clear liquid and add a little isinglass, boil this to half its bulk, set in clear jars—store cool and use for jellies, or flavour to serve with lemon rind, fruit.

Cladonia (e.g. *C. rangifer*)

Reindeer, feeding on lichens, often provide direct vegetable meals from their stomachs, for Eskimo and Arctic Indians alike. The stomachs are also used to ferment mixtures of lichens and meats, being hung over smoking fires to ferment, and may then be buried or cached for winter. This mix of lichens and meats is eaten raw or cooked, and is pleasantly acidic. (Mowat, c1973, *Tundra*).

CHAPTER THREE

The Grains

3.1 INTRODUCTION

Grains form almost a fourth of the world's diet. Here are not only included rye, oats, barley, triticale, wheat, maize, and rice, but the sorghums, millets, and the minor grains such as Amerindian rice grass and wild rice, amaranth, quinoa, buckwheat, etc.

There are many ways of preparing grains. They can be heated, soaked and sprouted, composted, fermented, or ground into flours. When soaked, grains expand; when heated in water or other liquid, they swell to their fullest capacity; this is called gelatinisation. This gelatinisation is evident in the thickening of cereals and in sauces to which flour is added.

3.2 RICE

Composted and Parboiled Paddy Rice

Rice that is still in the husk, either as cut heads or on the stem, is referred to as "paddy rice" or simply paddy. Home-milled rice has a higher recovery rate and is more nutritious (bran not removed beyond 5%) than machine-milled rice. Milled rice does not keep as well as rice in the husk which has been carefully dried to 9% moisture or less. Thiamine loss in milling causes beriberi in people on rice diets, and can cause sudden death in breast-fed infants. Washing of milled rice before cooking results in severe nutrient losses (55% of the thiamine in milled raw rice, 9% in parboiled rice).

Composting rice or wheat before storage (in the husk):

- Fixes the water-soluble vitamins and minerals in the starch.
- Improves the protein content in the gelatinised seed.
- Inactivates enzymes that may cause rancidity.
- Kills off mould spores and insect larvae.
- Lessens cooking time.
- Greatly increases storage life (by preventing germination and creating a dense, glassy starch difficult for insects to penetrate).

Composted grains are more easily digested and of enhanced food value when they are finally hulled and polished. Best of all, of course, is to compost the paddy and use as unpolished rice.

So much good protein, minerals, vitamins, and fats are removed by milling and polishing rice (and thus removing the embryo) that in some areas where nutrition is vital, the paddy is parboiled by a hot water dip plus steaming, so that some water-soluble embryonic vitamins (B complex) are leached into the endosperm. Dried again, the milling is now unable to remove the absorbed vitamins even though the embryo (germ) is removed.

Composted Rice Paddy (hakuwa) Nepal

This system is used with rice just short of ripening or unlikely to ripen in an early winter. The rice stalks are cut and stacked, unthreshed, heads-in to build a mound or haystack, layer by layer until four to six feet high. All seeds are in the centre. If the paddy is dry, water is sprinkled on each layer to equal the weight of the straw.

These khunyuns dot the fields and are often built in backyards in urban areas of Nepal. The top is domed, and clay turfs laid on to prevent wind damage; if rain is expected, they are rough-thatched with bundles of straw, and are often crowned with a pot, or a fancy straw twist-spiral; this top taper is regarded as a deity or shrine by the farmers.

The rice is left to compost for ten to twelve days, during which time anaerobic fermentation takes place and the pile heats to 60°C (140°F). Hot vapours come from the heaps on cold mornings. At the end of the ferment, the deity is worshipped.

After ten to twelve days, the heap is broken down; the husks and heads are now partly-boiled, or gelatinised. They are then dried for five to six days before threshing. The composted and dried paddy is now threshed out.

We now know that thiamine and other vitamins are absorbed from the germ and husk by the seed grain, and remain in this rice after threshing. Poor farmers eat a lot of this rice, which in part is a compost like chiura. One great advantage of hakuwa rice is the extended, pest-free storage life, high vitamin content, and light, digestible nature.

Not long ago most rice was eaten this way, and a hundred years ago, most wheat was also harvested unripe and sweated in stooks in the field, where it swelled and soaked up nutrients before being threshed. The technique should be tried for other grains.

Arroz Amarillo — Ecuador, Peru

Paddy is spread out in beds, sprinkled to dampen, and covered with cloth or canvas. Over the next ten to fifteen days this heats to 50°C (122°F). The beds are turned once and recovered. As with parboiled or composted methods used in India, thiamine is absorbed into the partly-gelatinised rice grains; riboflavin is also produced in the ferment, which involves both moulds and bacilli. The rice is then husked, dried, and milled as usual. Protein is about 9% of dry weight (Campbell-Platt, 1987).

Grey or Parboiled Rice (Ursina) — Nepal

Unhusked rice is soaked in cold water for two to three days (or less time if water is heated). Time in hot water is three hours at 70°C (158°F), which minimises fungal growth in the pre-soaked period. For large quantities of rice, hot water is poured over the wet rice in a large pit; smaller quantities are soaked in cauldrons. Just before total gelatinisation of the grains, the rice is removed, dried, husked, and parboiled rice is extracted. Raw paddy rice has 13% water, parboiled rice 35% (6% water in the dried product). Vitamins enter the gelatinous rice.

This part-cooked rice, dried, is a staple for many people. It is probably a little inferior to composted rice, but detailed analysis remains to be reported. Energy costs are much higher than for composting.

Parboiled Rice — India

Paddy and water are heated to simmer or boil a few minutes, then given a cold water steep for twenty-four to thirty-six hours. To prevent off-flavours forming during the steeping process, the water should be changed. After another brief boil, the rice is dried slowly (in the shade) and husked. In the Punjab, the paddy is given a slight roasting in sand for a pleasing fragrance.

The parboiled rice is incompletely gelatinised and has an opaque core. Husking is much easier with parboiled rice. It cooks better, and remains fresher when cooked than does raw rice.

Kheel (parched paddy in the husk) — India

This is a parboiled method for smaller quantities of rice. Paddy is sun-dried, then placed in pots; hot water is poured into the pot and let sit for one to two minutes. The water is then poured off and the pot inverted overnight. The next day, the paddy is briefly sun-dried and kneaded while moist, then parched. After the rice puffs, the husks are winnowed. I like this recipe—it avoids milling, hence nutrient loss.

Commercially, rice is heated in sealed containers at 288°C (550°F) for an hour. When the pressure is suddenly released, the rice puffs by steam release in the grains. (*Wealth of India VII* p. 176) Obviously industrial heat energy is the replacement of compost skills and incurs heavy energy use.

Chiura (flaked puffed rice) — India

Unhusked rice is soaked for two to three days in hot or warm water, boiled a few minutes, and the drained paddy heated in an iron pan or shallow fire-proof earthenware tray until it puffs.

It is then flattened by a wooden mallet or a pounder to remove the husk, or put through steel rollers to get very thin flakes. It is as good as parboiled rice for nutrition, and can be stored for months without deterioration.

This is a very economical and beneficial way to prepare rice at home, and to reduce cooking energy in rice consumption.

Chiura contains some extra nutrients from the germ and husk and is very popular as a stored food in Nepal. The grain are stored for later cooking, are eaten as is, are popped in oil, or can be cooked to a hot porridge, eaten as a cold porridge with yoghurt, and sugared or salted as snacks in a little oil.

Puffed Rice

Beaten flaked parboiled grain (chiura) is "puffed" in deep oil. Put a few tbsp of chiura in a perforated spoon; these puff immediately and are taken out and drained. Puffed rice is eaten with omelettes, or mixed with sugar, fried, and stored as a snack.

Murmura (parched parboiled rice) — India

Parboiled rice is thrown into an iron pan into which hot sand has been placed. It is done a few handfuls at a time, stirred rapidly, and as it crackles and swells it is sieved out. Using two pans enables the operator to pour sand from one to the other through the sieve, without losing much heat.

Sometimes the parboiled rice is pre-heated or fried in a pan, and sprinkled with salt water before parching. Or the rice is steeped several hours in dilute salt brine, dried, and parched. Unexpanded grains are usually ground by mortar. Puffed rice is ready to serve with milk, buttermilk, honey.

Sushi Rice — Japan

Sushi is raw or pickled fish, vegetables and so on, placed on cakes of vinegared rice. The rice is well washed and drained one hour in a colander. This latter technique leaches a great many nutrients.

Bring to a boil: 2 cups rice, 2¼ cups water, and 4 cm (1½") of kombu (optional), removing the seaweed just before the water boils. Put the lid on the pot, turn heat to low and steam for fifteen minutes. Take off the stove and let sit, covered, for ten minutes.

Sushi Rice Mixture

Make a solution of ¼ cup rice vinegar or cider vinegar, 1 tsp salt, and 1½ tbsp sugar. Spread hot rice on a tray and sprinkle on vinegar solution. Using a wooden paddle, scoop the rice up repeatedly, cooling it quickly and coating it with the vinegar solution. It is now ready to mould in little cakes for sushi.

Tapai Pulut — Malaysia

This is a rice ferment with the mould *Amylomyces rouxii* and a yeast such as *Endomycopsis burtonii* (Indonesia). The mould culture is spread over cooled boiled rice. Glutinous rice is washed, then cooked, after an overnight soak. When cool, it is mixed with ragi (see yeasts and moulds) and fermented at 28–30°C (82–86°F) for one to three days. At 30–37°C (86–99°F), incubation of forty-five hours is enough. Lysine, thiamine, and tyrosine increase, and protein may double to 16% (after Merican and Quee-lan in Steinkraus, 1989).

Lao-chao — China

This produces a sweet, low-alcohol (0.05 to 1.58%) rice which is widely used in Malaysia for sweets or as a base for sweets. At home, the fermenting cakes are stored wrapped in banana leaf, or in plastic press-top containers.

Tape Ketan — Indonesia

In Indonesia, rice tape is incubated five to seven days, and then pressed. This liquid is saved and set aside. Water is added to the press cake and re-pressed. Both liquids are then concentrated by boiling. The press cake and the reduced liquid are combined, spread out 1 cm (½") thick on trays, and sun-dried six to seven hours as little round cakes 4–6 cm (1½" –2") across. This dry "yeast" keeps for many months as brem, used to make liquors or as an inoculant for rice to increase ragi or yeasts for tape.

Rice-Legume Ferments

Idli—Variation One — India

Idli is a common food throughout India and parts of Africa, or any areas settled by Indians. This is a very good basic short ferment suited to most homes. It uses 1 part of black gram (legume seeds) to 3 parts rice, but the method works for lentils, peas, mung beans, small beans or wheat, steel-cut oats, and barley, as well as rice.

The rice and grams are washed, soaked four to eight hours, and then ground or electric-blended to a batter. The rice can be coarsely ground; steel-cut oats or bulgur wheat needs no grinding. Salty water (1% salt by weight) is used to help blend the grams, and the blending water should be twice the weight of the soaked grams.

Leave the batter overnight, and use the ferment for breakfast. Now it is possible to add an egg, a chilli, some crushed garlic or grated ginger at this stage for the final cooking (blend briefly). Some people add bicarbonate of soda or lime (1 tsp).

I usually cook on a medium to low heat in a covered oiled pan, but the mix can be steamed in covered pans or even baked. It is an excellent breakfast. In India, idlis are served with either sweet or vegetable meals, with stews and so on, or by themselves with syrup or gravy.

Idli—Variation Two — Korea

Make the idli batter as above. Additions to the batter can be, for example, chopped spring onions, green peppers, fine-cut onion, roast sesame seed, sesame oil, soy sauce, salt, and minced squid, octopus, or fish. It is these additions that make Korean food more tasty and interesting than idli alone.

Brush a pan with oil and heat on medium heat. Ladle some batter into the centre of the pan and spiral gently out with the back of the spoon to make a 1.5 cm (½") thick pancake. Cook two to three and a half minutes in a covered pan; turn pancake, and cook uncovered two to three minutes, dribbling oil around the edge. Cook all pancakes this way until the batter is finished.

Dosa (pancake ferment) — India

Wash, drain and soak separately 200 gm (7 oz) long grain rice and 85 gm (3 oz) dahl (i.e. split red lentils or split peas), each in three cups of water for eight hours. Grind drained dahl to a thick batter, adding 170 mL (5¾ fl oz) water, achieving a light fluffy texture.

Now blend the drained rice using 170 mL (5¾ fl oz) water and making a moist paste—pour this over the dahl batter. Mix the two and ferment sixteen to twenty hours. Add ¾ tsp salt, and some ground cumin seed. Cook as a pancake in lightly oiled pan, about two minutes per side. Pancakes can be filled with cooked vegetables, or meat and vegetables, and folded over.

Note that very sour yoghurt or yoghurt whey can be blended with grains if it is available, and replaces water.

Ang-kak (red rice, red colouring) — South East Asia

Washed, hulled rice is soaked a full day, drained and steam-cooked. After it has cooled, it is inoculated with *Monascus purpureus,* a mould producing a strong red purple colour soluble in water. The dried and powdered grains are used to colour food such as wines, pickles, sauces and pastes. Protein is about 20% of dry weight; powders or grain, dried, are kept as colouring.

Nuka-zuke (rice bran ferment) — Japan

Nuka = rice bran, Zu or Zuke = pickles

Nuka boxes are a feature of every Japanese kitchen. The data following on nuka-zuke (and minor uses) is from Arthur Getz, our "man in Japan".

Analysis of rice bran:

Per 100 gm (3½ oz) of bran:	**Minerals:**	**Vitamins:**
Protein 13.2 gm (½ oz)	Calcium 46 mg	A (carotene) 6 mg
Lipids (oils) 18.3 gm (¾ oz)	Phosphorus 1.5 mg	B1 2.5 mg
Glucide 38.3 mg	Iron 6 mg	B2 0.5 mg
Fibre 7.8 mg		Niacin 0.25 mg
Ash 8.9 mg		Vitamin C 0 mg

In the rice grain, the content of vitamins and minerals is: embryo 66%, bran 29%, starchy albuminous grain (polished rice) 5%.

Nuka boxes are used to remove alkaloids and harsh tastes from bamboo shoots and bracken fronds. Bran is used (5%) in jinda (traditional miso) of soybean, malted rice, and sake mash after spirits are removed; the bran comes from sake rice used for ferment.

Rice bran, rich in nutrients, is used in nuka ferment. A nuka box is invaluable for ongoing pickles and may be used for years with only the addition of salt brine and bran to replace that used. To start this lactic acid pickle base, combine together 454 gm (1 lb) sea salt, 3 cups water for a brine and 1 leaf (dried or fresh) seaweed.

Fill a wooden box or small barrel with 2 kg (4½ lb) rice bran or oat bran, and pour over it a pot of cooled brine. Stir the bran with a wooden spoon (this mixing can be in a large basin and then be transferred to the box). Add boiling water until the bran is damp, but when you squeeze it only a little water can be expressed. Set aside the box for a few days, during which time a preliminary ferment heats the box. It doesn't hurt to use some unsterilised miso in the brine to start with, or some already-fermenting nuka from a friend. This daily-tended "bed" (nuka-doko) is now ready to pickle crop vegetables or brined fish and squid.

Now add to the bran: 454 gm (1 lb) cabbage, radish or carrot leaves, a handful of wheat flour, 230 gm (8 oz) washed ginger root, 2 slices of cut-up bread, 280 gm (10 oz) dried chilli, 2 tbsp mustard, 2 tbsp turmeric powder, rind of 1 lemon, 3 peeled garlic cloves, ½ cup of stale beer.

Each day, for four days, discard the vegetable leaf and add fresh leaf; on the fifth day the pickle medium should be ready. Replace garlic every few weeks and every week add bread, a little bran, and 10% salt by weight.

After a few days, take large radish, turnip, carrot, scrub lightly (do not peel) cut into strips or long chunks, or quarter lengthways and wilt in the sun for a day or two. Now push these down into the damp bran, and leave for four to ten days, until sufficiently acidic. Cabbage can be part-quartered with stalks attached, and all root crops and greens wilted or slightly wilted before immersion. These vegetables ferment and become "pickled"; they are taken from the box and either refrigerated or sliced and eaten on the spot.

Bran and (if necessary) cooled boiled water can be added to the box as necessary. Bran should always be 60% of the total mass. I also add a little turmeric for colour in the bran. After using the nuka box for a few months, you will be able to judge times of pickling of vegetables in hot or cold weather.

Nuka boxes are traditionally made from cedar, cypress, or some such durable softwood, often tanned inside before use. They are **not** airtight, and if using glass jars, the lids need an airhole. Boxes should be aerated, once established, by a good stir daily.

It is usual to keep the nuka in a dark place in the kitchen, and to tend it by adding beer if dry, bran if low, or a couple of slices of chopped bread if not yeasty-smelling.

Keep the bran surface flat and the sides clean inside and out. If the nuka gets too acidic, add crushed eggshells, a little lime, or ⅓ cup of mustard powder. If it is too wet on top take a clean sponge and mop up surplus liquid. If a white mould develops, scrape it off. In hot weather, add a little mustard powder every ten days. To revive dry nuka, add bran, brine, bread, and leave for four days (stir daily).

If going on holiday, bag the lot and put it in a refrigerator, or lend it to a trustworthy neighbour. If no refrigerator is available, cover the surface with an inch of coarse salt and fit a paper on tightly. Store cool and remove the salt on return. Freeze the nuka only if you leave for a long time. Make sure no vegetables are left in stored nuka.

In abnormally acidic bran, no *lactobacilli* occur, and yeast flavours are lost, thus pH should be adjusted with limewater if necessary to 5.5–6.5 and a little sake added to deter bacilli.

Nuka boxes can be used for years if bran is kept topped up and pH adjusted.

Non-Food Uses of Rice

Rice Husks—can be used for making active carbon for filters, also milled-roasted to produce Portland cement, to make furnace bricks if used with clay, or insulating bricks for furnaces, and insulating boards of high quality, roof tiles, and cement insulation blocks for houses.

Rice Straw—is used for rope, thatch, and litter for poultry, then for growing straw mushrooms. Lye is made from the straw ash which is also used for preserving eggs.

3.3 MAIZE

For the purpose of this book, we are using the term "corn" to mean sweet corn and "maize" to mean field corn. Field corn is most often dried for later reconstitution, or dried and ground to make cornmeal, cornflour, or cornstarch.

Sweet corns are best eaten fresh, as the kernels shrivel when dry. Maize dries as plump grains for flours; wherever possible use unhusked kernels with germ (whole dried corn). If roughly ground to cornmeal, keep refrigerated and use in two to three weeks once ground. In maize it is the yellow corn which is high in vitamin A; the white corn is very low in this vitamin.

Storing Sweet Corn — Hopi

Boil firm sweet corn on the cob, three to ten minutes, until cooked, then dry thoroughly and store. Can be revived by soaking and warmed to eat. (Thomas Banyaca, *pers. comm.*).

Maize Porridges

All boiled maize or maize porridges have ash, lime, or limewater added to release protein and vitamins.

Ting (soured maize porridge) — Botswana

In Southern Africa, maize is a staple food, and it is most frequently consumed as fresh-ground meal made into a liquid or stiff porridge. This is often consumed thrice daily, with whatever side dishes or additions can be found at that season. The sour porridges can be heated for meals, and are served with fermented milk and sugar, or sauces.

Fresh porridge, or yesterday's cold porridge, may simply be mixed with water to ferment. A starter from a previous sour porridge aids fast ferment (to pH 3.5 in only twenty-four hours). More recently, culture starters plus calcium phosphate as a buffer are also added, or a little lime is substituted. People use agricultural phosphate and lime for this in Africa.

It is essential to use fresh maize meal; older meal gives butyric acid at too high a level.

The main flavour is from lactates, with a little alcohol produced by the yeasts.

Bubbling commences at 30°C (86°F) from twenty-four until fifty-two hours, when the ting is ready for consumption. The village ferment is complex, with eight groups of organisms (eleven species of *lactobacilli*; three species of *Saccharomyces* yeasts and three other yeasts; three aerobic bacilli; two species of *Staphylococcus*; two species of bacillus; two species of *Flavobacterium*; five species of moulds or slime moulds).

Care must be taken to keep the ferment at around 28–38°C (82–100°F). Off-flavours develop at low temperatures of 10–15°C (50–59°F).

Ferment Organisms Found in Nuka Boxes

Yeasts:	**Film yeasts:**	**Lactobacilli:**
Zygosaccharomyces nukamiso	*Hansula anomala*	*Lactobacillus cucurmeris fermenti*
Saccharomyces spp.	*Mycoderma* spp. (2)	*Lactobacillus lindneri*
Torula nukamiso		*Lactobacillus helormensis*

Mageu (sour porridge or field drink) — Southern Africa

Mageu is a common, almost daily, drink on farms, and resembles Nepalese field drinks. This is a very simple sour porridge, prepared in a few days. First a thin maize porridge is cooked, then cooled to 25°C (77°F) and a handful of wheat flour stirred in. This is allowed to ferment in the sun or a warm place for two days. It can be fortified with sour milk, whey, yeast powder, fish flour, pounded dry meats, or soy flour. Some cooked and pounded beans or bean meal would greatly add to amino acids.

Eight percent maize solids is considered optimum (the rest water and a little wheat flour), 1–2% sugar, malt, malt flour, or ground sprouted wheat will improve the ferment, but the porridge is low alcohol (1%) so sugars and honey are added as it is served. Starters can be carried over from one porridge to the next. Banana, pineapple, and guava are added as available, as flavours. The final pH should be about 3.1–3.9 (summer-winter).

The starters are kept as broth, and added at 1–2%. Buffering with calcium and phosphate salts has a beneficial effect (ash helps).

Ogi West Africa, Nigeria

Ogi is sold as a wet fermented starchy cake, based on soaked maize, millet, or sorghum ground in a wet state (never rice or wheat, traditionally). The cake is made into a slurry with cold water, and added slowly to boiling water as porridge, cooked with constant stirring for three minutes. This is eaten with fried cowpea flour, sliced fried banana, or steamed balls or cakes of cowpea flour (moin-moin). Meat, eggs, or fish are also added as available. Thick ogi can be moulded and cooled, thin ogi is used as a drink, especially for children or for field workers.

The grains are washed, soaked in water for one to two days, then wet-milled or ground (blended). Double sieving and grinding ensures a better product. Ferment one day for maize, longer for other grains, in a covered bowl. The excess water is poured off and the floury paste sold or used. It has a short shelf-life (two days). Any residues are fed to chickens.

Kenkey Africa

A fermented and cooked maize product of Africa. The best nutrition results from the whole grain, not husked grain. It closely resembles the Italian polenta or the Mexican tamales (see nixtamal in this chapter).

Combine 4 cups cornmeal with 1 tsp of lime and add to 4 cups boiling water, stirring rapidly into the boiling water to prevent lumps. Cool mixture to lukewarm, and stir in 2 tbsp yoghurt or 1 cup yeast barm (see also Worts in chapter two). Cover and leave overnight. Next day add 2 tsp of salt, then stir dough into 2 cups of rapidly boiling water, stirring constantly until the mixture thickens. Spoon into corn husks or squares of aluminium, fold over. Steam on a rack in a kettle for two hours. Cool slowly and eat hot or cold.

Raadi India

Soaked overnight and then ground and blended, maize kernels are well mixed with buttermilk or whey to ferment overnight, in the manner of idli (see idli in this chapter) and cooked as pancakes.

Sam!oti (! equals vocal click in indigenous language) Swaziland

Husked and shelled maize is stored below ground in large red clay pits; over the year, seeds near the soil become moulded and crusted. A hard cake a few inches thick lines the earth bins. This is broken out when the silo is empty, and is soaked, cooked, and served as a rather evil-smelling porridge, yet tests are aflatoxin-free. It is basically an emergency food, not all that popular today, but with no known ill-effects (C. Pemberton-Piggot, *pers. comm.*). Curiously, this mould layer is known to form a barrier to other pests of stored grains in earth pits.

Popped Maize India

Pour dry maize kernels (a little at a time) into an iron or earthenware pot containing hot sand. Use a sieve to remove the corn as it pops. Chickpeas are roasted in the same way, and soybeans are roasted in wood ash, as it helps in shelling the roasted beans.

Lotus seed and seed of *Euryhale ferox* (a prickly water lily) are also popped in sand; these latter are very popular and used for invalids.

Maize Porridge Relishes or Sauces from Africa

To add relish to maize-meal porridges, Africans serve small accessory bowls of the following. These relishes add protein, vitamins, or minerals to all porridges:

- Biltong or dried meats of domestic or wild animals, pounded to a paste.
- Roast powdered peanuts
  —with green spinaches (often cucurbit or bean leaf tips)
  —with the male flowers of cucurbits or of certain trees, boiled to a syrup.
- Powdered melon or cucurbit seeds (protein, zinc).
- Peanuts cooked with sweet potato.
- Tomato, onion and peppers.
- Dried caterpillars, crickets, termites, locusts.
- Honey and sugar syrup from soaked flowers, reduced by boiling.

- Prickly pear fruit boiled to a syrup.
- Pounded dried berries of *Grewia* spp.
- Oily seeds of nut trees pounded to pastes, or cooked and fermented.

Blue Corn Mush — Native American

In a large deep pan, bring 5 cups of water to the boil, and add few tsp bone powder*, 2 tsp baking powder, and 2 cups cornmeal. Stir continuously, adding more cornmeal (if necessary) until thick and smooth. Mush can be served spiced with onion, chilli peppers, or sweetened with corn sugar, honey, or syrup and milk. *Bone powder is made of the dried bones of deer, or fish, ground or pounded to a floury consistency.

Tortillas (an unleavened bread) — Mexico

Sift 2 cups fine maize flour and 1 tsp salt. Gradually mix in 1–1¼ cups warm water (with a little lime or wood ash) until dough is in a ball. Rest the dough for twenty minutes, pinch off walnut sizes and roll (well floured) very thinly. Rub a skillet with oil and cook for two minutes on medium heat, wrap in a cloth and keep warm in an open oven. Can be re-heated, a few seconds each side.

Kaanga — Maori

Mature maize cobs are steeped in running water in stone-weighted sacks until partially decomposed or fermented, then the seeds cut off and made into a porridge with a seasoning of ash. Alternatively, shelled corn is placed in a bag and soaked for up to two months (the water changed daily if in a pot or drum). When the corn is soft, it is ready. Mash 2 cups of this corn and add, stirring, to 6 cups boiling water; simmer gently for an hour, add cream and sugar to taste (kaanga wai).

Kernels may be mixed with grated sweet potato, and steamed in muslin or corn leaves for about an hour (salt and pepper; or butter, sugar and milk can be added to taste). The grains may be fried with salt and pork fat, or made into a gruel.

Maize cobs can be kept in water for weeks for this ferment.

Kaanga Wai (dried)

The "cured" (soaked and fermented) kernels can be minced or pounded, all juices pressed out, and dried over ovens or in the sun as cakes. (Most corn is cooked by the Maori with 4 cups of wood ash to 2 cups of white or yellow corn in 6–8 cups of water.) The hulls can be removed when loose, and the kernels cooked until soft and swollen.

Blue Corn Bread

Traditionally, save the ashes from juniper wood and sieve through a basket or colander to remove large particles. Mix together ½ cup juniper ash and 2½ cups boiling water then add 3 cups of cornmeal. Mix until you have a soft, firm dough. Shape into a loaf and bake in an oven at 180°C (356°F) for one hour. Traditional mud "beehive" ovens are used in the pueblos of Central America and south western USA.

Navajo Paper Bread

Mix 1 cup juniper ashes with 1 cup boiling water and strain through a cloth or colander (or grass wad). Stir this liquid into 3 cups of boiling water and add 1 cup cornmeal. Mix to make a thin batter, cool. Heat an oiled griddle iron or smooth stone and pour on by hand as thin a layer of batter as possible. Do not turn, but use a spatula to flatten and remove this bread. Continue until all batter is cooked.

Nixtamal (lime-soaked hominy)

Boil together 1 cup of hydrated lime, 15 quarts of water, and 5 quarts white maize grains until maize hulls loosen on the kernels (about an hour). Wash, rubbing kernels between the hands, in cold running water until water is clear then drain.

This mix is stored frozen in plastic bags, but kernels can also be fully dried, coarse-ground and re-heated in water for a breakfast porridge (hominy grits), or stored dry. Thawed nixtamal is ground in a corn grinder for tamales (steamed cornmeal in corn husks) with a variety of fillings (meat, spices).

Ground soaked corn can also be refrigerated or frozen to make tamale masa (corn mush with salt, some lard, meat stock). After cooking, the dough is ready when a test lump floats in cold water. Tamale dough is formed around a chilli-meat mix, tied tight in the husk, and steamed. Tamales are often re-heated to serve in green or soaked (dry) cornhusks.

In southern USA, steamed hominy grits (coarse-ground nixtamal) are often served as a "volcano" of grits, the crater filled with maple syrup and two fried eggs, and the base circled with links of sausages, or Mexican chorizo sausage. It looks very startling on first encounter, but this is a substantial southern American breakfast which I enjoy at traditional cafes in South Carolina.

Hominy Muffins

Mix together 2 cups cold boiled hominy (beaten smooth), 3 cups yoghurt or sour milk, 2 tsp salt, ½ cup melted butter, 3 well-beaten eggs, 2 tbsp sugar, 1 tsp baking soda in a little hot water, and 1 large cup wholemeal flour with a little bran. Beat all ingredients well and cook quickly in a muffin tray in a medium-hot oven. Blueberries or small fruits in season can be mixed into the dough.

3.4 WHEAT

Wheat Sprouts

Wheat sprouts are often used in baking bread:

- Sprout for three to four days, grind and mix 3:1 with flour for sweet doughs.
- Sprout one to two days, dry, roast and grind for sweet flour. Use 1 part sprouted wheat for 3 parts of plain flour. Such flours are used for sweet breads and biscuits.

Bulgur — Middle East

Remove any dark or shrivelled grains or stones from 2 cups clean wheat and sauté briefly in oil. Add 9 litres (19 pints) water and a little salt. Boil for forty-five minutes or until kernels are tender. Drain off any liquid, but save for soup stock. Spread kernels on sheets of metal and either bake at 120°C (248°F), sun-dry, or stir in a wide fry pan until completely dry (about sixty minutes). Grind or steel-cut coarsely and store.

To reconstitute, just soak and serve cold with parsley, chopped peppers, onion, tomato, chives, 4 tbsp olive oil, lemon juice, pepper, heat briefly in soups, or make into pancakes (soaked overnight) with blended pulse pastes. Note that cooking a large amount of wheat conserves cooking energy in subsequent dishes, and as bulgur keeps well, it is a good idea to cook a lot at once.

Couscous — North Africa

Aubert (1986) records the preparation of couscous (after Gast and Sigaut, 1979) as follows:

On a large brass tray, women place a large amount of coarse-ground grain, spray or sprinkle it with salty water and stir this over with their hands. The plate is inclined towards them at about thirty degrees, and the women scrape the bottom of the tray as though raking it in order to evenly spread the water throughout. After a while, they continue to stir the mass with one hand and with the other add a powdering of cowpea flour, then a sprinkle of water, until the mass reaches the bulk required, and the coarse flour adheres in particles two to three mm (⅒") diameter.

Now the couscous is passed several times through a sieve until the clots or particles of wheat flour are evenly sized. The coarse grain is at the centre of the particles, then flour the exterior of the little balls, which are now spread out on sheets and put to dry in the sun six to seven days. They are hung in linen or cotton bags in the sun for several days. This material can be kept several months, even years, stored in jars or metal boxes. To cook, they are lightly sprinkled with water, and steamed until cooked.

See also under grasshoppers in chapter 9 for what is probably the original recipe.

Frumenty (whole wheat grains) England

Boil 1 cup wheat grains briefly in 3 cups boiling water, then put in a haybox cooker or leave in a warmed, closed oven for twenty-four hours. By then the grains should have burst, and the wheat set as a creamy jelly. Milk and eggs, honey and lemon rind, spices, fruit and cream, etc. are variously added. Frumenty is normally eaten as a breakfast or supper.

Hopper South India

Combine 2 cups wheat or rice flour, 1 cup coconut water, and yeast into a dough, then cover and let ferment over twelve to fifteen hours. This dough is thinned to a batter with coconut milk, a little salt and sodium bicarbonate. The batter is cooked in a covered round-bottomed oiled pan for three to five minutes, giving a soft centre and crisp edges, and a steamed top. Yeasts provide vitamin B (Campbell-Platt, 1987).

Billos Australia

Soak overnight (or better, for twenty-four hours), any mixture of split peas or lentils (1 part) and grains (2 parts), especially rice, buckwheat, bulgur wheat, barley. In the soak water put a spoon of lime or bicarbonate of soda. In the morning, blend the mixture (in an electric blender) with 1–2 chillies, 1 egg (shell and all) and cook in a lightly oiled pan on medium heat. These puff up and are very nutritious pancakes. Turn and brown each pancake, keeping covered before turning. For main meals, minced squid, flaked fish, or chopped vegetable fillings can be added.

Sweated Wheat

In 1888, a contributor to the *Farmer's and Housekeeper's Cyclopaedia* (republished facsimile by the Crossing Press, Trumansberg, NY) speaks of the harvesting of wheat "in the dough state", when basal leaves are green and the heads unripe. Many farmers stacked and "sweated" such wheat before threshing, when the damp stacks heated up and were left to cool and dry out. The sweated grain was of good colour and plump, having absorbed nutrient matter from the stalk and husk.

This is the same method used in the rice-composts of Nepal (see Hakuwa) to increase vitamin content. It would certainly be worth trials today to duplicate this technique and analyse the vitamin content of sweated versus ripe dry wheat.

Kishk (after El-Gendy) Egypt, or known as Trahana in Greece

Cracked steamed wheat is boiled with sour milk (yoghurt or curds) and sun-dried. They are "...extremely nutritious stable foods which can be stored for years without deterioration." (Natural Science, 1979).

Whole clean wheat grains are boiled until soft, then sun-dried and ground. Bowls of this wheat are then dampened with boiling salt water (3–5% salt). Then soured milk, or diluted yoghurt is mixed in to make a creamy mix or paste, which is fermented under thick cloth for twenty-four to forty-eight hours, formed into balls or cakes, and sun-dried two to three days on mats. Ground pepper may be added to the dough, and the dried cakes heated in an oven to improve storage. Kishk is stored, like grain, in large jars or silos.

For strict vegetarians, the sour milk in kishk is replaced by liquids pressed from fermented gourd pulp *(Lagenaria vulgaris)*. The squash flesh is boiled, fermented for a month in closed jars, and filtered to provide the sour liquid for kishk.

The dried cakes keep indefinitely and are used after soaking in soup, stews, or for voyages.

Shamsy (after El-Gendy) Egypt

A starter (also known as yeast) is made by mixing a small amount of yeast dough with about 2 kg (4½ lb) of flour and 454 gm (1 lb) of water, left overnight in a warm place to rise.

Next day, the main dough is made from wheaten flour, a little salt, water, and the starter. This is mixed (about twenty-five minutes), left to rise in a warm place (forty-five minutes), divided into small loaves or rounds of about 454 gm (1 lb), then placed on bran or flour over boards or mats for sixty minutes in a sunny or warm place. These are then flattened into rounds about 20 cm (8") across and left to rise for sixty minutes in the sun, then turned over for thirty minutes in warm indoors. Beehive ovens or mud brick ovens are used to cook these.

Wheat Flakes

Cream together 1 cup whole wheat flour, 3 tbsp of peanut butter, and ½ tsp salt. Add ¼ cup of cold water a little at a time, continuing to mix until dough is springy and leaves the bowl. Divide, roll out paper-thin, bake fifteen to twenty minutes at 120°C (248°F). Turn the oven off and leave to cool and dry until very crisp. Crumble down and store in airtight containers. Serve as a cold porridge with milk, fruit, or yoghurt.

Noodles

Beat 1 egg slightly, then add ⅔ cup flour and ¼ tsp salt, and mix to make a stiff dough. Knead, stand covered for ½ hour. Roll out very thin and spread on a cloth to dry. Correct moisture is not a bit sticky and not so dry that the dough will crack or be brittle. The best flour to use is a hard durum wheat.

Roll pastry in a tight roll and cut into thin rounds. Unroll, place each strip on top of the next in a stack. Cut down through this stack in thin strips. Toss lightly to separate and spread on a board to dry. When thoroughly dry, store in sealed jars for future use.

To cook noodles, drop into boiling soup five minutes before serving.

Matzos — Israel

Make a stiff dough from flour, a little salt, and water. Roll out thin and perforate with a spike wheel. Bake in a brick oven until crisp. Store in a dry place. Used as crumbs or in egg dishes, as thickening, or with spreads.

Breads and Sourdoughs

Gluten (wheat protein)

Mix 4 cups wholegrain or unbleached flour with 1½ to 3 cups lukewarm water and form into a stiff ball. Cover with water for two hours. Then, still keeping the dough under water, knead it to work the starch out. Pour off the starchy water and replace several times, kneading the ball until the water is almost clear.

The lump of spongy gluten remaining is broken up, fried, boiled or baked. The cooked balls can be minced, cut into strips, or cut into halves and quarters; gluten is added to vegetarian dishes for protein. Gluten balls deep-freeze very well, and can be added to soups and stocks deep-frozen. Frozen gluten can be sliced, dried, and powdered. Waste starches can be added to livestock feed.

Bread

Holloway (1981) speaks of the soft, low-gluten content of European wheats, and the necessity to use a gluten flour to achieve a high volume, light, firm dough. Yeast expands the dough because the gluten in flour stretches. Well-moulded glutinous bread produces soft flakes when broken open.

Mrs Kander (1945) notes that light risen doughs are made in four ways:

- By using yeasts or sponges (liquid, cake, or dried yeast granules).
- Using baking powder (sodium bicarbonate).
- By using baking powder plus molasses or sour milk (yoghurt or whey).
- By beating until the mixture is aerated.

The Long Dough

The very best yeast breads have an eight to fourteen hour first rising (overnight). This long ferment gives the "bread" flavour to the product which is totally missing from short doughs (one to four hours), and thus few people have tasted real bakers' bread today. Before 1960 it was illegal to make bread with a dough less than eight hours fermented. It was also illegal to sell hot-baked bread, so that all bread was cooled overnight. As a small town baker, I would catch my horse to deliver the bread just after dawn. When I got back to the bakery; I would then set the dough in the morning, and (by nightfall) leave the new bread to cool. Then I usually went to a party!

Salt is added to all bread sponges, but sugar is never used unless to catch or increase a yeast. Malt, however, can be used to increase yeasts, and malt or molasses is often added to wholemeal breads, together with bran. Such breads can be 30% white flour or unbleached flour to improve the dough consistency. Added gluten flour assists ferment and texture. A long dough bread is given next.

Scotch Batch Bread (a long dough)

Prepare ingredients:

32 kg (70 lb) flour, 45 gm (1½ oz) yeast, 0.75 kg (1¾ lb) salt, 15 litres (4 gal) water, and 50 gm (2 oz) malt.

- Stage one—"Sponge" of yeast. All of the yeast is mixed with a quarter of the flour and water and 50 gm (2 oz) of the salt; the water is at 21°C (70°F). This sponge is left for fourteen hours (overnight) in a warm place. Traditionally, it is left covered with flour bags in a wooden trough, with a wooden lid, the trough about one third full at the start.
- Stage two—The sponge is broken down with 10 litres (21 pints) of water, 12 kg (26½ lb) of flour, 0.25 kg (½ lb) of salt and the malt. This produces a batter sponge; it is left to ferment for one hour.
- Stage three—The remaining flour, salt, and water are added and this dough left at 26°C (79°F) for an hour, then kneaded and moulded, and put in tins or deep trays to rise for final "proof". It is then off to the oven at 230°C (450°F), falling to 148°C (300°F) over one hour.

Yeast Dough Rolls

To make dough, mix together 1 litre (2 pints) hot water, ½ cup of milk, 1 cup mashed potato, and 1 tsp salt in a large bowl, stirring in 60 gm (2 oz) compressed yeast when lukewarm. Stir in 3 litres (6½ pints) flour (white, wholemeal, rye or mixed) to make a thick batter. Let this rise, then add flour and mix until elastic and light. Pinch off dough, let them rise slowly, bake at 220°C (428°F) for ten to twenty minutes or until done. Makes four to five dozen rolls.

Optional: beaten egg, sugar, or shortening (butter or lard) can be rolled into the dough with a little extra flour for special breads.

A risen dough can be kept in a covered crock in a refrigerator for a week. Bits can be pinched out to start bread "sponges" or to cook for rolls as steamed, fried, or baked rolls as required.

Nan (fast bread) India

Nan or naan is flat bread made with yoghurt. If no brick oven is available, use grill or an iron frying pan, very hot.

Mix together and sift 450 gm (15¾ oz) of white flour, 1 tsp baking powder, and ¼ tsp salt. Slowly add 425 mL (14½ fl oz) of yoghurt and mix well to a soft dough. When a soft dough is reached, it is left to rise in a warm place for two hours then is re-kneaded and made into ten balls, lightly rolled 5–6 mm (⅕") thick.

Each "pancake" is slapped into a hot pan and turned, or placed under a grill for two minutes.

Remove nan and brush with melted butter; stack in a clean cloth and serve, or briefly re-warm in an oven. Nans puff up in four to five minutes in the pan, when they are turned or grilled until the surface shows brown spots. (after Jaffrey,1983).

Breading or Crumbling

The only reason I keep corn or wheat flakes in my cupboard is to crumb cutlets.

However, bread is the best crumbing agent. Dry bread scraps in a cool oven to just brown-edged, then blend or mince. This can be stored in jars.

Lean meats are often juicier when lightly floured, dipped in beaten egg with a little soy sauce, and pressed into breadcrumbs, rolled oats, or crushed breakfast cereal. The essential in all such casings is to have very hot oil in the pan, and to cook well on one side before turning over.

3.5 SORGHUM

Sorghum lacks gluten and is not suited to yeast breads; use soda for breads made from sorghum flours. Some sorghum varieties are essentially perennial, and can produce a ratoon crop for many years in succession. If milled, sorghum flours are often mixed with dried cassava or cooked and dried sweet potatoes, and with flours from less bitter grains, although to me the purely sorghum porridges are not at all bitter, and are in any case eaten with soured milk and sugar.

Sorghum Porridge

In the Kalahari, water is scarce, and around homesteads a large quantity of Marotse melons, some very large, is grown. The pulp of this melon is used about 2:1 of pulp to sorghum meal, and makes a nutritious porridge if a handful of pounded melon seeds is added (Dorothy Ndaba, Gao Clark, *pers. comm.*). Served with sour milks and a little sugar, this is my favourite breakfast porridge.

Main Sorghums

- Sorgos—sweet stalks have luxuriant foliage for cattle or silage. Sugar can be extracted from this sorghum by crushing, much like sugar cane.
- Grass sorghums (Johnson and Sudan grass) for hays.
- Broom sorghums are common in south west Asia for brooms and grains.
- Grain sorghums (milo, churra, hundreds of local varieties). Only the dried leaf is free from prussic acid, so these are used for forage, while the stalks are left in the field, standing, to dry. The heads are first dried, then threshed for grains.

Furah — North Nigeria

Sorghum meal is sieved, wetted, and formed into 4 cm (1½") balls. It is then steamed until gelatinised, rolled in a little flour to part-dry, and dried.

These balls are crumbled into sour milk for a porridge, which gives it a refreshing taste.

3.6 BARLEY

Barley was, in early Japan, the main grain used for miso, hence also probably for "soy" sauce. It still is used today but is falling out of favour as white rice misos, sweetened and of short ferment, predominate. This is probably due to modern advertising, and the "white is higher status" or "white is right" idea that produces white sugar, white flour etc. to our loss in nutrition, but for a better profit by multinationals. White, in this sense, is wrong. Two of my friends at university law school (John Charles White and Aloysius Wong) have, or had, the famous law firm White and Wong. They could well debate the case of white and wong foods.

3.7 RYE

Kvass

Ryebread, often toasted, is crushed into water or cold tea to ferment with lactic acid bacteria (a little whey from yoghurt can trigger the process). Some sugar can be added. Once kvass is made, some can be saved as a culture. This is the field drink of Russia and Poland. Kvass can be used in sourdough breads.

Sourdoughs of Rye Bread

Prepare ingredients:

4 cups rye flour, 1 pint hot potato water*, 43 gm (1½ oz) compressed yeast or 1½ cups of kvass, 1 tsp caraway seed, 2 cups wheat flour, 1 tbsp salt, 1 cup riced (grated) cooked potatoes, drained and mashed.

* Potato water—Peel or scrub 2 potatoes (to six cups of flour), cook well, drain off water; mash the potatoes and add to water.

Cool potato water to blood heat, dissolve yeast (or add kvass) then stir in the flours to make a smooth, elastic dough. Let rise until doubled, then punch down and form into loaves. Bake at 190°C (375°F).

Keep one cup of the dough in an earthenware crock and in subsequent batches use ½ cup of this fermented dough to the lukewarm potato water, instead of yeast. Whey can be added to cooled potato water for *lactobacilli*.

3.8 OATS

Porridge

Porridge is made from rolled or pounded grains, chiefly oats, maize, wheat, sorghum, or millet, boiled in 2–2½ times its bulk of water, with a tsp of salt per 250 gm (9 oz). Porridges are often served with yoghurt and honey, or sugar and milk, or (in colder and hardier climates) with salt, pepper and butter. Today "fast oats" are steamed before rolling and cook in a few minutes.

Oatmeal Jelly

Cover a cup of coarse-ground or steel cut oats with water and stand for a few hours. Simmer for thirty minutes, with a pinch of salt, strain, add honey and lemon juice, to taste, and a pinch of cloves. Pour into moulds and cool, serve cold with small fruit, yoghurt, cream. (See also frumenty.)

3.9 THE MINOR GRAINS

Millet (Injera) Ethiopia

Millet or teff (*Eragrostis tef*) flour is made into a light batter and fermented. Each batch gets a starter from the previous batch, so start with a yoghurt-yeast mix and go from there.

Dissolve 1 tbsp yeast and 1 tsp honey in a cup of warm water. When this is well blended, mix in 4 cups of water then stir in (briskly) 3 cups of millet or teff flour. Ferment twenty-four to forty hours and stir in ¼ tsp soda just before cooking.

Test the skillet to cook without browning. The batter is poured on a griddle in a closing spiral, and gives a flexible pancake bread which is used in the same way as tortillas or chapattis. Cover pan while cooking, about a minute a side.

Millet Malt

Pearl millet grains are steeped in running or renewed water for twenty-four hours then soaked for seventy-two hours at room temperature. The germinated seed is dried in the sun or in air at 40–45°C (104–112°F), and then powdered. The enzyme diastase is extracted with distilled water. Either the ground seed or the powder is used as a malt in beers or sour porridges.

Shoshone Rice-Grass (*Oryzopsis hymenoides*)

Rice-grass is roasted as it is gathered, by flaming the dampened, dry stems over a deerskin or metal sheet. Then the ash is sieved out before rice grain is stored in baskets. This method can be used for other small seeds. Remarkably, the seed can be buried 20–30 cm (8"–12") in dry sand dunes to germinate, or covered in gardens with compost. Rice-grass is a nutritious forage grass and arid-land grain, quite perennial and the seeds very nutritious. Their "milk" or gruel is used to feed orphaned babies and invalids. (Gus James, *pers. comm.*)

Wild Rice

Wild rice (*Zizania lacustris*) is not a "rice" but rather a grain which grows in ponds and lakes in cold climates. It is native to Canada and north central USA, but grows well in Tasmania.

Storage is two-fold, by first fermenting in the husk and then drying and roasting for food storage in a large pan or revolving drum, until the seeds are dark-brown to black. These seeds are then suitable for eating, not for growing. For seed intended for next year's crop the fresh seed is bagged in hessian or cotton, and weighted to sink below the icy waters of the winter ponds, or in a cold water container in your refrigerator, or at least in a shaded or cool cellar. Seed is broadcast-sown in early spring in shallow flooded fields.

As seeds ripen over three to four weeks, heads can be tied together or even bagged over this period, and cut when all are ripe. Traditionally, canoes are driven into the stand and the seeds beaten off every three to four days. In machine harvesting most of the rice will be lost, but a "best guess" is made of the stage where a lot of seed is ripe and ready to shatter. Wild rice, roasted to keep, is cooked as for rice. Unhusked rice is fermented in covered heaps until dark-brown or black before threshing and roasting or drying for sale, like composted and parched rice or wheat.

At home, grow each grain in a tall pot stood in shallow water. Those yield about 300 gm of rice per pot, and 1100 pots would provide a staple for your family, with little waste due to machine harvesting.

The sheaves are harvested at 30–60% moisture and before husking are fermented six to twelve days, when the grain darkens. The paddy is then dried to 2–10% moisture in the sun or in kilns, and the grain husked for sale. Protein is 15–20% of dry weight. Wild rice is our most recently domesticated grain and only a very few selected varieties exist. It has great promise for shallow ponds 30 cm (12") deep in home gardens in cold climates.

Sesame

Tahina (sesame seed) — Greece, Arabia

Crush 1–3 garlic cloves into a little salt and mix with the juice of 3 lemons. Separately blend ½ cup sesame seed with olive oil to make a paste (tahina paste) and add the lemon juice and garlic mixture, gradually adding cold water to make a thick cream. Add ¼ tsp cumin, and taste to adjust with lemon and salt.

This cream is the basic tahina, sprinkled with parsley. Serve flatbread to dip out tahina. Some people add a little mustard, or 5 tbsp of ground almonds. It is a good variation, as in the Sudan, to replace the water with ½ cup yoghurt, and in Turkey 113 gm (4 oz) of walnuts are ground with the salt.

Tahina is served with fish, mussels, and salads or plain steamed vegetables, or just as is.

Sesame-Honey Preserves (also almonds, walnuts, rolled oats, dried fruits, pistachio, etc.)

- Caramelised sugar, syrup or honey, reduced by forty-five minutes boiling and thickening is poured over grain or nut clusters on an oiled plate, a marble bench, or on tiles, and allowed to set. These sweets are stored in jars.
- Grain or toasted grains are mixed in to reduced honey or sugar and poured onto oiled slabs, rolled out, marked into squares when warm and broken up when cool.

Bullrush Cumbungi (*Typha angustifolia*) — Maori

Bullrushes are harvested twice, once at the pollen stage, and once when the seed is ripe and the fluffy heads break up. The fluff is pressed into baskets and baked to remove insects, then used for insulation, quilting, pillows, etc. Fresh shoots are used as salad vegetables, and roots grown for winter range of pigs, in drained fields. A cornflour or starch can be made from pounded, washed roots.

Pollen (pua) is shaken from the heads into baskets. For every pound of pollen, use about ½ cup of cold water, and steam this dough in a greased bowl for about two hours. This bread can be eaten hot, or sliced and dried, and stored.

Traditionally, the seed heads (raupo) are gathered into sheds in the early morning, before they blow about, over several days. They are sun-dried to cure the seed, then stripped, bagged in fine-meshed bags, and shaken or beaten periodically to remove the fluff from the seed. The seed is poured into baskets or basins from a small hole in the bottom corner of the bag, leaf-wrapped, and cooked in earth ovens as needed over twenty-four hours. The seeds are then in a mass, usually cooked as cakes 10–20 cm (8") across and 2–4 cm (¾"–1½") thick, which were again dried as supplies for war or for hard times.

Seagrass (*Zostera*)

In Baja California, *Zostera* seed heads float in great drifts, and the Seri Indians gather, dry, and thresh them on deerskins, roast to separate husks, and winnow the seed after roasting and crushing. The flour, sometimes combined with cardon seed from the giant cactus groves there, is cooked with sea turtle oil or honey. Dried and part-composted *Zostera* stems have long been used as house insulation, as have the pithy stems of rushes.

Sprouts

Aubert (1986) notes that dry lucerne and wheat sprouts are superior to fresh soy sprouts, and records a doubling of available minerals and more than that in vitamins, lucerne again having a higher content of vitamin C (176 mg per 100 gm). Both wheat and lucerne sprouts are dehydrated (to 10–12% water). While legume sprouts reach their highest vitamin A content in three and a half days, cereals take longer (six and a half to seven days) to reach peak vitamin A content.

CHAPTER FOUR

The Legumes

4.1 INTRODUCTION

Several classical ferments have all or part of their mass as legume seeds. Miso, natto, soy sauces, dosas, idlis, and root crop ferments are all products of legume and seed mixes with other starchy bases.

Many of them are fermented with a mould (koji), salted, then aged with other starches. A few ferments are completed in only twenty-four hours; after an overnight soak, but all represent an increase in food value, as amino acids from the legume seeds are combined with starches from grains or roots. The longer-term ferments supply yeast, lactic acid, and amino acids.

Soybeans are just one of a vast array of possible legume bases. Miso, for example, can be made with peanuts, chick peas (garbanzos), broad beans, common beans, cowpeas (black-eyed beans), lima beans and so on. In temperate areas, soybeans are out-yielded by field peas and comptesse beans. Korea, for example, now imports large quantities of white lupin seed from Western Australia for making miso, tempeh, and "soy" sauces.

Soybeans and their products must be thoroughly cooked after soaking (four to six hours) or pressure cooked (twenty to thirty minutes) to destroy the trypsin inhibitors (TI) that prevent the trypsin enzyme (from the pancreas) digesting protein. While cooking alone may reduce TI 80–90%, the ground, cooked bean meal, curded, further loses TI in the whey.

Thus, soybeans need a lot of cooking (hence fuel) and perhaps should only be cooked in pressure cookers or in solar cookers. It also makes economic sense to cook soybeans in larger processing centres for many households (small factories) so that fuel is conserved. It is a good idea, when designing bean and grain croplands, to add firewood plantations at the margins so that there is enough fuel wood to cook the food.

Dried legumes are best processed to tofu, miso, tempeh, etc., as large batch quantities suitable for a village or region, just as a district bakehouse, walk-in freezer, washing machine or food processing centre is most economical in energy terms.

Legumes are the ideal complementary food for grains. Sprouting legume seeds also makes an edible salad addition, and is a way of helping detoxify legume seeds. Sprouted pulses are very common in eastern foods. They are soaked for twelve hours, drained, tied tightly in a muslin cloth, and hung for twenty-four hours before adding to salads, soups, stir-fries. If not tied in muslin, they are lightly pressed in a jar or box to thicken the sprouts.

Well-dried legumes, stored in ash or diatomaceous earth, are viable as seed for several years, and are often kept by households in gourds or paper bags hanging from rafters to avoid rats and mice. Parched seed, miso, or in extracts such as soy sauce are good long storage foods.

Garden and Crop Legumes (Family *Fabaceae*)

African Yam Bean—(*Sphenostylis stenocarpa*) Seeds to ferment, roots raw or cooked. Annual climber.

Alfalfa—(*Medicago sativa*) Sprouts, leaves powdered for adding to soups, sauces. Perennial herb.

Australian Pea—(*Dolichos lignosus*) Peas and pods eaten at all stages, split for pulse. Perennial climber.

Azuki Bean—(*Vigna angularis*) Beans to sweet pastes (cooked); sprouted, young pods as vegetables.

Bambara Groundnut—(*Voandezeia subterranea*) Seeds fresh, boiled, or dried roasted, ground, popped. Annual herb.

Black Gram—(*Vigna mungo*) Flour for papadoms, idli, dosa. Annual field crop, herb.

Broad Bean—(*Faba vulgaris*) Unripe seeds as vegetables, ripe seed to flours, soy sauce, tempeh. Sprouted. Annual herb.

Chick Pea—(*Cicer arietum*) Bengal gram, channa dahl; flour is besan. India. Parched, pastes, sprouted. Many cultivars. Herb, annual.

Cluster Bean—(*Cyamopsis tetragonolobus*) Green pods, dry seed, sprouts.

Cowpea—(*Vligna unguiculata*) Green pods and seeds cooked. Dried seed to flours, fermented, sprouted. Climber. Cultivars include yard long bean.

Horse Gram—(*Macrotyla uniflorum*) Seeds parched, boiled, fried, flours.

Hyacinth Bean—(*Lab-lab purpureus*) Green pods, shoots eaten cooked. Dried beans to tofu, ferment.

Jackbean—(*Canavalia ensiformis*) Immature pods eaten, seeds fermented to tempeh. Unripe seeds always cooked. Climber.

Jicama—(*Pachyrhizus erosus*) Grown for crisp tubers, stir-fries, starches. Annual herb.

Kidney Bean—(*Phaseolus vulgaris*) Young pods raw or steamed. Dried bean used in every way. Sprouts. Many forms and cultivars. Green beans, bush and climbing types.

Lentil—(*Lens culinaris*) Young pods fresh or cooked, seeds cooked, dry seeds boiled, pastes. Roasted, to flours. Annual herb.

Lima Bean—(*Phaseolus lunatus*) Immature or dried beans cooked, fermented, sprouted; young pods cooked. Vigorous climber.

Moth Bean—(*Vigna acontifolia*) Young pods cooked, dried seeds as a split bean. Drylands. Annual.

Mung Bean—(*Vigna radiata*) Young pods and seed cooked. Seeds a well-known sprout; parched and ground, fermented, flour starch to cellophane noodles. Annual.

Peanut—(*Arachis hypogaea*) Nuts raw, boiled, roasted, pounded as "butter", oils, annual herb.

Peas—(*Pisum sativum*) Eaten green, dried; as flours, sprouted, soups. Many cultivars. Bush and climbing varieties.

Pigeon Pea—(*Cajanus cajan*) Green pods, tips of shoots, dried seed eaten; sprouted. Perennial shrub forms.

Rice Bean—(*Vigna umbellata*) Green pods. Dry seed as a dahl or sprouted. Annual field crop, after rice.

Scarlet Runner Bean—(*Phaseolus coccineus*) Most favoured green bean of cool areas. Semi-perennial climber.

Siberian Pea Tree—(*Caragana arborescens*) Young pods and dried seeds cooked. Shrub.

Soybean—(*Glycine max*) Green, dried, fermented, cooked, sprouted; numerous ferments. Roasted, as flours. Annual herb.

Sword Bean—(*Canavalia gladiata*) Young pods eaten, pickled. Seeds always cooked. Climber.

Tepary Bean—(*Phaseolus acutifolius*) Arid area bean, dried beans well-cooked, parched, ground. Vine, can climb in shrubs; sprawls.

Velvet Bean—(*Mucuna pruriens*) Young pods as green vegetables. Seeds to miso, tempeh. Vigorous climber on trees.

White Lupin—(*Lupinus alba*) Sweet cultivars. Pods and green seeds boiled, seeds to flours, various ferments, green manure. Annual herb.

Wild Mung—(*Vigna vexillata*) Tubers eaten raw, cooked; 15% protein.

Winged Bean—(*Psophocarpus tetragonolobus*) Green pods raw or cooked, dried beans in all ways: roasted, flours, ferments. Roots raw or cooked, 20% protein. Vigorous climber.

Roasting Pulses

Wash and soak large legume seeds (chick peas, soybeans) overnight, then dry and leave in the sun for two hours to lose all external moisture. Deep-fry to a golden brown. Drain the oil off thoroughly, at first via a sieve and then on paper towels, and dust the cooked grains with salt, chilli, pepper, cumin etc., for flavour. Seal in packages, plastic, cellophane, or in jars, to eat as needed.

Parching

Throughout South East Asia, barley, chick peas, groundnuts, and nuts have long been roasted by stirring them in hot sand. The sand is first carefully washed, sieved, and graded so that it is easily sieved from the grains to be roasted. Then it is heated in a shallow iron pan or wok, and the grains or nuts stirred in until evenly browned and roasted. This prevents burning by contact with the pan. The roasted material is easily sieved from the sand, and this is done in many wayside stalls or cafes in India using a brass-wire sieve with a wooden handle.

Husking Hard Pulses—Variation One

To husk and split pulses: seeds are sun-dried for three days, cracked in a stone mill, soaked or sprinkled with sesame oil, then finally split in a stone mill. Hulls are winnowed to clean. 18 kg of split seeds (dahl) is recovered from 40 kg (88 lb) seeds.

Husking Hard Pulses—Variation Two

Seeds are soaked six hours, drained, mixed with fine clay (sieved), heaped at night and dried by day (two to three days), then cracked in a stone mill, yielding 31 kg (68 lb) dahl from 40 kg (88 lb) seeds.

Dahl is stored dry, keeps well. If mixed with oil, it keeps one to two months.

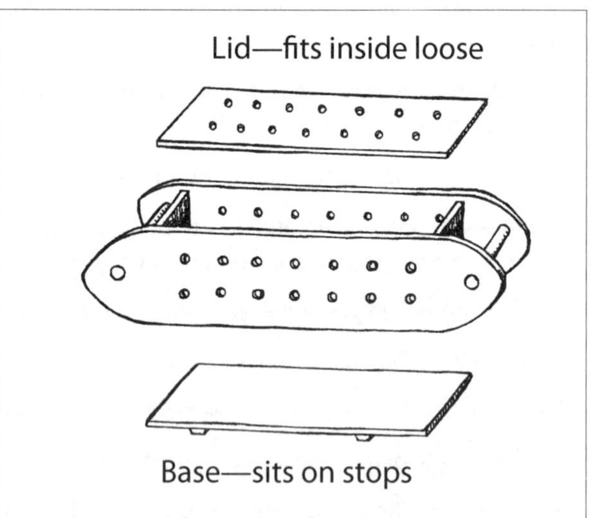

Figure 4.1—Tofu curd box for draining and water immersion. Turn box upside down to deposit drained curd in saline water.

4.2 SOYBEANS

Soybean Products

Soy Milk

Soy milk is used as a milk substitute, and is also the prelude to making tofu (bean curd).

Soak 454 gm (1 lb) of soybeans for ten hours, then blend, in an electric blender with an equal volume of water, to a smooth paste. Do a little at a time to avoid overheating the blender.

Pour the purée into twice the amount of boiling water and bring the purée to the boil.

Take off the heat and pour into a clean old pillowcase in a colander. Press out as much milk as possible (use heavy plastic gloves so you don't burn your hands!) into a big pot. Twist and press the fabric until the bean meats are fairly dry (this okara can be used in breads and casseroles, or fed to chickens and pigs).

Pour all strained milk into a pot, bring to boil and simmer ten minutes to take away the bean taste. Cool the soy milk and refrigerate. A little salt and sugar can be added for taste, if necessary.

Tofu

Make soy milk, as above, without any added flavouring. Bring soy milk to 93°C (200°F), then take off the heat and stir in a solidifier.

You can use any one of the following to cause curd formation in the milk: (quantities for each 454 gm [1 lb] of beans):

1½ tbsp nigari, available from a natural foods shop; 1 tbsp gypsum (calcium sulphate), which is the most popular curd-forming salt used today; 1 tbsp Epsom salts (magnesium chloride) or calcium chloride; 6 tbsp of lemon juice or vinegar; sour soy milk; vegetable rennet (e.g. fig juice or nettle juice) or rennet.

Nigari and gypsum will give the most tofu.

Stir in one-third of the solidifier while gently stirring, then stop stirring, hold spoon upright for a moment, and remove spoon from the still liquid. Then sprinkle one-third of solidifier over the surface of the milk. Watch curds form for three to five minutes, and then gently cut (with a spoon or knife) down into the pot and look to see if all the milk has formed curds. If not, gently stir the remaining one-third of the solidifier into the surface of the milk. The milk should now be in curds floating in a clear whey. Ladle out most of the whey and keep to use in soups.

Wet a cheesecloth in whey and line a large colander or perforated wooden boxes (see Figure 4.1) with cheesecloth and, very gently, ladle curds into these. Then fold cheesecloth across the top and weigh down gently.

Either gently press out the remaining whey from the curds, or leave six to eight hours to drain. Then in a large laundry tub or baby's bath, invert the box or colander and gently fill bath with cold water, then remove mould box (with the lid now under the tofu). Gently remove the cheesecloth, lift out the block of curd, and store in the refrigerator in slightly salted water in covered containers.

Use fresh or within the week.

Tou-Fu-Ru China

Wang (in Hesseltine, 1986) notes that tofu (Tou-fu) curds have been made since 179 BC in China. To further ferment tofu to Tou-fu-ru the tofu cakes are steamed, then covered with straw in incubation trays at 10–15°C (50–60°F). Three or four days suffice to produce a white mould, when the inoculated cubes are kept in salt and water to age. To this can be added rice wine, red rice (see yeasts and moulds), chilli, peppers, or ginger for flavour. *Actinomucor* and *Mucor* are found in this ferment, provided by the straw. I would presume that 16–20% salt solution would be used, as for cheeses, and that the fermented tofu is kept in sealed pots.

Miso (Bean-Grain Ferments)

Miso has an average 15% complete protein, and is the basis for most Japanese soup stocks. Traditionally, barley, nowadays rice or wheat, are variously mixed with steamed whole soybeans to make miso. Beans may be from 30% to 60% of the mass, grains the rest.

To start the miso, *Aspergillus oryzae* is cultivated on steamed rice or barley (see koji) and this is later mixed with whole cooked soybeans and salt. A salty paste results, with micro-organisms similar to those in soy sauce.

Blended temperature-controlled miso, with low salt and high koji levels, may be fermented for as little as one to four weeks (the so-called "white" misos). The red or dark brown misos contain high salt (13%) and low koji (20%) levels, and may be matured over one to four years. Most misos will keep two to three months refrigerated, but the well-aged varieties, kept cool and sealed, will keep many months and may improve with age. It is these latter groups that are most frequently made on farms, while the "white" misos are sold in cities.

While some misos are smooth-blended, traditional misos have up to 50% of the beans still whole or chunky.

Miso is sealed in plastic or cloth. To prevent storage jars or plastic pouches bursting, miso is heat-sterilised and sometimes a little alcohol is added to stop ferment when marketed. Unsterilised miso is also sold, and the latter can be used in new ferments as a starter.

The traditional use of barley is preserved in the name of the grain culture koji, with symbols (ideograms) for "barley" and "chrysanthemum" combined. One wonders whether the flower or its nectar was part of the original mould culture, as flowers are today in Nepal *(madhuca* flowers). Traditionally, barley miso is started in autumn, and is slow to ferment over winter; at least one summer is needed to mature the culture.

Miso is usually weighted with rocks on a board cover, or over a plastic sheet today, and the weights equal or exceed the weight of the miso. About 20 kg (44 lb) is minimal weight, even for small tubs or pots of miso during maturation.

Many types of pickles are added to both white and dark misos for short (three week) and long (three year) maturation. Foods pickled in miso should be salted and press-dried a few days to a few months (for fish), and never usually exceed 30% of the total mass for fear of diluting the miso and encouraging off-flavours.

Those wishing to make classical miso or tamari (miso soy sauce) should carefully read Shurtlieff and Aoyagi (1976), and devise their own recipes by trials. The Japanese themselves are always trying new cultures and mixes for special market niches, or to speed maturation; but of late there is a trend to natural and traditional long-ferment products without bleaches or preservatives, which in any case are not necessary in 13% salt misos.

Home-Made Miso

Assemble the following ingredients: 2 cups dried beans or peanuts; 2¼ cups water plus water added to open pots (as below); 9 tbsp natural salt; 1 tbsp unsterilised miso (optional, as you can later supply this from a mature batch); 2 cups mixing liquid in which most of the salt and miso is dissolved, 3½ cups of purchased or dried koji or 2½ cups of fresh koji.

This should make about 12 cups of miso, the minimum batch for maturing; any multiples of this can be made to fill the vat to four-fifths full.

You can use 20% legumes or soybeans, and the rest can be grain or root starches with koji.

Soak dried beans overnight; wash and remove any floating husks or beans before cooking. Pressure-cook beans for thirty minutes using 3½ measures of water to 2 measures of beans. Turn off heat and stand for ten to fifteen minutes. If using an open pot, cook for two hours, add hot water, and cook three hours longer, keeping water up to cover the beans and skimming off hulls periodically. If using a haybox cooker (see Figure 1.9), boil beans one and a half to two hours then immediately transfer to the box for overnight cooking.

Beans should be soft enough to crush with finger pressure. Drain and mash beans but leave half of them whole; beans can all be blended to a smooth paste but lose texture.

Save 2 cups of cooking water into which 1 tbsp of salt is added, and in this warm water dissolve 1 tbsp of any unsterilised miso you have purchased. Make sure the bean paste is at 41°C (106°F) and add water, a little at a time; stir in well. Crumble koji of rice, wheat, millet, barley, etc., into this from koji trays, of a weight equal to that of dried beans, mixing well with a wooden spoon. If using potato, yam, or cassava, cook and mash well and use 50:50 with grain koji.

Dry a vat or large-mouthed jar that will hold your miso, but leave about one-fifth of its depth unfilled, for the weights. Rub the pot with salt, place a tsp of salt at the base, and fill with miso. Pack firmly, sprinkle and press 1 tsp of salt on top. Cover with a sheet of waxed paper, foil, plastic, clean muslin, or tough paper. Top with a close-fitting board leaving about 3 mm (1⁄10") space between sides of jar and this board and weight with a washed clean stone or stones, pressing gently to expel air bubbles.

The first batch can be added to over several days, if it is a large container. Each layer is salted, pressed down with the weight, and covered. Now tie strong paper over the container and write on the date.

Place the vat in a cool room to mature. Examine at one month, and if no tamari (liquid) covers the miso, increase the weight. The best flavour is reached after a summer (say twelve months). Do not stir or allow air to the miso. If acidic, sour, or too alcoholic, it has to be discarded. For this reason, make up relatively small batches until the process is automatic.

To use miso, mix it well, scraping off any moulds and pouring off any tamari for sterilisation and bottling. Take out a smaller amount to use and refrigerate or keep cool, then smooth the surface, recover, re-weight. Lasts one to three years.

Mysore Miso

(From Shurtlieff and Aoyagi; developed by Dr. T.N. Rao)

Prepare the following ingredients: 20 parts (dry weight) chick peas, 4 parts soybeans, 3 parts chopped peanuts, 20 parts rice koji, 6 parts water.

Soak and de-hull chick peas for fifteen hours, soybeans twelve hours, and pressure-cook for twenty-five minutes at 15 psi. Blend some of this with chopped roasted peanuts, mix all ingredients with enough liquid to give a stiffish consistency (47–50% water). Ferment at 28°C (83°F) for five to ten days. Refrigerate to keep, if necessary under oil. Chilli and ginger can be added if this miso is to be added to curries, or a masala spice.

Miso Pickles

Miso is a basic environment for further pickling of fish, vegetables, eggs, and so on. In homes, ten to one hundred litres (2½–26 gal) pans, boxes and barrels are used, and food is put into the miso in layers to mature. Root vegetables (daikon radish, turnip, ginger, garlic), fruits (cucumber, melons, eggplant) and boiled eggs are common foods to be pickled. Tofu is frequently layered in miso for added flavour.

Fruits and vegetables undergo a preliminary slicing, salting, and withering process. They are sprinkled with salt, pressed for a few days, drained, and dried. All except the larger roots (quartered) are left whole, then packed with at least 2 cm (¾ ") of miso above and below them. In "sweet" white miso, pickles are removed in a few days.

Fish (previously salted or vinegared and lightly press-dried) are also layered in miso; vegetables and other pickled materials may be layered unwrapped, or wrapped in oiled paper, muslin, or foil to enable easy access and light washing after removal from the miso. The miso, flavoured with pickles, is used alone for soups, gravies, and rice dishes.

Sake, mustard, honey, chilli, turmeric, ginger, and garlic can be added to the miso, hence the pickles, for flavour.

White (sweet) miso processes pickles in one week; boiled egg or yolks, leaf, shoots. Red and barley miso is used for pickling over several weeks; roots, fish or squid.

Each pickle can be placed in its own bowl, pan or box. Like chiangs, there may be dozens of variations.

Miso-Limu Soup

Soak and boil ¼ cup of dried shrimp until tender, in 5 cups of water. Add ¼ cup of miso, dissolved in water, beat an egg and slowly pour it into the hot soup while swirling. Remove from heat, drop in sliced nori and sliced green onions or chives.

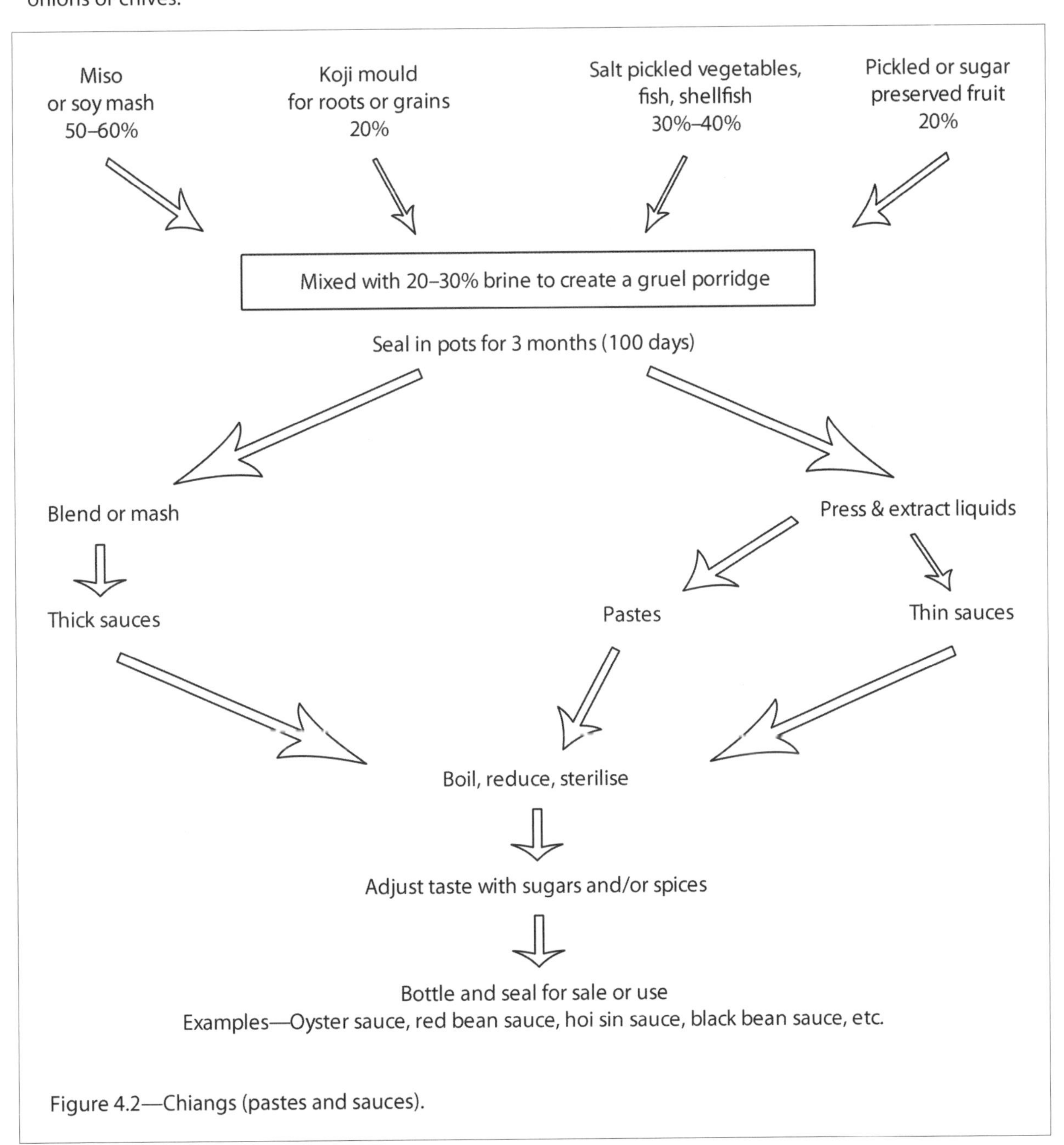

Figure 4.2—Chiangs (pastes and sauces).

Chiang

China

Chiangs (salted sauces) have been made in China since 500 BC. At first, many chopped or minced animal tissues were fermented, and later the bulk of the chiang was made of grains and soybeans, with salted or pickled meats added. Moulds can be used to create chiangs out of elm seeds, legume seeds, grains, and seed mixes ground or whole. Today, pure cultures of *Aspergillus oryzae* are often used, but see Chu for details of cultures.

Chiang is a starch/pickle fermented sauce in various combinations. It uses a process like miso, which is capable of hundreds of variations. Rice, potato, millet, or barley moulds are combined with salted, drained and pressed fish, vegetables, fruits, (plums, melons, peaches, apricots) to add specific flavours to the ferment.

Typical spicy flavours are added by (briefly salted and pressed) ginger, garlic, onions, turmeric, radish, turnip, greens, oysters, black beans, plums, etc. Special sauces of this type are well-known to Chinese cooks and their customers as black bean sauce, hoi sin sauce and plum sauce. Plums are part-salted and dried then added.

These chiangs have more brine than miso or fish pastes (16–20%), and a much higher water content than the 45–55% found in miso, which is mainly water taken up in the steaming or cooking of the substrate.

A typical combination for the hundreds of possible chiangs is to first separately make a koji. Add to it 10–40% of a steamed grain, cooled flour paste, or cooked beans. In the meantime make a pickle from preserved fish, pork, game, shrimp, crab, fish viscera, mushrooms, or fruits and spices pickled in a wine (mirin or sweet sherry) and brine preserve. The pickles are simply added, after briefly draining, to the koji mixture, and mixed to a medium porridge consistency, then cultured or fermented in a cool place in a sealed jar, pot, or wooden container for "one hundred days" or longer.

When it is "done", it is heat-sterilised at 80°C (176°F) for twenty minutes and bottled as a thickish paste; some mirin, honey, or any ethanol such as vodka, brandy, rum, etc. can be added (at 2% of bulk) to preserve the sterilised paste. Add a little olive or mustard oil floated on top to keep preserved. Usually, one sort of pickle (plus spices to taste) is added to a chiang of beans, rice, etc.

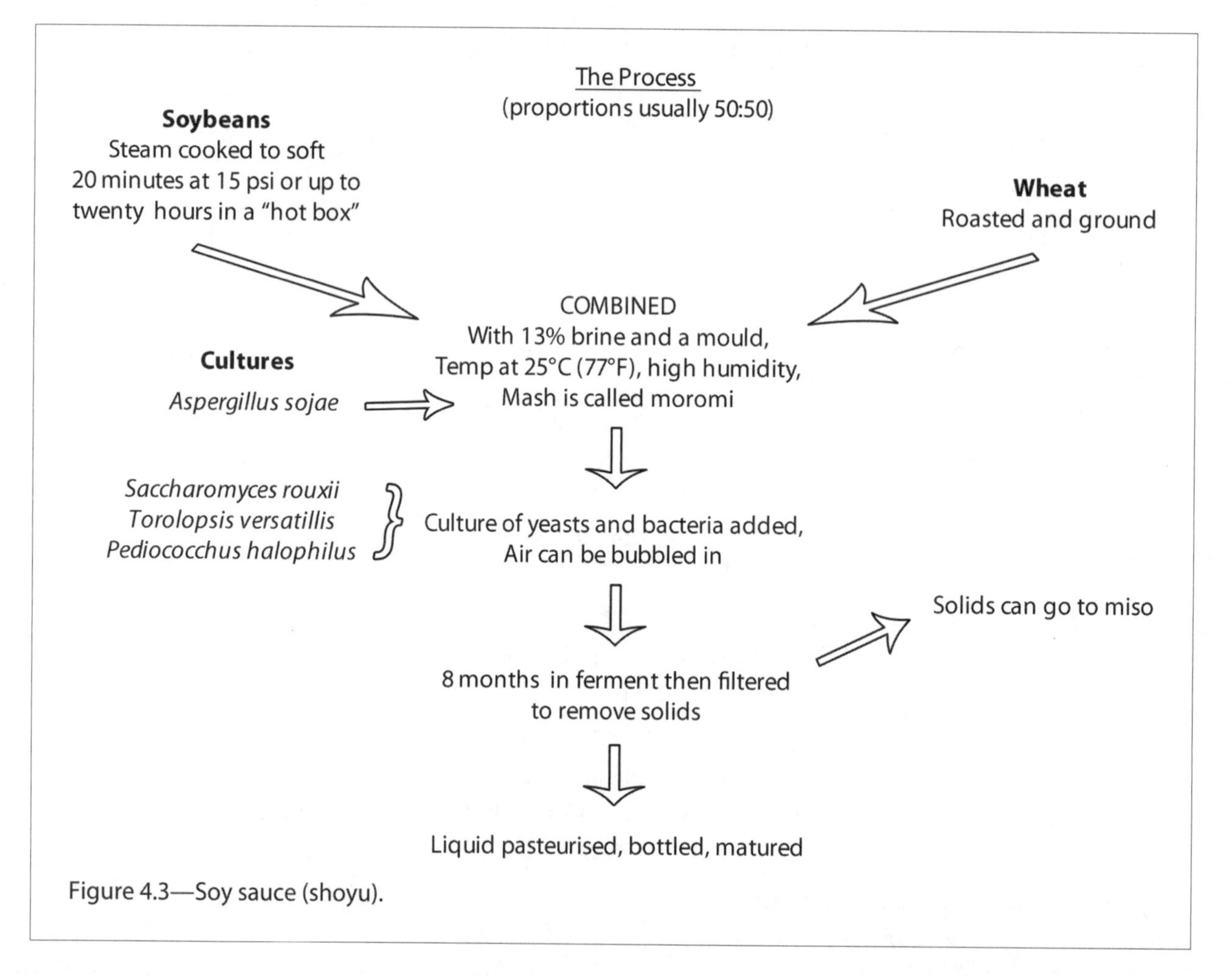

Figure 4.3—Soy sauce (shoyu).

Depending on tastes and the local culture, sambals (Indonesia), mustards (China and Europe), chilli, ginger, and garlic (China and Japan) can be added to the chiang. Koji is added as 60% or so of the grain ferment. The carbohydrates added at 40% can be fresh cooked, from potatoes, cassava, chestnuts, acorns (with the tannin washed out), millet, barley. Also used is jaggery, palm sugar, and dark brown sugar or malt (as in Indonesia), which gives a sweeter ferment. The chiangs can be briefly strained or soy sauces poured off, and separately sterilised as pastes and sauces, or bottled and served as a porridge poured over rice dishes or added to stir-fries as sauces, mainly to meats or fish "just fried", then combined with stir-fried vegetables on the plate. See Figure 4.2.

Soy Sauce—Variation One

Wash 10 kg (22 lb) of soybeans or legume seed, soak for twelve hours, drain thoroughly and steam at 15 psi pressure for seventy-five minutes or in an unpressured pot for five to six hours.

Meanwhile, roast 10 kg (22 lb) of whole wheat on an open dry pan until golden, and grind coarsely. Mix the soybeans and wheat together, then allow to cool to body heat.

Mix 2 tbsp of the spores of *Aspergillus soyae* or *Aspergillus oryzae* with 2 cups roasted flour, then sprinkle 1 cup of this at a time into the bean and wheat mixture, stirring thoroughly.

Spread the mixture on shallow trays and incubate at 25–30°C (77–86°F) for sixty to seventy-two hours (stir twice while incubating and remove any pink, blue-green or black spots).

Dissolve 3.6 kg (8 lb) salt into 25 kg (55 lb) water. Now add the soy and rice mixture and stir well. Do not cover; store in a cool place for twelve months. Stir daily for three days, then once a week for a year, to prevent surface mould growing.

The soy should be slightly thickish, red-brown, and have a pleasant aroma. Now ladle or pour this moromi (mash) into a clean coarse sack over a large container or bowl, on a rack of four to six strong tool handles and press down well using two people and a heavy plank or an hydraulic press; press several times, twisting down the neck of the sack as it empties.

Reserve residue as a seasoning in jars, refrigerated. Store the soy sauce in bottles in a cool room, periodically removing any white mould or oil from the top. A little alcohol can be added to preserve, or the bottles sterilised and well-sealed.

Soy Sauce—Variation Two Japan

- Almost equal volumes of cooked soybeans and roasted rough-crushed wheat are used as the substrate. The soybeans are first soaked for a day, then pressure-cooked at 10–13 psi for an hour (important). This mixture is then cooled to 40°C (104°F) and mixed with 1% volume of starter (koji).
- The whole mass is now mixed with an equal volume of salt water (23% salt) and forms a mash. (Some of this mash is reserved as a koji for the next batch.) The mash (moromi) is kept four to eight months in a large, deep tank and undergoes:
    - Lactic acid ferment by bacteria (*Pediococcus halophilus*).
    - An alcoholic ferment by yeasts.
    - Ageing to maturity.
- After tank maturation, the mash is pressed and the liquid separated. At decantation the oily layer is removed, and the sauce pasteurised at 80°C (176°F), then settled at room temperature before further decantation or filtering, and bottling. A pH range of 4–5 is necessary in the mash for yeasts to grow. The glutamic acid produced by the mould enzyme gives a good flavour. Miso provides yeasts for this maturation, and can be dissolved in the salt water.

Saccharomyces rouxii is present at vigorous ferment, *Torulopsis* spp. at late stages, utilising galactose (*Torulopsis nodaensis*), maltose (*Torulopsis versetilis*, *T. etchellsii*), saccarose (*Hansenula subpelliculosa* and *Torulopsis sphaerica*). Miso pastes are a source of fresh yeasts. Soy sauce contains amino acids and peptides from soybean protein, sugars from wheat starch, salt as a brine preservative, and 3% ethanol from yeasts. Although modern production plants produce shoyu in three weeks, this product is inferior to the traditional long-term ferment (after *New Scientist*. 2 October,. 1986). See Figure 4.3.

Soy Sauce (Kecap Asin)—Sweet Soy (Kecap Manis) Indonesia

In Indonesia, cooked, cooled, and strained soybeans are sprinkled with 5% wheat flour and incubated on trays for three to four days indoors and three to four days in the sun. Excess mycelia is then cleaned off and brine (at 20% salt) is added (60% of the volume) and exposed to the sun for thirty days. The mash is filtered, spices added (coriander seed, aniseed, cinnamon) and boiled. After filtering again, this high-quality sauce is pasteurised and bottled.

A lower-quality sauce is made for use in cooking. To all residues, the same quantity of brine is added and exposed to the sun for thirty days, filtered, and spices and jaggery added. The paste can be re-fermented with wheat flour, 2% sugar, shrimp paste, fermented seven to twelve days, boiled, and packed in sealed bottles. All keep about a year.

Sweet soy sauce *(kecap manis)* has 50–60% brown sugar added with spices. For this, add garlic, ginger, galangal to the coriander, aniseed, and cinnamon. Sugar solution is 38–50% of paste.

Black Bean Sauce China

Only black soybeans are used. The cooked beans are cooled, sprinkled with wheat flour, and a starter of *Aspergillus oryzae* added. This mixture is left two to three days at room temperature on trays, then placed in earthenware jars to which brine (15–20%) is added. The mixture is incubated for two months.

It is used as a condiment, and is one of the hundreds of chiangs or sauces that can be made, adding shrimp, fish, fruit, spices, etc. It is like a gruel in texture, with soft whole beans. Such sauces can be blended, sterilised, and bottled for sale and keep well (six to twelve months) if well-sealed. Wheat flour may be 1–5% of the soybean mass.

Soy Taotjo Indonesia

Soaked, dehulled and cooked soybeans are mixed with the roasted meal of rice or *Arenga* palm nuts; spread on trays covered with *Panicum palmifolium* leaves; sprinkled with ragi (*Aspergillus*) spores; and the trays covered with banana leaves and stacked to mould thoroughly.

The cakes are then sun-dried, brine-soaked, then mixed with sugar-palm syrup and rice ragi, and placed outdoors until the soybeans are soft (three to four weeks) before eating. Taotjo is boiled to consume, and used over vegetables, meat or fish dishes, or in sambals. Boiling reduces the strong smell of this paste.

For soy sauce, the mould cakes are layered in a basin or pot with fresh cooked whole black soybeans; and a brine-palm syrup poured over and left for a few days to assume an even brown colour (molasses may help here). This is the mix on which Indonesian shoyu is based.

Natto (fermented whole beans) Japan

About one-fourth of the soybean crop of Japan is made into natto. Soybeans are washed, steamed until soft, cooled to 104°C (220°F) and kept at about that temperature when inoculated with *Bacillus natto*. They are fermented eighteen to twenty-four hours and are very sticky and somewhat slippery; they should have long strands of slime when properly fermented. Natto is sweetened for deserts, or made into thick soups.

Thua-Nao (natto chips) Thailand

Clean whole soybeans are soaked for twelve hours, cooked one to three hours until soft, drained and spread on a banana leaf over a bamboo tray, and covered with banana leaf at 30°C (86°F) for three to five days. They should be shiny and soft, and can be eaten as is, or blended into a smooth paste, moulded into thin chips, sun-dried, and kept as thua-nao chips. The paste, heated and cooked, can be kept two days. Chip protein is 36–38%. The ferment organism is *Bacillus subtilis,* probably derived from banana leaves.

Tauco (soybean gruels) — Indonesia

Taucos use soybeans, rice or cassava flour, salt, and jaggery (palm sugar). Beans are cooked one to two hours, drained, cooled, and either naturally fermented two to five days or inoculated as for tempeh just after the flour (1–5%) is sprinkled on. The mass is sun-dried in trays, excess mycelia cleaned off, packed in pots, and brine (18–20% of weight) poured over. The pot of ferment is placed in the sun occasionally to heat, the containers covered at night. Then 20–30% jaggery or brown sugar is added, the gruel boiled, and the paste bottled. For tauco padat (with cassava flour or tapioca flour) it is again sun-dried two days and packed in sealed plastic sachets. These keep for a few months to a year.

Kochojang (a type of chiang) — Korea

Blend together: 5 kg (11 lb) steamed soybeans, 5 kg (11 lb) cooked glutinous rice, 2 kg (4½ lb) hot chilli peppers, 1 kg (2¼ lb) dried beef, 2 kg (4½ lb) honey.

Double the mass with 4–6% brine. This is matured in covered earthenware vats or barrels for four to thirty-six months, when vinegar can be added, and the mass well-mixed and strained to remove pepper skins before bottling. The pH is adjusted to 3.0–3.8 by the addition of vinegar. Kochojang contains B vitamins, carotene, and vitamin C (from the peppers).

4.3 SOME NOTES ON PARTICULAR LEGUMES

Peanuts

Ground peanuts are an essential ingredient in many African menus. Dig peanuts from the ground and dry them in the sun for a week (shelter from rain), or hang on the plant under a dry warm roof. Shelled, or in the shell, they can be roasted in burning straw or in a revolving drum over a coke fire, or in an oven at 70–82°C (160–180°F). Cook in small lots until the nuts are roasted but not burnt or raw. Store cool and dry. Reject all mouldy peanuts, as dangerous aflatoxins can build up.

Peanut Oil Seed Cakes

The oil seed cakes left after oil is expressed are first pulverised and then soaked twenty-four hours in water, when the remaining oil rises and is poured off. The cake is well washed, then steamed and pounded into shallow 2–3 cm (¾" -1") rectangular moulds placed over a bamboo frame, and covered with banana leaves to drain. They are then sprinkled with ragi and turned over on another frame of bamboo, and left in the shade to develop fungal hyphae. They are eaten fried, ash-cooked, steamed, etc.

Peanut seed is also germinated in a cool dark place, then pounded or eaten whole as toge, as are pigeon pea seeds, mung bean seeds, etc.

For dage, the seed skin is removed and they are boiled soft, then placed in a bucket, covered with a banana leaf, and weighted with a stone for two days in a dark place to sour as a form of natto (*Bacillus subtilis*).

Peanut butter

Roast peanuts to just dry for twenty minutes at 94°C (200°F) and rub the skins off. Blend with 1 tsp salt, adding ½ cup of edible oil slowly until the spread is at the preferred smoothness.

Bambara Groundnut (*Voandezeia subterranea*)

Bambara groundnut is a staple in Africa, cropped twice a year, and intercropped with sugar cane and turmeric; only cowpeas and peanuts are more plentifully grown. The immature seed is pounded to a pulp and the meal used for stiff porridges; the mature seeds are roasted before pressing to oils. The seeds are eaten fresh, roasted, boiled green in the husk, ground to flour, or canned cooked.

Dried seed pastes are used as hard rations, and keep well; pastes also added to sauces as a condiment, used as for peanuts in sauces with spinaches, chilli, ginger. Pastes are cooked in banana leaves and either steamed, cooked in ashes, baked, or as porridges. (Duke, 1981).

Marama Bean (*Tylosema esculentum*) Kalahari desert

An important food of the !xo people of the Kalahari. The young roots, rather like a large tapered beet, are eaten cooked in the ashes. Older (enormous) roots are dug by wildebeest, for their water content. The vining, butterfly-leaved plants die off in winter, then spread widely after spring rains. The beans and pods are cooked young, or the dried and roasted beans stored, and later ground as a flour or eaten as a nut with other food.

White Lupin (*Lupinus alba*)

A good semi-arid, sandy soil legume (annual), old in cultivation. The seeds are boiled to remove bitter principles (six to eight hours), the flours are used in breads (33% protein), or tofu made from the milk. There are sweet strains with low alkaloids (recommended for food), which can be used in the same way as soybeans, chick peas, etc. White lupin is a good honey plant and, as a green crop, a natural herbicide.

Polynesian Chestnut (*Inocarpus edulis, I. fagifer*) Palau, Tahiti

The large round pods carry single seeds. These are gathered and sometimes stored in earth pits. Collected seed is boiled for two hours, and then threaded and dried for sale. Cooked husked nuts are ground to a paste with grated coconut and cooked in bamboo sections or as leaf-wrapped cakes in ovens, or mixed in pits with breadfruit for long storage.

or

Pods are gathered, and can be buried to keep until enough are accumulated. Then they are boiled or earth-oven cooked two to four hours, cooled, and mashed. This mash can be eaten, used as any flour paste, dried, or pit-fermented with banana, breadfruit, or alone. As the tree is a legume, soy or miso sauces can include this paste.

Toge (green mung bean sprouts) Indonesia

Of all sprouted seeds (fenugreek, sunflower, alfalfa-lucerne, adzuki and so on) the green mung is most used. Sprouting activates enzymes and increases vitamin content many times over, as it does with grains.

The dried seed is spread, picked over, and soaked twenty-four hours (floaters removed). They can then be spread in a box and weighted with a 2 cm (¾") plank. Pour cool water over them twice a day for four days. These fat white sprouts are served with all salads, added to soups just before serving, and to stir-fry dishes a minute or two at the finish.

Mung meal is made by pounding or blending the beans to a powder. This is used to make pappadoms, or is mixed with banana and steamed in leaf baskets. Alternatively, it is mixed with eggs, fish paste and fried to serve with rice.

The seeds are also boiled with ginger and palm sugar to a porridge, which with milk is used to combat the nutritional disease beriberi. The leaves and pods are steamed when young as a vegetable.

Mesquite Bean Flour (*Prosopis* spp.)

Gather beans when dry, and further dry them to brittle, or parch at 107°C (225°F) for thirty minutes, a few days in a dryer or spread in the sun. When dry, break in a mortar, and grind in a heavy-duty hand flour mill. Sift and regrind if necessary. Add 1 part of this sweet flour to 2 parts wheat flour—the latter provides gluten, use for pancakes, rolls. (*Nature Seed Search*, Catalog, 1992, Tucson, Arizona).

Black-Eyed Bean Balls Nigeria

Soak 2 cups of black-eyed beans or cowpeas overnight and partially de-hull, then blend with: 1 crisp rasher bacon, 1 tsp salt, 1 tbsp cornflour, 1 egg, ½ cup vinegar, ¾ cup water (if necessary), 2 cloves garlic, ½ tsp chilli powder.

Form into 1½" balls and deep fry until crisp brown outside. These are delicious.

Velvet Beans (*Mucuna urens*)

The hairy pods are dipped in heavy syrup and the hairs scraped off into the syrup. When the syrup is as thick as honey with hairs, it is used to remove intestinal worms (1 tsp to 1 tbsp) in children and adults. It is safe to take and needs only one dose.

Mucuna Beans (*Mucuna utilis*)

Seeds of *mucuna* are cooked, the skins rubbed off, inoculated with ragi, wrapped in banana leaf and left to ferment to kedebel (Indonesia). For dage (Indonesia) the young seeds are cooked and the seed coats removed; they are then soaked two days in fresh or running water, covered with banana leaves and stored indoors, shaded.

Boiled again, they are ready to eat.

Tahoo Indonesia

Soy or peanut milk is mixed with salt, vinegar, lime juice, or burnt gypsum and left to form curds. The curds are pressed in muslin and dipped in turmeric or a decoction of turmeric to help keep them, as takoah, which is used over a few days.

Ontjom—Variation One Indonesia

Various ontjoms use washed and drained press-cake residues of peanut or coconut. With these, soybean residue, tapioca, bulgur, are combined at 10–15% each. The orange or red ontjoms are of *Neurospora*, the black ontjoms with *Mucor* and *Rhizoporus* spp.

Oil press-cake of peanuts or coconut is water-covered for about eight hours, then pressed to dry out the water, steamed, and inoculated with a previous ferment paste, or koji. It is left for one to two days in banana leaf wrap to ferment.

There is some danger of aflatoxins if fresh material is not used and cleanliness not observed. The inoculant is a mould *(Neurospora intermedia)* which may also be contaminated. Good peanut storage is essential.

White or Orange Ontjom—Variation Two

Ferment of groundnut (peanut) is inoculated with *Rhizoporus*, orange ontjom is inoculated with *Monilia*. The seed press-cake after oil extraction is used, proceeding as for (Variation One) above. The ferment is in a cool place.

The ferment can be spiced with fine-cut onion, ginger, chilli, salt, sugar; and roasted in ashes wrapped in banana leaf. Before fermentation, the meats of coconut or peanuts can be steamed.

Ontjom—Variation Three (Orange/red ferment of soy residue)

The residue left after straining soybeans for soy milk is used. The residue is pressed, mixed with cassava pulp, mixed thoroughly, steamed for one hour, cooled, inoculated with *Neurospora*, incubated thirty-six to forty-eight hours at room temperature. Keeps three days.

Hummous

Blend together 2 cups of cooked chick peas (garbanzos), 2 tbsp sesame oil, 1 tsp salt, 1 tbsp raw sesame seeds, and lemon juice, to a smooth paste. Covered with olive oil, it will keep for several days. Used as a spread with soft pan bread and salads.

Note that chick peas, like soybeans, need several hours of cooking until soft (when easily pressed flat with fingers).

Salting Green Beans and Corn for Storage

Blanch beans for three minutes in boiling water, or husked fresh corn for two minutes, and cut corn from the cob. Pack down in a jar at 454 gm (1 lb) salt for every 1.8 kg (4 lb) beans or corn. Weight well and fill with brine at 454 gm (1 lb) salt to 2¼ litres (4¾ pints) water. Cover with a cloth, a plate, and weigh down as for sauerkraut. Clean as the scum rises.

To use, soak overnight and cook briefly as for fresh beans or corn (heat to boiling for two minutes). When soaking, change water two to four times and taste to free vegetables of too much salt.

Papadoms

Mill mung bean, chick peas or black gram to a very fine flour, then sift. Mix in salt, pepper, chilli, and asafoetida and make the flour into dough, using a little water blended with a little mustard oil, kneaded, and pounded in a pestle.

Dip the dough frequently in mustard and olive oil to make it pliable, then form into small oiled balls and roll out into very thin wafers. Sun-dried, these keep several months and are a popular deep-fried snack. Baking soda can be added to the dough if required. They puff up rapidly in hot oil, and are sometimes flavoured with dried shrimp.

Vadias (dried pulse mash balls) Punjab

Vadias are small dried balls of ground dahl with other ingredients (spices) that are stored for adding to curries, soups, stews etc.

Moong (*Vigna* ssp.)

Black gram (moong)

Grind 1 kg (2¼ lb) black gram and 250 gm (9 oz) green gram coarsely and soak for a day in water. Drain and blend to a paste with ½ tsp of asafoetida.

Let stand all night, then add: Salt to taste; 25 gm (1 oz) peppercorns; 20 gm (¾ oz) chilli powder; 25 gm (1 oz) coriander seed; 454 gm (1 lb) cauliflower, minced finely; 5 cloves; and 5 cardamom seeds.

Blend well and roll into small balls. Dry these out on a clean cloth or old bedsheet before storing in an airtight jar as vadias, to be added to soups and curries.

Green gram

Soak 1 kg (2¼ lb) moong dal (*Vigna* seeds dry-ground to flour) overnight, drain, and grind to a paste.

Add: 1 tbsp milk, ½ tsp asafoetida, 1 tbsp coriander seed, 1 tbsp cumin seed, 10 peppercorns, 1 tsp chilli powder and salt to taste.

Mix thoroughly, form small balls of the mix and let dry on an old clean bedsheet in the hot sun. Store in airtight containers as vadias, to be added to soups and curries.

4.4 TEMPEH

Tempeh is produced from a mould of *Rhizoporus oligosporus*. It is certainly one of the most effective culture foods for nutrition and in saving cooking energy. (Steinkraus, 1989). Tempeh contains 47% protein, with double the riboflavin and seven times the niacin of soybeans. Even vitamin B12 occurs, from the bacterium *Leuconostoc mesenteroides,* while pantothenate and thiamine are decreased. Wheat and other cereals are also used for tempeh.

Tempeh inoculant is available from various sources (see Appendix D—Resources). To keep it, scrape off the dark surface spores and store in a sterile water paste (boiled and cooled) under refrigeration.

To start tempeh, soak beans overnight. Vinegar or lactic acid can be added to assist the removal of the bean coat; most of the bean coat is then rubbed off and floated away during the eighteen hour soaking process.

Bring 2 cups soaked beans and 8 cups water to a boil and simmer for one hour.

Drain, rinse, and allow beans to cool to body heat before adding the starter mould (½ tsp per cup of dry beans). Mix well, divide into packages wrapped in leaf, foil, wax paper or in small moulds. Incubate for twenty-four to thirty hours, at 31°C (88°F) or 26°C (78°F) respectively until mould binds them to a cake. Serve as soon as possible.

Tempeh is fried as slices or chips. It can be marinated with garlic and spices, or simmered in soy-sugar mixes, then fried. Or stir-fried with spices, chilli, soy, sugar, water. It is cooked (sliced) in soups (thirty minutes) or roasted and baked.

Hibiscus or Tempeh Starter

Aubert (1983) records that Dutch prisoners-of-war extracted the moulds for tempeh from the withered petals of *hibiscus* flowers, and so made nutritious food from soybeans. Flower yeasts and moulds are often mentioned in the "secret recipes" of traditional peoples. Tempeh provides 60 mg of niacin per 100 gm, a vitamin lacking in polished rice. Soy ferments contain 10 mg thiamine per 100 gm, riboflavin. Tempeh contains vitamin B12, probably from *Bacillus subtilis* or other bacterial sources.

Kecipir (winged bean tempe) Indonesia

The washed seed is boiled one hour, de-hulled, soaked overnight, steamed forty-five minutes, cut small, washed, cooled and drained, inoculated with ragi. Wrap in banana leaf, incubate at room temperature thirty to forty-eight hours. It is eaten within two days.

Tempe Koro Pedang (jack bean tempe) Indonesia

Boil, soak, de-hull, wash, cut up beans into small pieces. Inoculate with ragi, wrap in banana leaf or press into perforated stainless steel trays, or even plastic perforated bags. Incubate at room temperature forty-eight hours. *Rhizoporus* spp. inoculant.

4.5 TREE LEGUME RECIPES AND FERMENTS

Carob Pods

Pick and wash ripe pods, boil in just enough water or steam until tender, or pressure cook for twenty minutes. Then split the pods, remove the hard seeds, cut the pods into small pieces and dry well. When dry, pound or blend to a powder and store for a sweetener or chocolate base.

Soumbala

In Chad, Aubert (1986) reports that pulses and the seeds from *Parkia biglobosa* are cooked and fermented with the addition of bile from sheep. This bile is stored in bottles for that purpose (possibly with the addition of salt) for a year or more. Bile is a reducer of surface tension, and makes emulsions of oils and water.

Such ferments are used as sauces, like fish sauce, over rice or grain dishes, or as a side dish. They may reach 50% protein, and keep for months as a "black gruel" or dried cake equivalent to fish sauce in Vietnam. It is used sparingly, as one uses a condiment.

Dawa Dawa or Soumbala

African locust beans (*Parkia biglobosa, P. filicoidea, P. clappertoniana*) are large tree legumes of the Sahel and savannah, and are widely planted for their seeds right across the southern Sahara.

The red ball-like flowers are followed by long pod bunches which yield a yellow flour (pulp) and black seeds when dry. The "locust beans" are the basis of many sauces served with porridges of grains, the seed pod flours used as a beverage with flavourings such as lime leaf or juice.

The dry pods are pounded, the seeds sieved out and washed clean, the pod pulp sieved as a flour. Then the seeds are boiled over the fire in a covered pot for eighteen to twenty-four hours to soften the hard seed coat, which is rubbed off.

The kernels are heaped in beds to 15 cm (6") deep and covered (sometimes with plastic sheet, or leaves) at 25–35°C (77–95°F) until mucilaginous and strong-smelling.

The mass is then uncovered, dried in a warm place or in the sun, sometimes pounded, and wood ash added to reduce the smell. 3% salt can be added, and then the mass is formed into balls or small pyramids to dry. Dry matter protein can reach 50%.

These balls are mixed with other legumes, spinaches, soups, stews, or sauces accompanying main dishes. The ferment is basically bacterial.

Parkia Africana (Nete tree)

The seeds in Benin are first roasted then pounded and fermented in water. When well fermented they are washed and wet-mashed, then dried paste cakes made into sauces. The arils are made into a drink, or sweetmeat with sugar. *Parkia uniglobosa* pulp is sweet and floury and is eaten or fermented to sauces with other beans or flavours added.

Sauces made from *Parkia* spp. are the African equivalent of soy sauce, and the dried ferment cakes, with other bean flours or spices, can be made into a variety of side dishes.

Atta Beans (Owala oil tree, *Pentaclethra macrophylla*) Nigeria, Ghana

This leguminous tree is used in road avenues. The expressed oils are semi-solid and are used for candles, soaps, cooking. Boiled twelve hours, the beans are then husked, and salted 3–6% to ferment three days. The ground fermented paste is used as a condiment. The seeds have 21% protein, and dried pastes can be used as flours or the basis of sauces.

Pod ash is used as a salt in cooking, and the seed cake as a fertiliser. Uncooked crushed seeds can be used as a fish and arrow poison (the alkaloid is pancine). (Duke, 1981).

Tempeh Lamtoro (Subabul, *Leucaena leucocephala*) Indonesia

Leucaena is a widespread legume tree with abundant seed which stores well. Seeds of subabul are washed, boiled sixty minutes, soaked overnight, the seed coats rubbed off, washed again, steamed thirty minutes; cooled, inoculated with ragi (*Rhizopus* spp. inoculant), wrapped in banana leaves, and incubated twenty-four to forty-eight hours at 30°C (86°F).

This tempeh, fried and dusted with turmeric, keeps refrigerated about a week.

Tamarind (*Tamarindus indica*)

Jugo de tamarindo is a cool, refreshing drink made from the pod pulp and is bottled as a drink in Latin America. Tamarind is also the usual (secret) ingredient in sweet and sour sauces, and chutneys. The seeds, boiled and peeled, are eaten for a mealy starch and can be fermented.

Pod pulp is also added to fish pickle brines in Sri Lanka.

Tamarind can also be added to water in which the potato or aerial yam (*Dioscorus bulbifera*) is cooked, to detoxify the yam (Africa). The pulp is used to deodorise "suspicious" fish dishes.

Seeds can be hulled by roasting or soaking, also boiling. Pulp is a common antiscorbutic (vitamin C). Kernel meal is 15–23% protein, hence a valuable food additive.

4.6 THE SPROUTING OF GRAINS AND GRAIN LEGUMES

Although malting or sprouting seeds is a widely used method of food preparation, and is used for plants as different as coconuts, acorns, wheat, and mung beans, alfalfa, and fenugreek, it is not a normal part of western food preparation.

Sprouting has these effects on seeds or embryos:

- Metabolic poisons are reduced.
- Enzymes, especially amylase, are produced.
- Fibre content is partially reduced.
- Vitamins are enhanced by many factors.
- Starches are converted to sugars.
- Flours made from sprouted grains thin out thick porridge for children, and act as enzymatic thinners; sprouts are also chewed to a paste and added to children's porridge.
- Mineral absorption is enhanced.

Large seeds like coconuts are sprouted by burial or partial burial, as are acorns. Small seeds are sprouted by soaking and sweating, or simply by an overnight soak, which is rinsed daily until sprouted.

To Make Sprout Flour

Soak cereal grains or seeds overnight, rinse and keep damp under a cloth for a day or two until sprouted. Then dry seeds thoroughly, mill in a hand grinder, sift and dry, store. Nutritious high-protein flour is made by mixing legume and wheat sprouts 1:3. Use with porridges, plain flours at 1:4, for sweet loaf or buns.

CHAPTER FIVE

Roots • Bulbs • Rhizomes

5.1 ROOTS—STORAGE

Sweet potato, turmeric, and ginger are best left in the ground or in stone-lined ground pits for long storage (three to four months). Yams (*Dioscorus*) are dug and kept in open wooden bins or on shelves in dark yam barns raised on capped pilings to exclude rats. Potatoes, carrots and parsnips, are stored whole in sand or straw layers earthed over to form mounds; or in pits with such layers in well-drained sands, and are removed as needed. Carrots last fresh for ten to twelve months if carefully layered. Processing as starchy flours, or ferment with grains, prolongs root-crop life as foods, as does slice-drying and reduction to starches by pounding and washing, or processing to sagos. Any roots can be beneficially dusted with ash or lime before live storage. Radish is fermented and dried, and often then pickled in oil, as are turnip crops. Fodder turnip is put into clamps, like potatoes, or dry-stored in hessian bags (burlap).

Potato

Chuño (also known as dried potato) — Peru, Bolivia

The dried potato is pre-Columbian in origin, and originated on the altiplano (high plains) of the Andes where conditions are ideal for natural freeze-drying. On mid-winter nights frosts are severe, with the minimum at -2°C to -14°C (28–6°F), and the sunny days 26–35°C (78–95°F), humidity is 15–39% at dawn, and 12–18% at midday. Papa seca (dried whole potato) is made by boiling, the skins rubbed off, cut up, and again boiled to a thin gruel, which is then sun-dried to a crisp brown sheet, ground or milled to flour. It is stored for months.

Bitter potatoes (frost resistant) are not eaten fresh. But you'll find the glycoalkaloids are reduced from 30 mg per 100 gm to only 4–16 mg. These bitter varieties yield chuño blanco and chuño negro, while sweeter potato varieties are used for papa seca. The bitter varieties are exposed to frost whole and uncooked; examined to make sure they have been fully frozen (when the cell walls separate and cell saps diffuse), they are then trampled to remove the skins and to squeeze out the cell water. Covered with straw by day to prevent blackening, they are then submerged in running water (straw-covered) for from one to three weeks to sweeten, and spread to dry in the sun. The white crust of starches, gives them their name. The black (negro) potatoes are not soaked, nor are their skins removed, but more carefully trampled and immediately sun-dried to dark brown or black.

Even dried and somewhat de-toxified, bitter potatoes are eaten with clay by the Aymara Indians of the Bolivian Andes; the clay being very absorbent at the low pH of the stomach, and preventing gastrointestinal absorption of the glycoalkaloids in the bitter varieties of potatoes. Thus, geophagy again demonstrates how humans have managed to survive on poisonous foods by also eating clays to absorb poisons, just as clays remove tannins before eating, and bind to poisonous or pathogenic life forms in water (*New Scientist*, 30 March 1991).

Tongosh or tokosh are potatoes stream-soaked for up to a month, thus omitting prior freezing. They are in an "advanced state of decay" before rescue for sun-drying and have a strong smell in stews or when boiled. (Wolfe, 1987).

Papa seca are pre-sliced before drying, and ground to flour with a grinder; a flour much used in the cities. The dried potatoes are energy foods, 2–8% protein, 72–80% carbohydrates, about 2% fibre. They lose almost all of their vitamin C but retain niacin, thiamine, and riboflavin. However, the value of these methods is that a basic storable carbohydrate is available for many months, and enables existence over poor seasons with accessory foods, and in Peru, existence without these foods is precarious.

Note—Potatoes used for papa seca are *Solanum tuberosum* and *Solanum juzepczuckii*. Potatoes used for chuño are *Solanum juzepczuckii, S. ajanhuivi, and S. curtilobum* (Wolfe, 1987).

Potato Cheese

In Saxony, potatoes boiled and beaten with sour milk are made into a sort of cheese, which packed in well-sealed vessels is kept many years (from an old book, the title page missing; it might be worth a try!) I imagine that 6–15% salt would be essential.

Dage Indonesia

Unpeeled potatoes are sliced, half-cooked, and soaked in fresh water for a few days (water changed at times). The slices are air-dried and fermented between banana leaves in a dark, cool place two to three days. When the slices are soft, they are used in sambal; the smell may be a bit off. Cassava is also used, or coconut press-cake, seed press-cake, or mealy seeds. See also velvet bean.

Sweet Potato

Storing Sweet Potato Zimbabwe

Roots are stored for up to three months in bell-shaped ground pits covered with stones. When dug, roots are washed, sun-dried, and stored in ash for a few weeks in a dark, dry, airy place. Only round whole roots are stored this way.

Dried Sweet Potato

Firm roots are cooked by boiling (not to the soft stage), sliced, and sun or oven dried until leathery. Pounded into flour, these dried slices are used as 30% of grain flours in recipes. Like yams and cassava, sweet potato can be left in the ground if rats or other pests do not attack them there.

Drying Sweet Potato Papua New Guinea

Use sound clean tubers sliced at 5 mm (⅕") thick and the slices spread on bamboo mats or trays covered with plastic. At night, and in case of showers, have sheets of good plastic and stones ready to cover the trays (black plastic heats better), or bring trays indoors to a dry room. When the slices snap cleanly they are sufficiently dry. Waste pieces can be cooked or fed to pigs.

As weevils do not attack this product it can be stored in bags hung from rafters in a dry barn for a year or more. Shredded sweet potato dries well if blanched in boiling water momentarily.

Sweet Potato Flour Papua New Guinea

This is ground from air-dried chips (as above), and is used mixed 10% with wheat flour for bread, or made into batters as for dosa with 20% legumes, ground with peanuts for pancakes, or made into hard biscuits. The best way to make flour is a hammer mill, as the dry material is tough to grind.

Sweet potatoes, cooked and pulped, with 0.2% of malt added at 15°C (60°F) yields an edible and fermentable syrup (maltose 43%, sucrose 7%, dextrins 14%).

Sweet Potato Cake

This is a light, nutritious cake. Boil 1 kg (2¼ lb) of yellow sweet potato until soft, grate and mix with ½ cup of coconut cream or lard: add ½ a cup of yeast wort, then mix to a dough with coconut milk, animal milk, or water; let it rise twice, knock down first time and roll into balls. Bake on a greased tray.

Sweet Potato (also known as kumara) Maori

Storage pits are used for the crop, and large rotted roots are skinned and squeezed out in a basket to a flour, into which butter and sugar is kneaded, then flat cakes made and baked for a half hour over medium heat. Alternatively, the kumara can be cooked soft, skinned and mashed, or roasted with bacon fat and sugar in an oven.

Also pastes are used mixed with minced fish, meat, fruits, vegetables, spices and baked hot as a roll for three quarters of an hour. You can also bake: 2 cups kumara to ½ cup of grated coconut and a little salt, at 180°C (350°F) for an hour.

Fermented sweet potato can be mixed with a little arrowroot, salt and pepper, formed into cakes and baked in hot ashes or fried (kotero). Sweet potatoes are often fermented in pits (rua).

Dried Cooked Kumara

Selected small kumara are scraped clean, washed, eyes removed, wrapped in leaves (their own or other leaves), and steamed in an earth oven (hangi) for twenty-four hours, then dried in the sun for two weeks. They are eaten when dried, or added to stews.

Yam (*Dioscorus)*

Yams are peeled, bad parts removed, cut into pieces, boiled until soft (like a potato). The cooked yams are mashed to a glutinous dough, and this dough sliced to 2 cm (¾") thick, or the paste spread out this thick and dried at 50–70°C (122–158°F) to 10% moisture. The dried cake can be ground to fine flour and used as for sweet potato flour, or packed into sacks and bags for storage.

Cassava

Cassava is best used as it is harvested, within a few days. Otherwise, prune all leaf and shorten stem three to four weeks before digging, and store undamaged roots in wet sawdust, packed tightly (up to two months). However, there may be vitamin A losses in storage. Pit ferment extends storage life to a year or more.

Microbial activity breaks down any bound cyanides, but a few days' water rinsing in baskets after ferment does reduce free cyanides in bitter varieties. "Sweet" varieties have almost no cyanide.

An efficient method of removing cyanide in cassava tubers is by a long soak followed by grating and sun-drying, when reduction is 48–99%. Prolonged soaking followed by cooking is also very effective, as is soaking plus ferment (soak to soft, then pit ferment).

Low-cyanide (sweet) cultivars are being developed, but it is unlikely that cyanide-free plants will resist insect and wildlife or domestic stock attack as well as the current varieties. Despite the cyanide content of cassava, poisonings are rare, due to traditional methods of preparation.

Remove not only the "bark" but the outer peel (cortex) from the cassava roots. Tubers are cut open, peeled, split and any fibre removed, then cut into strips 2.5 x 10 cm (1" x 4"). These are soaked and fermented under water in a trough or in a heap covered to keep it moist for several days, then thoroughly sun-dried for storage or to be kept for a porridge-grain addition.

Although good sweet varieties can be dug in six to twelve months, the older, bitter cassava varieties are often left as long as four to six years, so that it suits "bush fallow with weeds" between crops, and it is also a good pioneer crop to clean new land. Thus, cassava serves as a food at either end of cropping cycles in poor soils.

Bitter cassava has more than 100 mg of hydrocyanic acid per kilogram of pulp and is considered poisonous. Moderate cassava has 50–100 mg of hydrocyanic acid per kilogram of pulp. "Sweet" cassava has less than 50 mg of hydrocyanic acid per kilogram of pulp.

Sweet cassava can be simply boiled, but the bitter varieties are used to make gari. Obviously, gari cakes can be sun-dried and ground to flour for long storage. Ferment detoxifies the cyanide with the enzyme linamarase.

Cassava Flour — Nigeria

Roots are pit-soaked and fermented one to two weeks, then peeled, grated, pressed (modern presses are hydraulic), sun-dried, milled and sieved to produce flour.

Grate the roots, squeeze out well in water, then place the pulp in a bag and squeeze thoroughly. Keep all the water in a basin overnight, then decant the clear water and spread the starchy residue on a flat rock or sheet of metal to dry. Crumble and sieve to store. Long storage is possible in a press-top can.

Fufu — Nigeria

Peeled roots of cassava are steeped a few days, fermented under wet sacks, mashed, pressed, sun-dried, milled, sieved, and sold as fufu (flour).

Cassava Starch — Papua New Guinea

Roots are washed and peeled and the skin and cortex removed, then ground or grated to a pulp or slurry. The pulp is sieved to remove fibres, and well washed into a tub, where it settles after six hours, when the surface of the starch mass is carefully scraped off with its impurities. This process is carried out two to three times, each time stirring the slurry until only a creamy-white mass remains.

The starch cake is sun or oven-dried, crushed to a powder and finally sieved.

Cassava, along with bananas, sweet potato, and taro forms a staple for millions of people in the humid tropics; sago is now almost entirely produced from this plant, replacing that from palms. Bitter (hydrocyanic acid) plants are grown in forests, where they mature slowly (ten to twelve months) but are not browsed, and these are used for sagos and starches. Sweeter, short-maturing (seven to eight months) varieties are used in home gardens for (peeled) root consumption. Sweet roots are boiled, baked, or steamed as for potatoes; the foliage of sweet varieties is used as cut fodder for pigs and cows.

Cassava — India, West Indies

Normal preservation of sweet cassava roots is by drying thin uncooked root slices, which store for months (the fresh tubers decay in four to six days). Fresh-peeled and grated, juices are expressed containing the enzyme phosphorylase, "Q" enzyme, and Beta-amylases. Reduced 50% by boiling, the expressed juices store indefinitely and are used in the preservation of fish and meat; this sauce is called cassareep in the West Indies.

The pulp can be fermented for about three days, then well pounded, sieved to remove fibre, and the meal so produced is well roasted in flat discs over a low fire for storage as an energy food. Locally, this is called farhina (flour). In Brazil, such flours are pan-roasted with a little butter and are on every table as a light condiment.

The flour (60 parts), ground from the very dry chips is mixed with peanut flour (15 parts) and wheat semolina (25 parts), made into a dough with boiling water, extruded through a mincer or die cutter, and dried at 49°C (120°F), or fanned with hot air. It contains double the protein of rice and cooks in five to six minutes. It is richer in minerals and vitamins than rice.

Tapai Ubi — Malaysia

The cooked and mashed or whole roots of sweet cassava are cooled, and inoculated with ragi at 28–30°C (82–86°F) for one to three days, and the sweet paste or roots used as a basis for sweets. Tubers for tapai should be only ten to twelve months old, freshly harvested, and scraped to make grooves after cooking, to allow the ragi to penetrate.

Covered but not totally airtight containers are ideal. Protein content goes from 1–2% to 4%. Ragi is sprinkled on the mash or tubers, which are leaf-wrapped or placed in press-top plastic containers (not fully airtight as the process needs some air).

Tape Ketela — Indonesia

Young sweet cassava is cleaned, peeled, washed and cut into 3–15 cm (1"–6") pieces, steamed thirty minutes, cooled on a tray, then every piece coated with ground-up ragi *(Rhizoporus* spp., *Chlamydomucor, Candida, Saccharomyces*). The coated pieces are piled in a banana leaf-lined basket, covered with banana leaf, and incubated at 27–30°C (80–86°F) for twenty-four to seventy-two hours. It is sweet, slightly sour, and keeps two to three days.

Gari—Version One — West Africa

Cassava root is washed, peeled, grated, and pressed tightly into cloth sacks or wooden troughs to press out extra water, then left for three to six days. The fermented pulp is then coarsely sieved and lightly grilled. Palm or cooking oil can be added before grilling or during the toasting process. Again sieved, the ferment is ready to be consumed.

This is a lactic acid ferment, with a final pH of about four, and while vitamin content rises (except for vitamin C), cyanic acid content is very much reduced.

Gari— Version Two — Nigeria, Ivory Coast

Cassava is peeled, the cortex removed with the hard core, grated (blended or pulped), the grated material well squeezed dry in a cloth bag under a weight while it ferments six to twelve days outside in the shade; the water content is now 55%. The pulp is then sieved to remove lumps and fibres, then roasted in a large pan at 120°C (248°F), constantly turned, in a little oil. It is then sold or used over the next two to four weeks.

Dried gari is soaked soft in hot water, made into balls, and eaten with meat and vegetable stews. Bitter varieties are mixed with yam and sweet potato, with fish, or with peanuts, coconut etc. (and sometimes sugar).

Finished moisture content should be 9–10% after roasting, and if well dried in a hydraulic press, gari has a shelf-life of up to two years.

Note—Hydraulic presses for dewatering bags are used horizontally over a trough, the bags (folded at the top) separated by thin steel sheets. See Onyekwere in Steinkraus, 1979.

Ginger

Dried Ginger

Papua New Guinea

Wash the roots, peel or scrape off the skin and slice 2.5 cm (1") thick or leave whole, then dip in 4% sodium benzoate solution (2–3 tsp per litre [2 pints] of water) for five to ten minutes. Spread slices on bamboo screens, but if sunlight fails, dry at 66–67°C (150–152°F) with a fanned draught. Ginger should weigh 9–11% of wet weight. Pack away in tight containers to deter insects. Grind to powders as needed, or soak and cook.

Pickled Ginger in Syrup—Version One

Use new "thin-skinned" plump roots from the early autumn crop, scraped or cleaned with a hard brush. Cut into smaller pieces. Use 225 gm (8 oz) green ginger, 568 mL (19 fl oz) water, and a 2.5 cm (1") cinnamon stick. Save the water and add 1 cup of sugar to 1 cup of water, boil on medium heat until the ginger is clear and syrup is reduced. Bottle and seal. This will keep indefinitely.

Ginger in syrup, when cool, can be sealed in a plastic bag in the jar, so that as pieces are removed, air can be excluded by folding down the plastic.

Pickled Ginger in Syrup—Version Two

Wash ginger well, mince and blend with some of the water, add sugar cup for cup and cook until syrup is thick, then strain through a fine sieve to remove fibres. Store the ginger pulp in a cool place. Used in hot sauces, drinks with orange juice and ice, as flavour in fruit salads, fruit frappé, and meat sauces etc.

Ginger and Garlic Paste

India

Equal quantities of garlic and ginger are pounded or blended to a paste. Sesame, cumin, fennel, black caraway and dried chilli are roasted and powdered. In a pan, the garlic and ginger paste, with a little oil, is fried to golden brown, stirring well. Then the powder spices are added and fried for a few minutes. The material is cooled, a little lime or lemon juice is added, and the paste is bottled. Used as a spice in curries.

Ginger and Garlic Salt

Very finely minced or blended garlic and ginger (50:50) and a few tsp of chilli or cayenne can be mixed with salt (1 part spice to 5 parts salt). Let stand overnight and then put the mixture in glass jars and seal. Use small amounts to flavour soups.

Turmeric

Dig rhizomes in the cool season and cover with water to 5 cm (2") in a large iron pan. Slowly boil until soft (two to four hours). Test regularly with a thin stick and do not overcook or undercook; just until tender.

Sun-dry cooked rhizomes for up to ten days (heap and cover at night). The dried rhizomes are now polished (peeled) by rotating in a drum mixed with smooth stones, usually in a wooden perforated drum (holes 0.5 cm [⅕"]) with a 20–25 cm (8"–10") door in one side. Turmeric is now grated, ground and stored. The dried material can be ground to a powder.

Taro

Taro is best "stored" in the ground. Although normally harvested at six months, it can be left for extremes of two to three years if the area has no serious root-rot problems, which is more common in wet terrace than in dryland (mulch) gardens. Harvested roots will last two to three months in leaf-lined and earth-covered pits. As poi ferment in sealed and chilled sachets, it will last longer; but duration is not known, whereas the flour of frozen, sliced, and dried poi can be kept for a year or more.

Giant Swamp Taro (*Cyrtosperma chamissona*) Solomon Islands

Lift the giant taro carefully, wash, and dry it for a day, peel carefully to remove all skin but do not wash once peeled. Grate on a wooden paddle with long thin nails or bicycle spokes driven through them, then mould grated root until very smooth. Mould the paste into football-size rounds and place these on oiled leaves of taro or banana. Pour more oil on the leaves and cover with more leaves.

These go in an earth-oven for three to four hours, then are pounded with a mallet in a wooden trough while hot, adding coconut cream (oil is expressed). Now roll into smaller (oily) balls, stack in a wooden trough and cover with the leaves from the oven. Inspect frequently, and if necessary wipe off mould and remix balls.

This lactic acid ferment is used over six months as "puddings" and can be re-heated to porridges with banana, coconut, or papaya added, or eaten as is. (Parkinson, 1991).

Poi—Version One

Poi is fermented taro. The peeled taro bulbs are first cooked and let cool, then pounded. Traditionally this is done with a stone pounder on a long oval board, between the legs (board 91 x 46 x 3.8 cm [35" x 18" x 1½"] thick).

The paste is frequently moistened with sprinkled water or yoghurt whey and packed into large stone or wooden vats or jars, and allowed to ferment three to four days until sour. Taste periodically to ascertain acid levels, and chill when ready.

Poi is served and used with all meals, especially fish and salads. It is sour, pinkish, paste-like, and is kept from spoiling today by refrigeration, so that large quantities can be made at a time. Refrigerated or frozen poi is kept in sealed plastic bags.

Poi—Version Two

Taro, peeled, boiled or baked, pounded, sprinkled with a little water (or whey), and sieved to remove fibres, is eaten fresh or allowed to ferment in pots two to three days, or at times longer for those who like very acid ferment.

Frozen for a day, the paste sets into cheese-like cakes, easily sliced, shredded, and dried to nutritious flour with good keeping characteristics. Taro bulbs, once lifted, rarely last two months.

Poi Drink

The Polynesian equivalent of lassi is to put in a blender ½ cup of poi, crushed ice or ice cubes, mint, vanilla (4–5 drops) and 2 tbsp of honey or syrup. Blend and serve with a dusting of nutmeg; dilute with chilled water or fruit juice.

Dried Poi (also known as popoi) Tahiti

Cooked, mashed and fermented taro paste is thoroughly pressed, dried, and wrapped in ti leaves (*Cordyline australis, C. fruticosa*). This desiccated product is said to keep for several months (Pecaro, 1968) Ti leaves are used because the ferment acids don't attack the leaf.

Taro, peeled and boiled, can be soaked, beaten or blended with any dal (split peas or beans soaked overnight and mashed) in a 5:1 ratio (taro:dal). A few days of ferment (to sufficiently acidic taste) produces a lactic acid paste of good food value, useful for sour blended drinks (with fruit or salt), pancakes or with steamed leaves. Poi is normally used as a cold dip at Hawaiian luaus. Luaus are feasts of pit-cooked wild pig, seafoods, and salads or salted salmon with tomato and onion. We use every excuse for a luau, as up to one hundred people contribute and consume!

Taro, peeled, sliced thinly, and fried makes very good chips, superior to potato chips, and marketed in much the same way as salted or spiced crisps sealed in cellophane or plastic. They are also a popular domestic item.

Tubers are stored in a dry place for two to three weeks to become blue-grey, then steamed. In Indonesia and New Guinea the tubers are then sliced and sun-dried.

Madrai Ni Viri (also known as bread of Fiji)

Parkinson and Lambert (1983) note that taro, bananas, breadfruit, and cassava were buried in pits to ferment until soft. The ferments, mixed with coconut cream, were wrapped in leaves, baked in earth-ovens, and preserved by drying in "times gone by".

Such foods could be stored for long periods. It is probable that the roots, at least, were cooked, and the bananas and breadfruit partly heat-ripened, (see also breadfruit). Nut flour pastes from large cooked legume seeds can be added to these pit ferments.

Taro, tree banana, cassava, potatoes, sweet potatoes and other root crops can be part of any grain, grain legume, or flour-based ferment of foods, including breads, buns, and lactic acid ferments. It is only necessary to cook and mash root crops and to make these a part of any ferment, or prepared food based on flours (add 5–15% of starchy fruit or root pulps).

Taro and Banana Loaf

8 cups of grated or chopped taro is thoroughly mixed with a cup of coconut cream and 4 cups of mashed ripe banana. About 3–4 cups of this dough is wrapped in banana leaf and oven-baked for an hour. Parcels left in the cooled oven, or chilled, keep for a week.

5.2 MINOR ROOT CROPS

Wapato — Mainly North & South America

The roots of wapato (*Sagittaria* spp), cooked in an earthenware oven and dried will last for years, and re-soaking will revive them, or they can be chewed dry.

Arrowroot Flours

These are made from peeled, ground, and well washed roots of *Maranta, Canna edulis,* and *Curcumia angustifolia* (a type of non-spice turmeric). The pulped roots are well washed and strained through muslin; the fibres used as mulch or as fodder and the milky starches allowed to settle. The clear water is decanted, and this process of stirring, settling, and decanting repeated three to five times to produce a starch cake, sun or low heat dried. It is most used as a thickener in gravies etc, or as biscuits, and is valued as infant or invalid food.

Arrowroot flour, like taro and breadfruit flour, can be mixed with sugar, grated coconut, banana pulp, water, and coconut cream to a dough, then leaf-wrapped and oven-cooked to a cake or bread. Or it can be baked as biscuits in an oven (if coconut toddy is used, a yeasty bread or scone results).

Arrowroot Loaves

The white flours from arrowroots (*Maranta; Tacca; Canna, Cucurmis.* spp.) is mixed 1:3 with sugar, and made into a dough with water or coconut water; to which is added about 25% by weight of mashed ripe banana, papaya, and a little lime or lemon juice. The mix is either cooked as a porridge, or combined with grated coconut and baked as leaf-wrapped loaves for one to two hours.

Pickled Beets

Boil, skin, cool and slice beets. Pack in a jar and cover with 600 mL (20 fl oz) mild vinegar and 1 tsp salt per quart of beet slices. Seal jar and keep in a cool dry place.

Camass Bulbs

Vast meadows of camass bulbs were developed by Amerindians in Oregon, and a few still exist today. The bulbs were dug, and cleaned in autumn, then pit-cooked over twenty-four hours. Then the caramelised bulbs were uncovered, cooled, squashed flat, and sun-dried as a sort of date or fig for long storage, or mixed with pemmicans. Grasses and bulb tops provided the layers for steaming and separating the bulbs when cooking.

Garlic in Pork Fat

To preserve garlic in pork fat, render 454 gm (1 lb) pork back fat and simmer this slowly until fat is thoroughly melted; strain, then add 300 gm (10½ oz) peeled garlic and cook until very tender (about one hour). Spoon hot into a small jar, seal and keep chilled. When needed, spoon out some garlic, melt briefly, pour fat back into jar, and serve cloves on meat.

Konnyak (*Amorphophallus konjal*) Japan

The ufi or elephant-ear is a taro-like plant with large, rough, creeping rhizomes. The roots are peeled, cooked and pounded as for taro (see poi) and are then strained, the fibres rejected, and the starches solidified with lime, much the same way as for tofu. Poured into moulds (with or without dry-crumbled nori or seaweed) it is again dried as 250 gm (7 oz) flat cakes. It can be freeze-dried to last indefinitely. An alkaline food, low in calories.

Radish and Turnips

Sinki (see also Kawal) Nepal

Radish roots are cleaned, quartered, and sun-wilted with the radish leaves hung in garlands for two to four days.

A pit 1.2 m (4') across and 1.8–2 m (6'–6½') deep is dug, (the wilted radish measured to fill this pit). The walls of the pit are mud plastered, and a thick straw layer spread on the floor, while sheaves of straw are stood close all around the walls and added to as the root and leaf layers rise. The pit can be dried out by fire before packing, or allowed to sun-dry.

Now about 30 cm (12") of wilted radish root is put in and pounded down with a heavy wood rammer, then 10–15 cm (4"–6") of straw or wilted leaf is added, and so it proceeds to the top, which is then covered with straw, then clay, and is well stamped. The ferment must be under pressure from stamping and earth.

At about a week, the odour of ferment can be smelt and the radish is dug out, sun-dried, and stored in close-woven bamboo baskets. Note that this method may omit radish leaves. However, when leaves are omitted, water is sprinkled on each layer before the straw is added. Freshly-fermented, sun-dried radish sold from the pit is called sinamini. High grade radish is used for this product.

Daikon Radish Nepal

Long strips of radish root and long strips of the leaves are exposed to the sun, wilting or drying over a couple of days until they feel leathery (during November-December). A pit 1.2–1.5 m (4'–5') deep is dug, its length and cubic capacity depending on the volume of dried radish. It is dried out by lighting a fire of dry sticks and leaves, and then lined with dry banana leaves. All sides are covered, and the pit must be dry for a successful ferment.

Smaller quantities are fermented in a tin or wooden container, with drain holes and also lined with leaves of banana or sal (*Shorea robusta*). In pits, roots and leaves are in alternate layers.

In tins, radish ferment takes about eight days (in pits twelve to fifteen days for the leaves alone). Fermented roots are separated, cut into pieces and dried out to sinki. The sinki can then be mixed with the oil spice "achar" which is made up of the following ingredients: roasted and ground cumin seeds, a few black peppers, red chilli paste, and fenugreek seed which have all been browned in a little mustard oil. This ferment is frequently used in soups and dishes or kept for later use.

Pits are covered with banana leaves and soil, and well trodden until compacted, then more soil heaped on top. Placed in tins or boxes, the radishes are compacted and straw covered, then pressed down with a heavy stone.

The pressed and fermented wilted leaves (grunduk) and the sinki are stored dry in airtight jars in the oil-spice mix, and will keep a full year. The usual store is an earthenware jar (gagro). Leaves can be of cauliflower, mustard, or spinaches for grunduk (fermented leaf).

To make a famous Nepalese soup, a handful of pickled leaves or root, sliced onion, tomatoes, oil, and salt to taste are used. Oil is heated in a deep pan, onions browned, then pickles added and fried crisp, then tomatoes and lastly water to cover. All are boiled to cook the tomatoes. Soaked, pre-cooked, or green soybean can be added (Gyanu Acharya).

Wasabi (leaf and root) Japan

Sliced stem, leaves, and root of *wasabi,* Japanese aquatic horseradish (*Eutrema wasabi*), are pickled in sake lees. Dried and powdered, the root powder is then mixed with water for eating with sushi in a soy sauce dip. The powdered root keeps well for long periods.

Horseradish Sauce (*Armoracia rusticana)* England

- Clean and wash white horseradish root and grate or blend. To each tbsp of horseradish, add 1 tsp mustard seed, 1 tsp fine sugar, 2 tsp salt, 1–2 tbsp vinegar, 1 tsp cream or melted butter, and 1 cm (¼") slice of ginger. Blend together, pack in a jar and keep cool until used.
- To a cup of grated horseradish, add 1 tbsp of mustard seed and 1 tbsp of salt. Blend with vinegar to a smooth paste. Bottle and keep cool for use.
- Grated horseradish, a hard-boiled egg-yolk mashed with a little cream, mustard and vinegar is blended cold, then heated before bottling and sealing.
- Wash roots, scrape and cut off peel; grate or blend with white vinegar to cover horseradish. Add 3 tsp of sugar per 454 gm (1 lb) and bottle, sealed. Blocks of this sauce can be frozen.

Radish Achar and Vegetable Achars

Fresh radish, turnip, chayote, or cucumber flesh, cut into rectangles, are dried for one day. Mustard seeds roasted and powdered, turmeric powder, and salt to taste are added. Pack all ingredients in a jar in mustard oil and stand in the sun for seven days. Carrots, cauliflower, squash, and cucumber are also pickled this way, and green (split) chilli added to all pickles. The whole pickle is oil-covered and used quickly as it cannot be stored for long (as it moulds). For sinki, the spices are fried and the hot oil poured on the dried fermented roots for long storage.

The following achars are made in Nepal:

- Vegetable: cauliflower, green tomatoes, chayote, chilli, cucumber, ginger, onions.
- Fruits: mangoes, Chinese dates (jujube), Indian plum, Natal plum, jakfruit, lapsi, Myrabolan plum.

Dongchimi Korea

Wilted root of long radish (daikon: *Raphanus sativus* var. *longipinnatus*) is pressed with a little green onion, ginger slices, chilli, garlic cloves, under a 3% salt brine (by weight). It ferments at 5–30°C (41–86°F) and when sufficiently soured (to taste) store six months at 5°C (41°F).

Liquorice (*Glycorrhiza glabra, G. echinata)*

The plant is sown in deep sandy soils, well-manured, and allowed to develop over three to four years, then trenched for the roots, which are washed clean, semi-dried in air, cut into pieces and thoroughly boiled in water. When the water is saturated, it is drawn off and reduced to a doughy consistency. The liquorice is rolled into 10 x 2 cm (4" x ¾") rolls, and wrapped in bay leaves after thorough air-drying. It is made into the confection by dissolving in warm water, straining, and evaporating. Sugar, flours, and mucilage (gum) are added for the confection, but pure liquorice is resin-like and brittle, and as children we would buy it like this from chemists.

Asafoetida (*Ferula assafoetida)* Persia, Afghanistan

An umbelliferous plant like a giant fennel. After four years of growth, when the leaves are yellowing, the top part of the root is uncovered and the stem twisted off, a slice taken off the root, and the cut root covered with the stems and leaves. Periodically, the top slice is scraped and a new slice removed. The yellow resin so obtained is fetid and musky in odour, forming a clear tincture in alcohol. It is a moderate stimulant, a mild laxative, and has been used as a condiment forever, the Persians believing it to be a food of the Gods.

Gum Arabic (*Acacia senegal, A. seyal*)

Although this is not a root or tuber crop, it has starchy qualities, so I am including it in this section. The trees are grown from seed on poor rocky soils, unfertilised, and tapped at four to six years by cross-cuts in the bark, removing a long strip of bark, or by gathering natural gum exudates. The gum is sun-dried (ripened), and is soluble in water (so it can be cleaned, sieved, and dried). It is used in making liquorice, confections, and in cough and dysentery medicines. It dissolves in an equal quantity of water. (Duke, 1981).

Tragacanth — South West Asia

The gum from *Astralagus gummifer* is used as an emulsifier, and to suspend insoluble powders in water, or mix oils in water-based medicines. It is also used in salad dressings, ice-cream, milk powders (as a stabiliser), cheeses, citrus and rose oil emulsions, toothpastes, lubricating jellies, or to emulsify fat-soluble vitamins.

Ti (*Cordyline terminalis*) — Cook Islands

The large roots of older plants 30 kg (66 lb) or more are dug, cleaned, and enclosed in earth ovens for two to three days. They are then at 20% sugar and can be dried. Grated, the sweet roots are added to ferments or used to sweeten porridges. A powerful alcohol can be made from the cooked and mashed roots, or the sugars added to sweet dishes and drinks.

Kapok (*Ceiba acuminata, C. parviflora, C. pentandra*)

All species produce large tuberous roots, harvested after a rainy season in Mexico by Pueblo and Yaqui peoples. The roots were a staple for thousands of years. (Camotes in Mexico).

A useful perennial dry-country crop, the trees grow from seed and cuttings or truncheons. Seed fibre is used as a floss or insulation; young leaves are used as a spinach, and seeds and roots roasted for food. The tree is long cultivated in deserts. (Nabhan in Wickaus, 1984).

Ortiguilla (*Cnidoscolus palmeri*) — Mexico

A shrub of granitic drylands producing up to 15 kg (33 lb) of tubers, eaten raw or roasted; an important desert staple. (Both the above woody perennials produce root crops for processing in drylands.)

5.3 A VARIETY OF PICKLED ROOT CROPS AND VEGETABLES

Various methods of pickling vegetables are used in Japan. The processes in brief are:

Tsuke Mono (also known as Japanese pickles)

1. Salt pickling

- Long white radish (daikon), gourds, beans, cucumber, or turnip, are washed, withered a few days to dehydrate, and layered in salt in a barrel, lidded and weighted. This is for shio-zuke, salt pickles. The pickles are removed and washed as needed.
- Small quantities, salt-sprinkled and with some warm water added, are weighted in a basin from a few hours to a few days.

2. Nuka box (bran) pickles

- Large quantities are laid up in the nuka box. Again, vegetables are withered (one to two weeks for daikon radish) and all are cut to a similar size. The vegetables are close-covered and heavily weighted. Greens are withered a few days. In large vats, these stay for six to thirty months, are washed, packaged, and sold. 6–12% brine is used to top up.
- Small quantities spend two to three days in domestic nuka boxes, usually unweighted.

3. Miso Pickles (miso-zuke)

- A sweet miso; vegetables are buried in the miso a few days.
- Large miso vats may have layers of fish or vegetables buried or layered in them for months.

4. Sake lees (kasu-zuke)

The lees produced from making sake may be used as for miso pickles, in large or small quantities.

5. Minor systems

- Thin sliced radish with sweet pepper powder are layered in kelp (senmai-zuke).
- Horseradish in sake (nara-zuke).
- Ginger in nuka box or sake.
- Green plums in salt (umeboshi). These "plums" are unripe green apricots!

And so on. Thousands of variations are possible! Other vegetables include: Chinese cabbage, eggplant-turnip-carrot mixes; cucumber; scallions; eggplant alone; gourd in sake (after Richie, 1985).

CHAPTER SIX

Fruits • Flowers • Nuts • Oils • Olives

6.1 VEGETABLE FRUITS

Cucurbits

Cucurbits, especially hard-shell squash and pumpkins, are stored in dark-dry areas in autumn. Often, they are slung in nets under the shady eaves of houses, in attics, or in cellars in cold areas. It is very important to keep all the stem on stored pumpkins, either by letting the vine dry off completely, or by carefully cutting the vine an inch or so above the stalk.

Smaller or immature pumpkins do not keep well. These are peeled, halved or cut into wedges, and hung to dry in the sun or food drier. Melons (in strips) are treated the same. Dried pumpkin or melons are used in pies or for thickening stews. In pies they can be cooked with dried fruits and nuts. Young squash are cut in rounds and dried on cloth on rooftops or in driers, as is chayote (choko).

Dried Pumpkin (and all squashes)

Fruits; peeled, seed removed, and thinly sliced, are steamed five to ten minutes to sterilise, then dried in hot sun or an oven over two to three days until brittle. They are kept cool, dark, and well-sealed and used in soups, stews, breads, or scones, after soaking or pounding to a flour.

Narra Melon (*Acanthosicyos horrida*) Africa

Flesh is boiled, seeds strained out, and the fruit pulp dried and solidified for storage. The seeds, as for other cucurbits, have good mineral content, including zinc.

Salted Pumpkin Seed Africa

Melon or cucurbit (cucumber, pumpkin, squash) seeds are field-dried, often in their own half-shells, and stored until plentiful. Some Middle Eastern varieties of seed melons, with only papery seed coats, are easily cleaned by rubbing. Most are, however, difficult to decorticate and it may help to soak them, then boil in a large shallow pan over medium heat, adding a light salt to the pan (3 gm [larger pinch] per litre [2 pints] of water). Boil and stir until dry, do not allow to burn. These lightly salted and dried seeds are stored and eaten, often pounded with sorghum as porridge.

These seeds are a valuable source of protein and zinc in deserts and the fruit pulp supplies vitamin A, while the growing vine tips are a spinach and a source of vitamin C.

Ogiri ("as is") Nigeria

If ogiri is to be made; boil watermelon or cucurbit seeds until they are soft and the husks easily removed, then the seeds are mixed with the cucurbit leaves and again boiled for two to three hours. The seed pulp is strained out and can be placed in buried earthenware jars with the cooked leaf, and covered with sacks, allowed to ferment naturally for five days. The resultant paste is dry-smoked over charcoal, and powdered once fully dry.

The high value of this paste is that it contains lysine, limited in cereals, and all cucurbit seeds contain zinc, limited in deserts. It is also protein-rich.

Eland or Gemsbok Cucumber

The African horned melon (*Cucurmis naudianus*) is a common vine fruit of the Kalahari, eaten by the !Kung people after ash-roasting, cutting off the top, mashing the contents, and spooning the hot flesh out, like a boiled egg. Sweet varieties, as grown in Australia, are eaten fresh. In New Zealand, this melon is called kiwana.

Cucumber Crock

In an earthenware crock, any vegetables of similar size, or cut to even size, can be pickled as follows:

- Layer base of crock with fresh or dried dill, then two to three layers of vegetables, a layer of dill, and so on.
- Boil together: 2 cups vinegar, 1½ cups salt, and 1 litre (2 pints) water. Skim and cool. Cover the vegetables with the brine, then weight a plate or a wooden plug to keep them under. Tie cheesecloth across the top of the crock and store at 20–26°C (70–80°F).
- After a few days, remove the scum daily and wash the plate, stone and the sides of the crock free of scum. Top up with brine to replace what has evaporated. In one to three months, the vegetables should be firm and fermentation will cease. Transfer pickles to glass jars, using a fresh brine, and seal tops. If glass jars are unavailable, the crock itself can be sealed with paraffin wax. Pickles are removed as needed and the crock re-sealed.

Pickled Cucumber

Scrub 25 cucumbers, wash and pack into a large glass jar with 4 garlic cloves, a few celery leaves, sprigs of dill or bruised dill seed, 10 peppercorns, and a saline vinegar solution (85 gm [3 oz] salt, 3 cups water, 1 cup vinegar). Pour solution over cucumbers and seal. Ready to eat in six weeks.

Dill Pickles

To one part vinegar add 3 parts water, and to 4 litres (8½ pints) of this add half a cup of salt. This is the brine for dill pickles, and will cover about 2.5 kg (5½ lb) of pickling cucumbers, whole or sliced. Finely sliced garlic and fresh dill is packed in jars with the cucumbers, the brine boiled and skimmed, and poured over them, still boiling. Usually, 3–4 one litre (2 pints) bottles are filled, stood in boiling water for twenty minutes, then sealed. Pickling spices are typically pepper, mustard, and chilli with seed removed.

The Japanese sprinkle cucumbers with salt, and roll them to soften before pickling.

If jars go cloudy, the brine is renewed and they are refrigerated to consume.

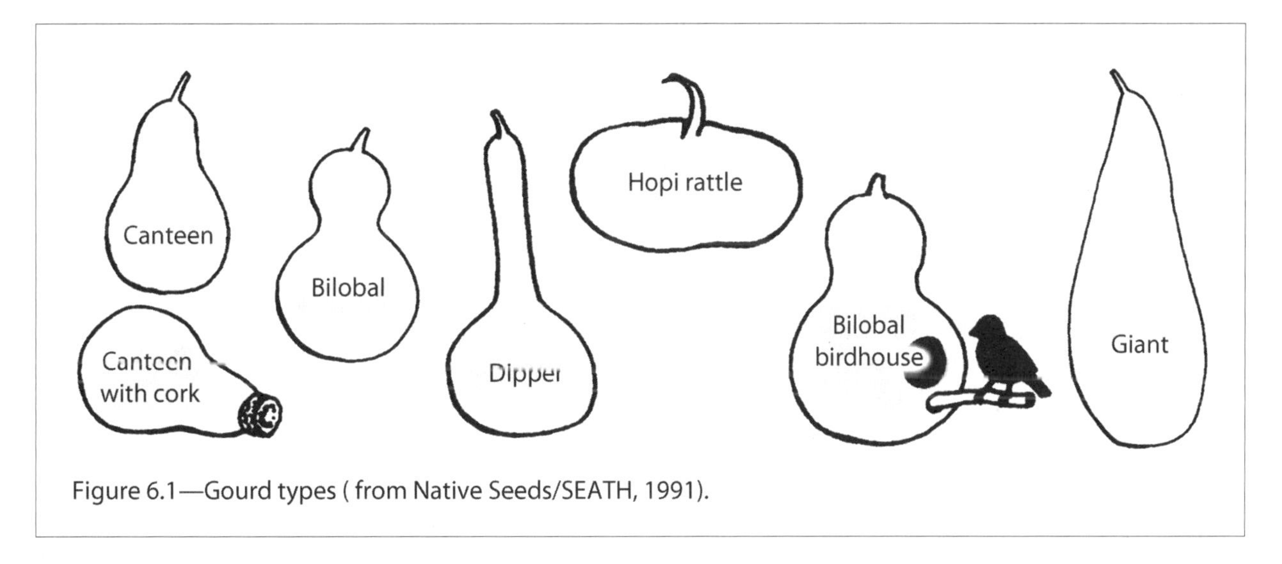

Figure 6.1—Gourd types (from Native Seeds/SEATH, 1991).

Gourds

Gourds are grown by every rural tribal society and largely supplant pottery in these cultures. They are in use world wide, in many forms, for such diverse functions as covering penises (New Guinea), souring milk (Africa), making drums and sounding boxes (Africa and India), and as utensils and storage jars.

Pick ripe gourds with hard skins and dry them on a shelf in a dry place, turning them occasionally. They may mould, but this can be wiped off; if they shrivel, throw them away. Fully dried, the seeds will rattle, and now you can soak them well in warm water and scrape or rub off the mould and skin with steel wool, or a loofah.

Dry and sandpaper lightly, and they can be cut, carved, and decorated. They dry better if frozen, when the skin loosens easily. Seeds taken from these are not so fertile.

A saw, sharp knife, and carving tools produce a wide range of products from birds nests to musical instruments.

Young green gourds can be eaten as vegetables, as can the vine tips. Figure 6.1 shows the shapes of various gourd types.

6.2 TOMATOES

Whole Tomatoes in Oil

Use medium-sized perfect tomatoes, whole and without blemish. Slice thickly and dry the tomatoes well (either sun-dried or in the oven), then place carefully in large-mouthed jars and cover with excess nut or olive oil to an inch deep. Herbs, spices, and vinegar are often added. Seal the jars hermetically (e.g. with cork and wax, or rubberband and plastic).

Italian Tomato Paste

The best sort of tomatoes to use are the Italian tomatoes. They are pear-shaped and do not have the juiciness of regular tomatoes, so they don't take as long to concentrate.

Use sound ripe Italian tomatoes. Wash, dry, cut into thick slices, sprinkle with salt at about 1 cup per 36 litres (8 gal). Drain for six to eight hours, then cook soft, cool, rub through a fine mesh and discard the skins and stalks. Cook pulp over a low heat, stirring all the time until thick. Spread the paste on a board and score to help drying. Dry in full sun or warm oven and work over.

When dry enough, keep in a pan about four days then roll into egg shapes; dip these in olive oil and place in a stone jar and cover with a heavy paper, plastic, or cloth dipped in oil and salt. Weight with a plate, and keep oil up to cover the plate. Will keep indefinitely. To use, dissolve some paste in boiling water.

Dried Tomato (in oil)

Small whole cherry tomatoes or slices of larger tomatoes sprinkled with a little sea salt are sun-dried three to four days, then bottled, and a few slices of garlic cloves and a little chopped chilli added. They are then covered with olive oil. In Australia, these are a very popular preserve, eaten as a side dish, added to salads, or as tapas (snack foods).

6.3 STONE FRUIT

Dried Apricots — India

- The cooked pulp is spread on cloth or oiled sheets and then dried for papad. The dehydrated pulp is pressed, and the sheets of pulp rolled, sometimes with a fine sugar powder, and stored in jars in a dark place. See fruit leathers in this chapter.
- Fresh ripe fruit is sun-dried or sulphured and dried (split and stones removed).

Dried Prunes

Prunes are a variety of plum which not only contain little moisture, but are very concentrated in sugars. Prunes are collected either on the ground or off the tree when almost dry, and further dried in the sun. In India, fully ripe (often fallen) prunes are first dipped in hot lye to remove skin wax, then rolled over a needle board to allow moisture to escape, and sun-dried ten to fifteen days. They can also be stored for long keeping in a medium syrup, or the pulp made into fruit leathers.

Date Palm

Ripe dates are eaten raw or dried, also boiled and fried. Syrups and date "butter" are also made. Pulp fermented in water or milk, will ferment to beer, wine, vinegar, or is distilled to spirit.

Drying Dates India

Perfect ripe dates are dried to 60% of weight over three to seven days in the sun or fruit ovens. Unripe dates can be boiled in water with a little oil, dried hard, and are called bhugrian.

Umeboshi Plums Japan

Half-grown plums or green apricots are first air-dried a few days, then packed in earthenware jars layered with coarse salt and the leaves of the shiso (beefsteak leaf, *Perilla frutescens*). The layered jar is then fitted with a wooden plug and weighted with a stone for ferment. The plums mature in six to eight months. They keep for years, and are pitted and minced to spice dishes.

Pickled Jujube (Chinese Date)

Jujubes (*Ziziphus*) are grown, like dates, in dry sunny climates. They look and taste similar to a date.

Usually sun-dried, jujubes can also be pickled. For pickling, fully ripe (yellowish) fruit are rolled on a pricking-board (to pierce the skin) and immersed in a 2% salt solution which is increased daily by 2% salt until the salt content is 8%. Then the fruits are transferred to a fresh 8% salt solution, with 0.2% (2 gm per litre volume) sodium metabisulphite for microbial control.

Jujubes can be stored for one to three months.

Salted fruits are washed in fresh water until tender, then transferred to a heavy syrup with some citric acid, for sale. The dried pulp of this fruit is powdered as a flour in Rajasthan. As there are many varieties, selection is possible for most semi-arid locations (*Wealth of India*).

Mango Indonesia

Fruits are peeled, sliced, salted, drained, and packed into jars sealed with ashes for a month, or in stoppered bottles for two weeks, and the pulp used as a spice with fish and rice.

The fruit kernels are first grated, then cooked in boiling water and sugar to a kind of porridge, cooled in moulds and eaten as a sweet.

Amchur

Green mangoes, sliced and dried crisp, are powdered to give a sharp resinous powder used in curries, achars, marinades, and in creamy gravies served with fish or poultry, or dusted onto grills.

Salted Green Mangoes

Wash and dry green whole mangoes 5–6 cm (2"–2¼"). Lay up in a large jar, layer by layer, with coarse salt and a sprinkle of chillies about 200 gm (7 oz) chilli to one hundred mangoes. Put a smaller jar of mangoes aside nearby. Every day, use mangoes from the small jar to fill the space left by shrunken mangoes in the main jar. Leave fifteen days to pickle, when it will be ready. Take out mangoes as needed, and slice thinly as a pickle with curries.

Ripe Mango Achar India

For 24 large, round, ripe, undamaged mangoes, use a large, thick pan and one 738 mL (25 fl oz) bottle (like a beer bottle) of oil. Heat the pan and very carefully fry all the mangoes on all sides, being careful not to break the skin. Remove carefully and cool for a half hour.

Meanwhile, combine: 375mL (12.6 fl oz) vinegar, 250 gm (9 oz) chillies, 250 gm (9 oz) garlic cloves, and 250 gm (9 oz) cumin seed.

First grinding each of the above spices separately, on a stone moistened with vinegar, or oil if necessary.

Beat together: 2 tbsp mustard and 2 tbsp turmeric with vinegar.

Add, already powdered: 25 gm (1 oz) cloves, 25 gm (1 oz) cinnamon, 25 gm (1 oz) cardamom and 10 gm (⅓ oz) white pepper and more vinegar.

Then beat all the spices with all the vinegar for three minutes. Add salt and sugar to taste (6 or so tbsp). Then add the oil from the pan.

Now, roll each mango in spiced oil and place carefully in a jar. Pour on all the spices and oils at the finish. Seal tightly for three weeks. Chilled, keeps one year. Remove mangoes whole, one at a time; do not damage remaining mangoes. Top up with oil if necessary. Serve in a small dish, sliced, with curries or rice dishes.

Green Mango Chutney

Prepare ingredients: 454 gm (1 lb) green mangoes, 454 gm (1 lb) sugar, 2 cups vinegar, 113 gm (4 oz) sultanas, salt, mustard seeds, 3–4 chillies, garlic, and sliced green ginger to taste.

Peel and slice mangoes, sprinkle with salt, and leave for twenty-four hours to drain. You may bruise the slices with a mallet before salting. Heat vinegar, dissolve sugar, and add all the other ingredients. Simmer until syrupy and add slices of mango until tender. Spoon into glass jars and seal.

Mango Chutney — New Caledonia

Cut up 1 kg (2¼ lb) peeled or unpeeled green mango and boil softly until thick and golden with: 454 gm (1 lb) sugar, 2 cups vinegar, 3 garlic kernels, 1 cup dried raisins and 4 pieces of ginger. Bottle and seal.

Fermented Mango Pickles — Papua New Guinea

Immature green mangoes are sliced thinly 5 mm (⅕") and placed in jars in 1% brine (by weight with water). They ferment over a week and should smell pleasantly. Stir daily.

Pour off brine, boil it, and remove the scum, if any. Rinse fruit slices with boiling water, pack into (new, sterilised) jars and cover with the boiled brine. Cap and seal immediately.

6.4 CITRUS

Storing Lemons

Pick with 1 cm (½") of stalk when just ripe. Rub all over with oil or petroleum jelly and roll in tissue paper. Pack in an airy box in a dry airy place. Check regularly for mould. Will keep fresh for months. These can also be stored in ash or dry sand (see Figure 6.2).

Whole Preserved Kumquats

Cut deep cross slits at outer ends of kumquats, wash and boil for one hour in 1½ cups of water to 1½ cups of sugar. Leave covered until cool, bottle and seal. A glass of vodka, brandy or gin can be added to jars before sealing and just before cooling the hot syrup.

Citrus Rinds

Thin-peeled (no pith) orange, citron, lemon, lime or *Fortunella* skins can be matured a year in gin or like spirits, or immersed in sugar in a crock. These are diced and added to dips or sauces, used to garnish fish or duck, or added to simmered miso sauces.

Chuk Amilo (concentrated lemon or lime) — Nepal

Large quantities of juice are boiled until viscous and red-brown in colour, the frothy foam carefully removed as it boils. This stores for decades and is used in souring just as is fresh juice, or a little added to pickles.

Nimkis — Nepal

In Nepal whole fresh limes are salted in barrels until soft, or placed in saturated brine; these keep indefinitely. They are used at the table, or as a therapy for nausea (Gyanu Acharya *pers. comm.*) Salted limes are stood in the sun for a week, then stored in a dark and cool place when softened.

Lemon Juice

As vitamin C is reduced by boiling, lemon juice can be concentrated by three to four episodes of part-freezing in a shallow bowl, and removing the ice until it is reduced to 30–50% of the original bulk, and 2% alcohol or salt added. It is then stored under a film of oil in a dark cool place, and removed by a pipette for use.

Pickled Lemons

Lebanon

Scrub and slice lemons, sprinkle with salt, leave for twenty-four hours to drain and go limp. Now stack in a jar, sprinkle with paprika and cover with a good oil (corn, nut, olive). Close the jar tightly. It will mature in three weeks. If the slices are first frozen, then sprinkled with salt, they may lose water within two hours, which speeds up the draining period and softens the lemon slices.

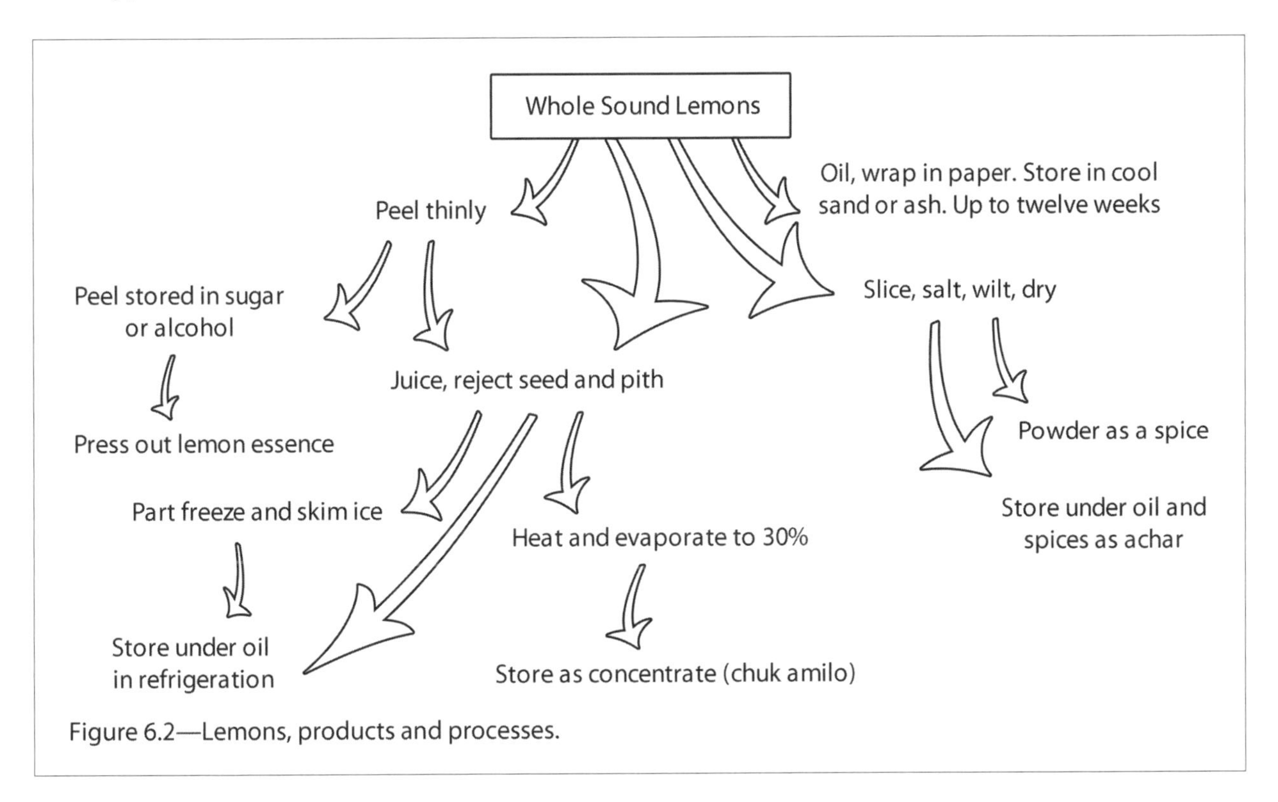

Figure 6.2—Lemons, products and processes.

Marmalades

Essentials with citrus marmalades are:

1. Soak the fruit overnight by covering with water. (Save the water for the next step, as this provides pectin). Fruit can be kumquats, oranges, limes, or grapefruit. A lemon or two can also be used.
2. Slice fruit thinly, including the pips and cover with the water from the first soak. Leave overnight. A few pips in the jam don't matter.
3. Cook the fruit for forty-five to sixty minutes at slow boil using 7 cups water to 2 kg (4½ lb) fruit.
4. Heat sugar in the oven (use 1 cup of warmed sugar to every cup of fruit and water) then add to the cooked fruit and stir until dissolved, boil rapidly until trials set. Bottle in sterilised jars and seal.

Pickled Limes—Variation One

Punjab

Use 16–20 larger green sour limes, and drop them into a large pot of boiling water, just removed from the fire. Let them cool there for half an hour, then dry and slit deeply so that each is almost cut into four lengthwise. Push 2 hot green or red chillies and a piece of ginger into each lime.

Now mix: 30 gm (1 oz) chilli powder, 30 gm (1 oz) turmeric, 75 gm (2½ oz) fenugreek seed, 75 gm (2½ oz) salt, and 75 gm (2½ oz) sweet oil (gingilly oil or sesame oil).

Pack the stuffed limes in a clean wide-mouthed jar and pour the oil and spices over them. Stand in the sun for up to ten days. Will keep for about six months under oil.

Pickled Limes—Variation Two

Cut 12 large limes into thick slices, remove pips, and drop into a boiling solution of: 1 cup of water, 3 tbsp salt, 2 cups vinegar, and 3 tbsp chilli powder.

Boil lime slices until soft, then add: 8 cloves garlic; 5 cm (2") piece of ginger; 5 cm (2") piece of cinnamon stick, broken up; and 6 cloves.

On a low fire melt: 2 cups of sugar and ½ cup of water.

Watch and stir until at the thread stage; then mix with limes. Usable after a few days, this lime pickle keeps very well.

Lemon Achar

Extract juice and pips. Salt skins and flesh well and leave overnight. Rinse thoroughly and lay in cold water, changing this water several times over twenty-four hours. Boil until soft, strain, dry the lemons and marinate in a sauce made by blending together a large onion, a bell pepper, a large piece of ginger, vinegar, and a tsp of turmeric. Blend these spices with oil (a good olive oil), and pack lemons well in jars. Cover completely with oil and spices, seal jars thoroughly to keep.

Dried Limes — Middle East

Halved large or whole small limes are boiled with about 1 tbsp salt per litre (2 pints) of water for three to five minutes, then drained and spread to sun-dry for a week, turning each day. On wet days transfer indoors to a warm place or put in an oven at 60°C (140°F), or in an electric dryer. Dehydrate until dark and the flesh is dry. Store whole or pound to a powder in a mortar. Use as for any spice, or on grilled and baked meats.

Pomegranate

Picked before maturity, these keep for months in a cool dry place. To extract seed juices, split the fruit and spoon out the seed and pulp under water. When the pith and skin float, they can be removed. Blend with water for one minute and strain through a cheesecloth to make juice, or use as a base for grenadine, jellies, and to marinade meats or fruits (Creasy, 1982).

Pomegranate Syrup

The sour red-brown sauce is made by reducing unripe pomegranate pulp and seed slowly until very concentrated. It is then strained and bottled hot, and is used in casseroles, game cooking, waterbirds, lamb etc. (Grigson, 1991).

Anardana (dried pomegranate seed) — India

Seed, with pulp removed from the ripe fruit and sun-dried ten to fifteen days to red-brown, is used as a spice.

Grenadine

Grenadine is a syrup made from pomegranates used in drinks and over sweets or ice-cream. The roughly-cleaned seeds of 3 large pomegranates are added to 2½ cups of sugar, ground and mixed well and left for twenty-four hours in the saturated sugar. Add a tbsp of water, heat, and using a strong wooden spoon, press thoroughly to express seed juices. Boil, press again when cooling, and store in sterilised jars.

Berries (small fruits) — Africa

Many are dried, especially those without large hard seeds, then pounded for addition to porridges, to form cakes, or to be cooked as a dry biscuit. Many fruits are pulped, seeds sieved out, and the pulp reduced to a syrup, dried, or fermented to wine and (later) vinegar. Some fresh pulps are mixed with milk as a drink. Many fruits are of course eaten fresh or made into jams with sugars.

Some acidic fruits or their pulps are used to curdle milk (lemon, tamarind, *Garcinia* pods).

Cacti (prickly pear)

There are twenty to thirty cultivars of the *Opuntia* species in Mexico (tuna rojo, tuna amarillo, etc.). Pick the fruits with a pair of tongs and a knife to avoid both the sharp spines and almost-invisible hairs. There are also spineless and seedless varieties. Remove hairs carefully with a small knife before eating.

The pads (nopales) from *Nopalea* are carefully de-spined (some varieties are spineless), sliced, and stir-fried or pickled as for any achar, in oil and spices, or as a vinegar pickle.

Drying Fruit

To prevent fruit oxidisation, fruit slices to be dried can first be dipped in a 25% lemon juice 75% water mix. This, like sulphuring, preserves fruit colour in sun-drying.

Dried Pineapple

Cut pineapple into 5–10 cm (2"–4") slices (cores removed) and dry in trays in the sun or a cool oven, then store in jars for addition to cakes, salads, pork dishes, stir fries, or stews.

Preserving Fruit in Sugar

A satisfactory proportion of sugar is 3 parts sugar to 4 parts fruit. Boil the fruit or juice first, until fruits are quite soft, and add sugar to the hot pulp off the stove, or (if sugar and fruit are boiled together, and the fruit acids contact the sugar) sugar will convert to glucose, caramelise, or invert.

Rumtopf (bachelor's jam) Germany

Using a syrup of 50:50 water and sugar (cup per cup), bring the syrup to the boil and hold for ten minutes, then in a sieve dip whole, firm fruit in for ten seconds (peaches, apricots, plums, nectarines and so on). The fruit skins should now slip off easily, or prick the fruit (e.g with kumquats) and the skin can be left on. Halve the stone fruit and leave kumquats whole. Pack carefully into glass-topped jars.

Pour on syrup and leave for twelve hours, then pour off a little over half the syrup and store in a large jar for later preserves. Fill jars with vodka or brandy and screw down lid. Store in a cool dark place. Fruit not halved can be pricked with a skewer to allow spirits to penetrate.

Dried prunes with slices of orange peel in brandy can be preserved this way. If fruit is added as the season progresses, add 454 gm (1 lb) sugar to each 454 gm (1 lb) of fruit and enough alcohol to cover.

Whole Fruit in Vinegar

Using wide-mouthed jars, fill with good white vinegar in which you can dissolve enough sugar to make an acid syrup. Over some weeks place in fruit at a perfect (dry and ripe) stage. The fruit is ready at six to seven months. Keep jars in a cool dark place, well sealed when full.

Fruit Mincemeat (from Mrs Beeton)

Prepare ingredients: 1.4 kg (3 lb) lemons; 1.4 kg (3 lb) apples; 454 gm (1 lb) raisins; 454 gm (1 lb) currants; 900 gm (2 lb) sugar; 454 gm (1 lb) suet (beef fat); 2 tbsp orange marmalade; 1 teacup brandy; 28 gm (1 oz) each of candied fruit of citron, orange, and lemon.

Grate rinds of lemons, boil lemons until tender and then chop. Bake apples and remove skins and cores; mix all ingredients in one by one, mixing thoroughly, and store in a stone jar with a good lid. Mature for fourteen days before using for pies. Also, mincemeat is a good energy food for walking or emergencies. It is not unlike the berry-fat-meat (salted and pounded) mixes of the native Americans, called pemmican.

Fruit Pastes and Cheeses

- Quinces, apricots, guavas, mangoes, dates etc., are made into pastes or "cheeses". Quartered fruits are boiled until soft, then pressed through a sieve; the resulting pulp can be lightly salted (**not** salty). Weigh the pulp and add an equal weight of sugar.

  Use a heavy pan to reduce the pulp, stirring gently all the while until the paste starts to candy and come away from the sides of the pan. Pour into 20 mm (¾") deep trays and let cool and dry in the sun or in warm ovens when convenient. Cut paste into squares, wrap in greaseproof paper, and store in tins.

  In Brazil, fruit pastes are served with cheese and fresh rolls for breakfast.

- Follow the previous recipe and instead of pouring pulp into trays, pour into jars or wooden moulds lined with wax paper. This is often sold as square block or further sun-dried and wrapped.

  Almonds, apricot kernels, or other nuts can be added to paste at the purée stage and are also preserved in the "cheese".

Kamaradin (dried fruit pulp leathers)

Overripe and blemished fruits can be used to make fruit pulp. Peaches and apricots are good and apricots dry as bright orange. Pears should be mixed with peaches to prevent cracking. Plums may tend to be sticky and need apple or fig pulp.

To all pulps, minced dried fruits or nuts can be added. Stones need to be removed, and the fruit minced or blended. To each 3 kg (6½ lb) of pulp, add 100 gm (3½ oz) sugar and 1 level tsp of sodium or potassium metabisulphite to prevent enzymatic and bacterial browning. This should be added as soon as fruit is blended or minced. Omit if you don't mind a brown colour.

Shallow trays are lined with plastic film, flat stones, or boards (oiled), and the pulp is poured in at 12–15 mm (½") thick to dry in the sun. After three to five days, the sheets of pulp can be peeled off and laid over racks or ropes to dry for eight to nine days.

Sticky slabs can be dusted with castor sugar and plain talc. Roll up the slabs, or cut them up and store in glass in a dark cool place. If sugar is increased to 300 gm (10 oz) to 3 kg (6½ lb), the product is softer and is used as for any dried fruit, or pieces soaked and blended in water for fruit drinks.

Try various mixes of fruits for flavour; ground ginger, pineapple, gooseberry, passionfruit, *Monstera deliciosa* pulp, and cinnamon or caraway seed give flavour to bland pulps such as pear. Mangoes make excellent dried pulp.

Titora (fruit leather) Nepal

Sour fruits such as green gooseberry (Amla) are cooked, any tough skin or pips removed, and the sour pulp mixed with salt, chilli, and asafoetida. A mat or piece of plastic is oiled and the part-dried and kneaded fruit pulp spread over it as a thin layer of paste. Sun-dried for four to five days, it is cut into strips and kept for snacks.

Sweet fruits are cooked briefly with sugar and spices such as cardamom, and made into a leather as above. Rolled strips are kept in tight jars and the sugar used is much less than for jam, and if fruit is very sweet, sugar may be omitted. Sweet fruit can be dried as small balls rolled in fine sugar for storage, but most drying fruits are preserved by natural sugars (grapes, peaches, apricots, bananas, mangoes, apples, pears, plums, dates, and figs).

Leathers are made from mangoes, apricots, pineapple, and mixes of the pulps of these fruits with sugar, tamarind pulp, and sometimes crushed nuts. Raw or cooked (fifteen to twenty minutes) fruit pulps are spread on oiled trays or mats in slabs 0.5–1 cm (¼" to ½") thick. These oblong sheets, a little larger than tea towel size, are sun-dried or cabinet-dried at 60–62°C (140–145°F). Once dried, they can be rolled or cut and pressed into blocks. Sugar and citric acid are sometimes added to preserve them at the cooking stage.

In peach, apricot, or plum jams, the seed can be cracked and the kernels cooked with the pulp and sugar to reduce or eliminate cyanides. To me, apricot jams without the kernels lacks flavour and texture, and I subject myself to hours of seed-cracking to ensure this special flavour.

In Afghanistan and Turkey, the fruit leather sheets are hung to finally dry like so many translucent golden towels, on ropes or reed racks. They darken with age. Cabinet dried sheets are stored wrapped in dark containers to keep the best colour.

Tamarind Pulp

Tamarind pulp is 8–14% tartaric acid and 30–40% sugar, and can be preserved by the addition of sugar or a little salt, the paste darkening with age. Flowers, leaves, and seed are also edible and can be dried or pounded to meals. The overripe pods, with sea salt, are pounded to a paste to clean silver!

Fruit Jelly

Use apples, apple and elderberry, red or black currants, quince, or gooseberry. Wash fruit first and cut up if the fruit is large. Just cover with cold water, and boil. When soft and tender, tip through a muslin bag and drain. Do not squeeze if clear jellies are wanted.

Measure juice as you return it to the boil, cup for cup with sugar. Boil juice and sugar together, testing frequently on a cold plate until it jells. Pour into sterilised bottles, pour on a little paraffin and seal. Many people prefer whole berries in cranberry jelly; the procedure is the same, omitting the strainer.

Guava Juice

Ripe firm guavas are sliced, the ends cut off and rejected for fodders. The slices are just covered with water, and boiled fifteen to twenty minutes until soft, then the pulp strained through muslin (do not squeeze the bag). Use the pulp for guava sauce, bottle the juice in sterilised jars, and keep in a dark place to preserve vitamin C. Mix with a little cold, boiled water to drink. Seal bottle corks with wax or use a sound top.

Guava Sauce

2 cups of pulp, from juicing, are cooked with a chopped onion, a garlic clove, a chilli, ⅓ cup of vinegar, 1 tsp each of cloves, allspice and cinnamon, ½ cup of sugar, and salt to taste.

First, boil the onion until soft, in a little water, then add all ingredients and simmer thirty to forty minutes before sealing in sterilised jars or bottles. Use on meats, especially pork.

Pectin

To extract pectin, you can use apples, quinces, or hard plums. Take 2 kg (4½ lb) hard, acid, firm apples, slice thinly and boil quickly for twenty minutes, strain through four thicknesses of muslin or cheesecloth, pressing pulp lightly with a spoon. Return pulp and weigh it, and add to it an equal weight of water. Boil again for twenty minutes and strain. Add second juice to first and measure by cup back into boiler; it should be about 12 cups. Using a wide pan, boil rapidly for thirty to forty-five minutes and reduce to 3 cups. Pour into sterilised bottles and seal (each bottle a ½ cup). Use with overripe fruit to set jellies and jams, or to set uncooked fruit and meat brawns in a jelly.

Durian (*Durio zibethinus*) Malaya

Some varieties have thick oily seeds, which are fermented until sour in bamboo joints or pots. After three to four days it is eaten with salt, fish, rice. The flesh of the durian is also fermented ten to fourteen days in buried pots or in pots hung in smoke. The ferment is then cooked and spiced with chilli and salt to make a sauce or sambal for rice, fish.

Jak Fruit

Jak fruit are so large (up to 20 kg [44 lb]) that they are often canned with mango, pineapple, or passionfruit pulp, or made into jam. Rody (1988) records that 1 cup of jak fruit pulp with 1 cup reduced milk and a little salt can be cooked (stirring for five minutes), blended until creamy and used cooled as a thick sauce with ice-cream.

Jak fruit seeds are either boiled or roasted as for chestnuts.

Papaya Indonesia

Grated and boiled, the bitterness of green unripe papaya is removed by adding a large lump of clay to the cooking water of the fruits, leaves, and flowers. The water is then strained off, and new water, salt, sugar, and guava leaves added when the papaya, the leaves, and unripe fruit lose their bitterness.

Green Papaya Thailand

The "male" fruits are peeled, grated, and served as a salad after salting, squeezing, and rinsing. This is a side dish for rice, meats, peanut paté, and sambals.

Green Papaya Pickle—Variation One

Peel and remove seeds and grate a green papaya. Add finely-chopped onion, 2 garlic cloves, a pinch of ginger, a de-seeded chilli, and a tbsp or 3–6% of salt by weight. Ferment at 28–30°C (82–86°F) for six to eight days. This pickle is vitamin-rich and is kept at 10°C (50°F) as a condiment with curry or salads. *Lactobacillus* spp., *Leuconostoc, Pediococcus, Streptococcus* are present.

Green Papaya Pickle—Variation Two

Peel, halve, remove seeds from green papaya, grate or slice thin, sprinkle with salt, drain ten minutes, rinse, and add chopped chilli. Serve as a salad or side dish. Proceed as above but then cover with lime juice or vinegar, mix well, salt to taste, and store in glass jars to mature. Long storage should be in a cool place.

6.5 MISCELLANEOUS FRUITS

Marula

This prolific tree (21–91,000 fruits per season) produces fruit of very high vitamin C value. The kernels are rich in protein, minerals, and vitamins.

The marula (*Sclerocarya birrea*) of southern Africa is one of several trees of traditional importance now being assessed and valued in product terms in Africa. For example, in the northern Transvaal, a commercial farmer (beef and game) today harvests 1,700 tons of fruit from his 8,500 marula trees. The product value exceeds that of his cattle or game (source: *The Farmers' Weekly*, September 1991, South Africa).

Fruits are removed by heat-softening and maceration, and fermented to beer. The kernels are extracted (each has three nuts); the nuts are sold to natural product shops, and nut oils pressed out and sold for culinary oil. This oil is slow to oxidise and has ten times the shelf life of olive oil.

Dry nuts from below trees gathered by the Ovambo peoples are ground and boiled to extract oils, and the nut cake used as a paste for sauces.

Note, however, that unhulled nuts from marinated fruits lying in sunlight in wet heaps quickly develop the pink mould *Fusarium roseum*, which generates the mycotoxins aflatoxin and zearalenone, the latter a hyper-oestrogen, causing spontaneous abortion in people and livestock, the former causing severe liver damage. Such toxins are fairly common in foods in Swaziland at 20 ppm (with 20 pp billion allowed in the USA), and will show up in milk products and meats (Crispin Pemberton-Pigott, *pers. comm.*).

Thus, enzyme-assisted maceration of fruits plus rapid drying of kernels is essential if these toxins are to be avoided. Peanut press-cake, coconut press-cake, and maize all develop similar toxins in wet heaps in sunlight.

Fruit Porridges

Even thin fruit pulps, such as the dried fruit skin on the manketti tree (*Ricinodendron rautanenii*), and dried berry fruits from *Grewia* bushes, are made into aromatic porridges in Africa. Macerated fruits or fruit pulps are also boiled down to storable syrups, and the syrups in turn can be fermented to beers, and of course the beers distilled to spirits and flavoured for liqueurs (amarula liqueur is made from the marula fruit).

6.6 STARCHY FRUITS

Bananas—Ripening Green Bananas

The banana bunch is always cut full-sized but green, and ripens best at 15–21°C (60–70°F) over a week, or (for ferment) smoke-heated in a room, for three to five weeks at 13°C (56°F). Dwarf Cavendish is stored twenty-two days at 11–13°C (52–56°F). At below 11°C (52°F), dark areas and off-flavours are developed. The cut stalk can be sealed with vaseline.

In Nepal, mature-size green bananas are cut, and the hands cut from the bunch; if the cut end is touched, they may spoil later. A large dry pot is straw-lined with random-packed straw, and the hands laid in and straw-covered. A pot on top, cemented on with a mud-manure mix, is filled with rice husk and shavings and top-lit. This heats the lower pot, and after five days it is opened to reveal ripe bananas.

Large quantities are placed in an earth-covered pit as bunch bananas and a fire is lit at one end, although straw baffles are fitted at both ends. When the bunches are ripe (in a few days) the fire is doused and the bananas are removed. City people ripen green bananas in closed plastic containers where they sweeten and ripen at room temperature. To ripen banana or jak fruit in Indonesia, they are buried in a pit filled with leaves of *Erythrina*, which generates compost heat. Bran is also used in these pits, a sort of earth-based nuka box!

Coconut

Coconut is the staple food and drink, fibre, wood, and thatch of coral atolls and tropical coasts to 25° north and south of the equator, also inland to 20° in frost-free monsoon climates. It is the world's most important tree crop for people, and provides over two hundred by-products, from activated charcoal to a base for plant-cell propagation, from "milk" to toddy, and as undersea log reef shelters for crayfish (all grown over with coral). Husks and shells provide fuels, fronds provide mulch. Life without coconuts on tropical islands is unthinkable.

Planted in dished and mulch-filled circles 3–5 m across, and 12–20 trees per circle, the trees yield as well or better than "on the square", and take up a fraction of the space.

A great many crops can be grown, from pepper, beans, and passionfruit on the stems, to rice, yams, taro, cassava, banana, coffee, guavas, citrus etc., below the crowns. The nuts provide oils and fats usually provided by dairy products in temperate lands.

There are five basic liquids derived from the coconut palm:

- Sap (aguamiel) tapped from the stem of the flowers. This is fermented to toddy, boiled and bottled as a sweet drink, or evaporated to a thick syrup. Toddy can be allowed to ferment further to vinegar.
- The water drained from the nuts. This is almost always consumed as a refreshing drink, and can be substituted for water in cooking.
- The "milk", expressed from squeezing fresh-grated coconut. Water can be added to the grated material until it is exhausted.
- The "cream", which is really the first squeezing of the coconut gratings. It is frequently used throughout Oceania.
- The oil, obtained by heating or boiling coconut cream and skimming off the oil as it rises. Copha can be derived from cream or oil.

Thus coconut water, cream, oil, toddy, and coconut vinegar are all obtained by a variety of techniques. Of all these, toddy is common to many palms, and is distilled to a variety of spirits or liqueurs. Only the coconut provides the milks and creams used so widely in cooking.

Coconut — Papua New Guinea

The split kernel flesh is grated and ½ cup hot water or hot coconut water added. After a few minutes the pulp is squeezed, then pressed or wrung out in cheesecloth. This is coconut cream. Twice the hot water is added to make coconut milk, and if this is cooled the cream rises and can be lifted off.

The grated coconut, squeezed or unsqueezed, can be sun-dried in a shallow pan, or more rarely in an oven. Store tightly sealed.

Coconut Oil

Use a small press to squeeze out the cream from shredded coconut, then rinse and squeeze the coconut meat again. Gently heat the liquids over a fire, and add 3% salt or seawater to settle the solids. The oil will begin to rise and can be skimmed off. It goes rancid quickly. Coconut oil is in most use as a cosmetic and massage oil. The grated residues can be dried and used for fodder. Half-meats, sun and oven-dried, are exported for the extraction of oils for cosmetics.

Coconut Chips — Papua New Guinea

Dry (mature) coconut is split and as large pieces as possible of flesh prised out, sliced thinly, salted, spread on a flat tray and roasted on low heat, in an oven, until crisp and brown, or fried crisp in a heavy pan. Keep sealed once crisp.

Fermented Coconut (kiora) — Fiji

Grated fresh or whole ripe coconuts are variously fermented to slightly sour as additions to fish and seaweed dishes in the Pacific. The palm sap or juices from the flower stem are first fermented, eight to nine hours, then boiled while still sweet to make a cool drink (kareve) at evening in Kiribati. Fermented toddy mixed with flour is used to make yeast breads, or let go several days to make vinegar for sousing raw fish.

Coconut Marinade

Fish in Tahiti are marinated with coconut milk, seawater, and lemon juice in which fish flesh is hydrolysed. This sauce is used as a marinade for raw fish slices which are eaten over a day or two.

Tajoro — Tahiti

Tajoro is a ferment of grated coconut, small freshwater shrimp, and clam flesh fermented in a calabash (hue). This strong-smelling paste is much appreciated in Tahiti. For some reason, the name of this fermented shrimp is also the love name of Tahitian girls for French sailors. (Information supplied by a friend who wishes to remain anonymous).

Sami Lobo — Samoa

The green coconut, with the soft albuminous flesh, is pierced and the water emptied. The nut is then filled with seawater and plugged with grass for one to two weeks. The resulting thickish gelatinous ferment is cooked with young taro leaves to make a nutritious paste or sauce, or used uncooked with starchy foods.

Miti Hue — Tahiti

The sliced albuminous flesh of the green coconut is combined with seawater in a gourd or bottle, and the heads of a few prawns added. The bottle is shaken and sunheated a few hours daily until the mix is thick and creamy.

As this is used, fresh seawater is added until the albumin is exhausted. Use as for sami lobo.

Nata de Coco — Philippines

Nata de coco is a white creamy solid, flavoured as needed with spices or juices for a flavoured jell. It is used as a snack or accessory to rice dishes.

Drain water from mature coconuts and add 10% sugar, then ammonium phosphate or sulphate (0.5%), strong vinegar, and ragi. The calyces of the rozelle (*Hibiscus sabdariffa*) may also be added for colour and flavour.

First the juice, sugar and ammonium phosphate are mixed and boiled ten to forty-five minutes. This is cooled, and the vinegar and 10% ragi starter by weight added. This mix is fermented at 28–31°C (82–88°F) for nine to fifteen days and can be stored for up to a year in sealed jars at room temperature.

The organism listed by Saono (1986) in nata is *Acetobacter xylinium*.

Sprouted Coconut

Coconuts are left to sprout on sand or soil then either halved by machete, or split on a stone by a vigorous overarm swing using the sprout as a handle. This exposes the coconut "apple", a sweet pithy centre which can be eaten out of one's hand.

This is a "fast food" of people on many atolls. All fallen nuts are placed in nurseries to sprout, and visited daily to test for edibility.

Breadfruit (*Artocarpus* spp.) — Papua New Guinea

Breadfruit is gathered in quantity in March, April, July and November when beads of latex exude on the surface. If the fruit falls, pigs fatten on the fruit.

The breadfruit are pit-cooked or oven-cooked just long enough to lose the green colour, then cooled and peeled. The fruit is halved and the walls of the fruit reduced to 5 cm (2") thick. Fruits are then washed in 3% brine or seawater, and dried on screens or mats, hollow side down over a fire pit, and the screen placed over the fire pit for four days, or oven-dried. Three days sun-drying (hollow side now turned up) gives a biscuit-hard breadfruit flesh.

So dried, and tightly bound in pandanus leaves, the preserved dry breadfruit will keep for several years.

Ripe fruits can be stored overnight under water, or a few days in a cool dark place. The skins are pricked before baking or roasting, and the fruit peeled before boiling or steaming.

Peeled and cored, slices of breadfruit are dried in a cool oven or in the sun, and stored dried and sealed in tins or dry leaves for additions to stews. It is also cooked, spread as a paste on oiled leaves, and dried. Well-dried breadfruit is pounded into flours, sifted, and stored as for any flour.

Raw, peeled and sliced breadfruit in pits, weighted, matures in two months but keeps for a year or longer. When used, this ferment can be washed in sacks in streams if too sour, then mixed with banana pulp and coconut cream for baking. If frozen, the paste can be sliced fine and again dried to flour, or fried. (South Pacific Community Food leaflet Nº 9.).

In Tahiti large quantities of breadfruit are cooked in earth-ovens for forty-eight hours. Families then take home enough of the cooked fruits for their needs. Air-dried, these remain edible for some weeks. Cooked breadfruit is grated, sun-dried, and used as is for porridges or is pounded to flour.

Mahi or Tioo — Marquesas

The main crops of ripe breadfruit are peeled and cut up. The heart is removed and thrown away. The cubed pieces are thrown into pits, lined with a weave of ti leaves (*Cordyline*). At about a metre (3') thick (of breadfruit pieces), they are trodden tight by feet, then covered with a layer of the dried leaves of the breadfruit, then a second and at times a third layer may be added (to 3 m deep) until the pit is crammed full. It then has a mattress of ti leaves put on, and the excavated earth is replaced and trodden down to cover these. The mass ferments to an acidic paste, which keeps for a year or more.

Such pastes are mixed with coconut cream or condensed milk and used as a cheese spread, or cooked and pounded with fresh breadfruit (also cooked), water, milk, or coconut cream.

In times of hunger, the people assemble and share out the preserve, which rolled into balls or made into cakes is folded in leaves and cooked. The preserve (a breadfruit poi, see taro) is not made much now, as the need for famine food is less frequent, but it presents an opportunity to make a ferment for good nutrition.

In the New Hebrides the cleaned breadfruit are leaf-wrapped and these bundles placed in stone heaps just below the high tide mark, where alternate saltwater flooding and sun-heat produce a similar ferment. The paste can be sun-dried and used as is (reconstituted) or pounded to a flour.

Breadfruit Paste — Kiribati

Methods of preserving breadfruit:

- Peeled, cored, and sliced breadfruits are oven-cooked two hours, pounded to a smooth paste, then sun-dried on oiled leaves, turning as it dries. Such paste sheets are rolled into "logs" and leaf-bound to keep or to carry on voyages. They keep for years.
- Grill-roasted over fires overnight, breadfruit pieces are sun-dried and stored in baskets for long storage.
- Where cooking fires are constant, pieces are left for two days above the fire for thorough drying, then stored, (Parkinson, 1991).

Dried Breadfruit — Tahiti

Mature breadfruit is roasted for an hour in hot ashes, then peeled, the seeds taken out, the flesh cut up and spread on a mat, about 4–5 cm (1½"–2") thick. This is now suspended over a smokeless fire or put to one side of a hot fire, the pieces turned from time to time until dry (about six hours). The product stores for about a year.

Pandanus — Tahiti

The fruits, grated with a clam shell, are made into fermented paste. The fruit pulps are eaten mixed with grated coconut. Fruit pulps are 25% sugar, and the cell pulp contains protein, thus fully mature fruit pulp ferments quickly, and can then be sun-dried. Varieties with oxalic acid crystals are first boiled, the pulp mixed with grated coconut, and sun-dried for indefinite storage (Petard, 1986).

Of the cultivated and edible species of *pandanus* (*Pandanus tectorius* and spp.), the ripe fruits are boiled or baked in an earth-oven, the outer rounded ends of the fruit segments cut off, or the pulp scraped off them. These can also be dried separately on dry banana leaves when they look like mashed dates, then stored in press-top tins as te tuae.

Otherwise, the pulp is mixed with grated coconut and toddy and larger pancakes of this set out to dry. When dry, these are heated on hot stones until crisp, then pounded to powder (te kabubu). This is also stored sealed, and used for puddings and drinks.

6.7 FLOWERS

Rose Water

Indian recipes frequently mention rose water as an ingredient. This is produced by hydrodistillation (steam distillation), but this method is limited. Further oils can be recovered by a solvent extraction, with benzene and petroleum ether, from which is obtained "oil of rose water", used to reconstitute rose water.

A strong solution is obtained by water distillation (separating the volatile oil) and can be diluted by an equal volume of water. A popular rose used in this process is cabbage rose—the Rose de Mai of France (*Rosa centifolia*). Dried petals are used in sachets and to scent tea. Bulgaria and India use (and cultivate) the damask rose (*Rosa damascena*) for 60–70% of production.

Rose petals, packed into a perforated-base pot are subjected to saturated low-pressure steam for two to two and a half hours, and later discarded. The rose water carries over in a coiled, cooled, worm condenser, and is collected in glass. The main content of rose water is phenyl ethyl alcohol. (*Wealth of India IX* p. 75).

Rose petals, layered in sugar in a glass jar, flavour the sugar which can be used in rose syrup.

Rose Petal Syrup

454 gm (1 lb) of petals, is simmered in 300 mL (10 fl oz) of water for only three minutes, then has 454 gm (1 lb) of sugar and the juice of 2 lemons added until sugar is dissolved. It is then boiled down to a thick syrup. Used in sweet dishes, chilled drinks, or ice-creams.

6.8 NUTS

Chestnuts

Chestnuts are, with acorns, a source of good-quality complex carbohydrates. They fall and are fairly easily husked by drying and lightly rolling. The shiny outer skin is removed by slitting slightly and boiling for twenty minutes. After a further soak, the inner brown skin can also be rubbed off, leaving a yellowish kernel. They can then be eaten fresh, or hard-dried for storage and grinding to flours.

Softened by soaking, they can be minced and combined with grain flours at a 1:3 ratio, chestnut to flour, for a sweetish bread (Tuscany).

For a meat stuffing, soaked kernels are cooked soft in boiling salted water, minced, blended, or pounded to a paste with sweet potatoes, parsley, an egg soup stock, fine chopped liver or giblets of poultry (to stuff birds), or in rolled veal and pork fillets as stuffing.

To make *marron glacé* (sugar-glazed chestnuts), chestnuts are boiled soft and repeatedly dipped in heavy syrup and dried until sugar-glazed, or stored in heavy syrup.

Chestnut water from boiling is best used to cook fresh corn, cut from the cob, as a nutritious mash.

Acorns—Variation One

Acorn meals and flours are stored as any meals, in jars, tins, or pots, but live acorns are reportedly stored to retain viability, in peat swamp soils. I have stored them over winter in boxes of sawdust, in the shade but not sealed, where they germinate very successfully in spring, as do chestnuts and walnuts. However, if they are well stored in acid peats, I see no reason not to attempt storage in nuka boxes (see chapter 3), where similar anaerobic and humic conditions prevail. In fact, nuka boxes may also be tried for wax-wraps of butter, cheese, and solidified fats.

Acorns— Variation Two

Make a small incision in the husk of the acorns, soak for four to six hours, then boil. Remove the husk, grind the kernels, and taste for tannin (tannin may make the acorns too acidic). Wash in warm water and drain water twice through a coarse cloth. Water dissolves most tannins. If the meal is still bitter (the black acorns), mix in a handful of fine red clay per 4.5 litres (9½ pints) of meal. Meal may now turn black, as insoluble ferric tannates form.

Now mix 50:50 acorns with flour and bake as a bread (a blue-black bread). Sweet (white, annual) acorns may be eaten or made sweet with the first two washings and will make a white bread without clay being used. Acorn meal can be used with grain meals in porridges or breads. I have made good breads with 30% acorn meal.

Just sprouted acorns can also be boiled and shelled in the spring or thaw period when enzymes are activated.

In India, tannins are removed after crushing and repeated warm water washes. Then limewater and potassium permanganate are added before drying and grinding to a paste or a flour.

Racahout (sweetened acorn meal) Turkey

Several species of acorns are buried to rid them of tannins, dried, washed, and pounded with sugar and aromatics. These were commonly fed to fatten seraglios, but as few women today need fattening, this dish is probably rare. The acorns of *Quercus ballota* (Morocco) are quite sweet and are eaten fresh.

Pine Nuts (*Pinus monophylla* and spp.)

Shoshone peoples buried greenish (not quite ripe) cones out of the reach of squirrels, in the field below the pine trees, for food reserves, as poor nut years are common. If they needed to be dug up, the cones were heated in pits, and the nuts extracted, roasted, and rolled under a cylindrical stone to extract the kernels. Caches of cones were stone-covered for protection from squirrels.

But "when the rose hips are red in the valleys, the pine nuts are ripe in the hills" (Paiute saying) everybody went harvesting the year's supply. Nuts were gathered, sieved clean, part roasted by shaking with hot coals, cracked to extract the kernels, and the latter ground to thick soups, with rabbit or fish, fresh or dried.

Pine nuts are well preserved in honey for many years. Jars of these were found, in good condition, in Pompeii in 1873. This is a winter energy food for combining with porridges or cake mixes.

6.9 MINOR NUTS

Karaka (*Corynocarpus laevigata*) Maori

Karaka is a large, glossy, dark green sea-front tree with prolific berries, but the kernels contain prussic acid (cyanide). To make them edible, the ripe orange fruits are trodden in water and the flesh removed. The kernels are cooked twenty-four hours in an earth-oven. Ferment releases the cyanide, so flesh is removed by treading or boiling three to four hours. If boiled to remove flesh, the kernels are further cold water steeped for a week, and can be sun-dried for storage. So treated, the nuts are used as flours, or in stews.

Ngali Nuts (*Canarium* spp.) Solomon Islands

The nuts may be buried in moist earth for the shells to soften, or dried over a fire before shelling.

The kernels are sun-dried, or dry-roasted in 1 m lengths of bamboo, leaf-plugged at the end and stored over the fireplace. Dried nuts are also stored over the fire, in cloth bags or baskets. Fresh or dried kernels are pounded and added to taro and fruit porridges or breads, added to doughs and pastes, or added to breadfruit ferments. (Parkinson, 1991).

Mongongo Nut (*Ricinodendron rautanenii*)

A staple food of the San (Kalahari), growing in groves in dunes. Both fruit and nut contain high food value, minerals, protein, energy, and vitamins. Both seed and truncheons produce slow-yielding trees. The species has sexes separated. The thin fruit rind is fermented or added to porridges.

Cocoa

Using a warm to hot pestle and mortar, place the cocoa nuts in it and pound them to an oily paste. If sweet chocolate is required, add half the weight in refined sugar or crystallised sugars. Pound, mix well, and keep it warm in a pan in the oven, and do this with each mortar load. Place another pan to warm, then take portions of the paste out, roll them out smooth with a stone roller on a marble or stone slab, place in the empty pan, and continue until all the mass is rolled and smooth. Then, divide up and press into oiled moulds. When cold, tap out the chocolate and use for drinks, cooking etc.

Variously combined with copha, butter, molasses, and cream to make sweets or drinks, these are flavoured with grated chocolate melted in, or chocolate with butter and egg white are beaten to a froth and baked on low heat.

6.10 OIL SEEDS AND OLIVES

All oil seeds are pre-heated, usually in a large iron pan, before processing (castor oil, rapeseed, mustard oils, sunflower, safflower etc.). They are then either placed in a close-woven sack and pressed between great wooden beams with a screw press, in an hydraulic press, or via a water-cooled tapered barrel screw press (rather like a mincer) with nozzles placed below the barrel at various lengths for cooler and hotter oil to fall out, at about three places along the barrel. Alternatively, seeds can be crushed and boiled, and the oil skimmed off.

All oil seed-cake is valued as a supplementary feedstock or, if too "burnt" in a barrel process, as a fertiliser. Dark (heated) oils are used in lighting or as lubricants, and cool yellow-green oils as salad and cooking oils.

Sesame Oil (*Sesamum indicum, S. orientale*)

Sweet oil

Sesame oil tastes sweet and keeps for years. It is used as a culinary oil in Japan and Egypt. The oil cake is mixed with honey and preserved citron as a sweet, or the cake used as bee fodder. The leaves form a mucilage in water and are used in clearing sugar syrups when making jaggery. The stems and leaves, or the plant itself, repels or destroys soil nematodes, and the pelleted leaf can be tilled into crops as a nematicide.

Olives

Olives are traditionally soaked a few days in ash water (lye) or soda ash to remove the bitter taste. Then they are washed and covered with a saline solution of 3% salt. Fermentation over a few months passes from bacterial species to mainly lactic acid bacteria in six to ten days or thereabouts, then to yeasts which also develop with the lactic acids at twelve to fifteen days. About five yeast genera can be identified. When fermented, olives can be sterilised by heating, and bottled or marinated or kept under olive oil.

Olive Oil

Ripe fruit, beginning to redden, is first milled by stones set far enough apart not to crush the seed. When the pulp is bruised it is pressed in bags, yielding a salad oil, the "virgin oil", greenish and transparent. The pulp is now water-soaked and again bruised to yield a cooking oil. Finally, the pulp is broken up, soaked, and fermented in large vats, then pressed to yield an oil for soaps and cosmetics, and the seeds pressed for a lighting oil.

Olives— Variation One — Greek method

Wash 454 gm (1 lb) olives thoroughly and put them in an earthenware crock large enough to easily hold them. Cover with a saline solution of 1 litre (2 pints) water to 1 cup salt. Weight a heavy plate on the olives and keep them submerged in a cool place. After several weeks, drain and rinse thoroughly, hit each olive with a mallet or hammer, or slit each olive. Fill a sterilised jar with the olives.

Then mix: ¾ cup olive oil; ½ cup wine vinegar; 1 large bay leaf; 1 clove of garlic, cut fine; and 1 tsp of fresh oregano.

Pour mixture over olives and gently shake the jars. Keep jars cool or refrigerated once they are marinated. They will be juicy and keep for months.

Olives— Variation Two — French method

To remove bitterness from olives, soak in water for ten to fourteen days, changing the water daily. Then pack the olives in a jar with a few chilli and some dill seed and cover with a saline solution of 30 gm (1¼ oz) salt per litre (2 pints) of boiled water. Cover and keep cool for two to four weeks to mature, then again add the same quantity of salt as before, now dissolved in very little water. Best eaten after two to three months. If olives are removed from the salt solution, just cover with olive oil to keep.

Salty Black Olives — Greece

Mix 4.5 kg (10 lb) olives and 2.25 kg (5 lb) salt well and spread into a wooden box; stir once a week, letting brine seep out. Leave salted four to six weeks, when the olives should be shrunken and less bitter. Turn onto a platter and add a little olive oil, just enough to coat each olive, put olives into jars and refrigerate. To serve, sprinkle with oregano and lemon juice.

Pickled Olives

Olives are green from March to June and black from April to July in the Southern Hemisphere.

To every 3 kg (6½ lb) of black or green olives, use ⅔ cup of sea salt, 8 cups of water, and 1 cup of olive oil. Note that black olives are ripe green olives, not a different variety.

Sort olives and discard blemished fruits for oil pressing. Cut two to three slits to the stone in the remaining olives, and place in a 5 litre (10 pint) drum until it is ⅔ full, then cover with water and weight with a wooden plug and a stone or a water filled plastic bag; a scum may form, but it is harmless.

The water is changed daily. After seven days taste black olives for bitterness (black olives take less time to leach). Taste green olives after fourteen to eighteen days. Continue changing water until olives are almost non-bitter.

Combine the ⅔ cup of salt with 8 cups of water, boil and cool, then replace water in olives with brine at this strength. Add the oil to make airtight. These will keep several months but need a fresh oil topping if the top olives and the oil "skin" are removed periodically. They are usually stored in a rubber-sealed jar. Start to eat after ten days in the brine solution. (*Australian Women's Weekly*, May 1992).

Marinated Olives (store for six months)

Use pickled olives from above:

- Thinly peel a lime and chop skin finely with a sprig of lemon grass, a small onion, and 2 red chillies. Combine in a glass jar with 1½ tsp cumin seeds, and 3 cups of olive oil. Take 600 gm (20 oz) of drained pickled olives, and pour the above marinade over them. Store at room temperature.

Or in the same quantity of olives, alternative marinades are:

- ¼ cup fresh grated ginger, 1 clove of garlic, ½ tsp sesame oil, and 3 cups of olive oil.
- Combine a handful of oregano, rosemary, thyme, 2 cloves of garlic, 2 slices of lemon, with 3 cups of olive oil. Proceed the same as in Variation One.

In Thai Markets

Regional varieties of shrimp paste are mounded for sale

Pickled fruits for sale at market

Dried squid are ready for rolling and roasting at point of sale

Soybean Processing

A soybean husker removes bean husks. Husked beans (in baskets) are placed in a wooden steamer (on floor) over a boiler for cooking. Once cooked, they are inoculated with koji

Husked beans await cooking

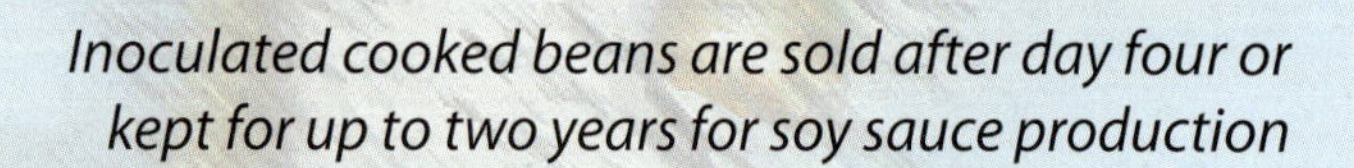

Inoculated cooked beans are sold after day four or kept for up to two years for soy sauce production

Soybean Processing

A monk at the Asoke Monastery presses down fermenting beans to reveal sauce at an early stage of the process

Beans are sold as a sour bean ferment or later pressed to extract soy sauce

Marine Products

Salted, dried and smoked fish are threaded on bamboo skewers

Cleaned squid are laid whole on a tray for artificial heat drying in humid weather

Dried seaweed in a Korean shop in Japan

Marine Products

Salted fish are presented in useful containers

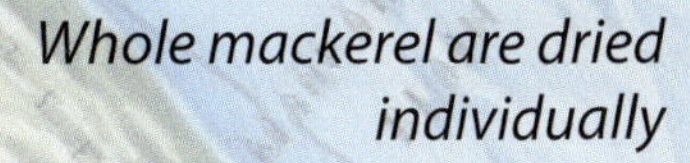

Whole mackerel are dried individually

Different sizes of shrimp, salted and dried

Marine Products

Various species of fish are differently prepared for drying

Small fish are split and headed

Thicker fish are split and headed with the backbone remaining

Trevally-type fish are headed and backbone is semi-detached

Aguamiels

Palm inflorescence is tapped and the sap collected in a container

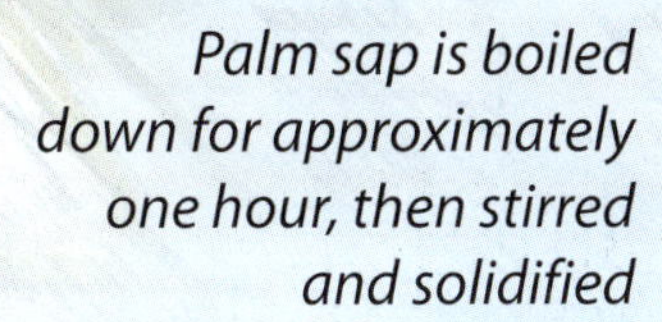

Palm sap is boiled down for approximately one hour, then stirred and solidified

Spirals of palm sugar packaged for sale in a market

Bamboo

Water-cured bamboo shoots from a soldered can will store for one year

Shredded bamboo under lactic acid ferment in a farm kitchen. Note the grass plugs

Rice and beans cooked in thin walled bamboo nodes at a market

At the Markets

Yeast bread cooked in a beehive oven in Egypt

Maize popped at a market in Thailand

Giant insects for sale in Thailand. Males have a vanilla-type flavour

At the Markets

Parkia pods and brine fermented seeds for sale in Thailand

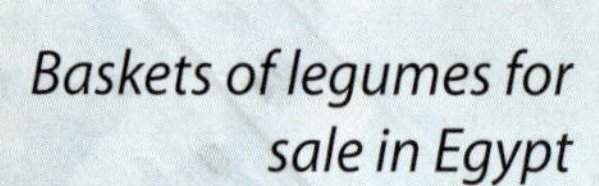

Baskets of legumes for sale in Egypt

Salt-pickled duck eggs for sale

Fungi

Filling pots with sawdust, lime and bran, for inoculation

Sterilised bags, fruiting with three varieties of edible mushrooms

Wild fungi from termite mounds, salted for sale at a market

Meat Products

Sour and dried fermented sausage for long-keeping

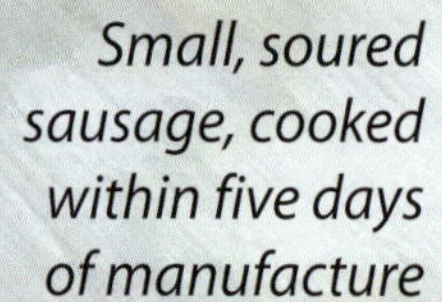

Small, soured sausage, cooked within five days of manufacture

Mealworms in Saigon, ready for a stir-fry

Asian Delights

Fifty chop meal, filled with fermented products

Bean paste fermenting in large jars

Bound seaweed waiting for its next treatment

Asian Delights

Seafood on sale in Korea

Daikon radish and cabbage ferments on sale

Blah-rah and bamboo in buckets at a market

At the Markets

Giant kelp bundled into blocks for sale in Chile

Kelp floats, threaded up and dried

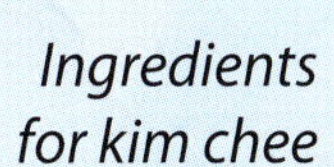

Ingredients for kim chee

At the Markets

Fermented sausages

Pots on sale, used for home fermenting

Thai greens

CHAPTER SEVEN

Leaf • Stem • Aguamiels

7.1 LEAF FERMENTS

Shredded leaf of the cabbage family, mixed at times with grated ginger root, garlic bulbs, or a few fine-chopped chilli pods, and tight packed with or (more rarely) without salt into water-sealed jars, is common to all Eurasia as sauerkraut (western Europe) and as kim chee (Korea). It is a lactic acid ferment and is very similar in nature to dill pickle ferment, preserved olives in brine and so on. Jars are left in cool cellars or refrigerators after they are sufficiently soured (three to ten days). Preliminary souring is in a warm place, or in the sun. A small bowl of this is served with most meals.

Sauerkraut—Variation one

Wash a head of cabbage, pulling off the older leaves. Quarter and remove stalk, then shred remaining young leaves finely. Wash the whole older leaves and line the bottom of the pot. Pack in shredded cabbage tightly, with 1 sliced garlic corm per cabbage, and 1 chopped chilli to 2 cabbages. Add a sprinkle of peppercorn and juniper berry (optional). Bring to within 5–8 cm (2"–3") of the top of the pot, weight down a lid or wooden cover, and let ferment. See Figure 7.1.

During ferment remove scum daily, wash the weight (usually a stone or iron weight), then weight again. When sauerkraut tastes okay, transfer to sterilised jars or sealed plastic bags and store in a cool place or a refrigerator.

Sauerkraut—Variation two

Prepare ingredients: 15 heads of cabbage, quartered and outer leaves and stem removed, 1 kg (2¼ lb) salt, and 24 green or tart apples.

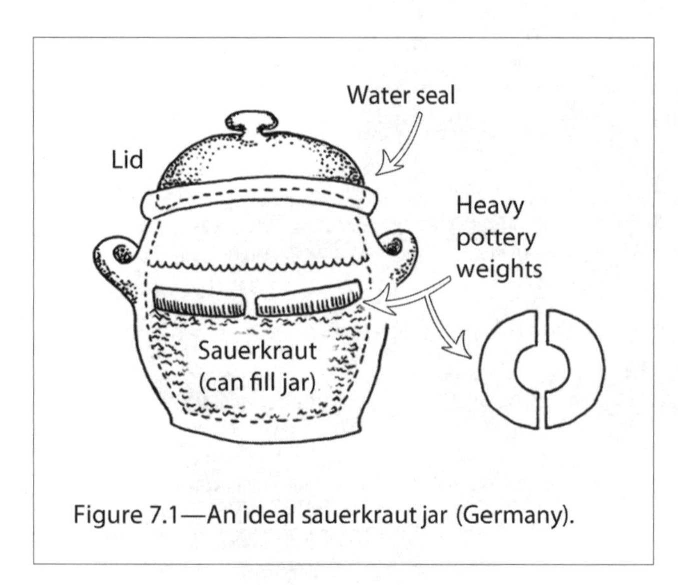

Figure 7.1—An ideal sauerkraut jar (Germany).

Slice young leaves of cabbage very fine. To every 2.25 kg (5 lb) shredded cabbages sprinkle ¼ cup of salt, mix well, and pack into a 36 litre (10 gal) crock. Add 1 cup of shredded apple and stamp down until the juices cover the cabbage. Continue to the end, always pounding tight to raise the brine.

Use cabbage leaves to cover, fit a round cloth, then a fitted round of wood, and a heavy stone. Leave space above board to allow swelling. Keep in a warm place for two weeks, examining and carefully lifting out any scum with the cloth; wash cloth, board and stone and upper side of the crock, cleaning off scum. Re-cover, move jar to a cool storage (can be part buried in shade) and examine for cleaning weekly. Pack sauerkraut in plastic bags or jars and heat seal, or store in a refrigerator. To top up jars use ¼ cup of salt to 1 qt of water. (Kander, 1945).

John Peceks' grandmother used him as a compacter for sauerkraut, washing his feet and putting him in a big vat to compact layers of cabbage and salt, then weighted it down with a wooden plug and a big stone. My grandmother simply kept me on a butter churn handle for hours! (Dr. Pecek is an agricultural scientist at Iowa State University, whom I met in Russia.)

Additions to sauerkraut are cumin seed, fine sliced apple, buttermilk (1 litre [2 pints] per 20 kg [44 lb] of cabbage), sliced radish, coriander and mustard seed, chilli, garlic, ginger and also chives can be used.

Swiss Sauerkraut—Cooking sauerkraut

Heat sauerkraut with black mustard seed and juniper berries, then mix with onion (which has been browned in a pan) and 2 glasses of champagne. Roast peanuts to crunchy stage, adding these on top when serving. (Heidi Ruef *pers. comm.*)

Kim Chee—Variation One Korea

Prepare ingredients: 454 gm (1 lb) Chinese cabbage, 454 gm (1 lb) white radish (daikon) cut in slices 1.3 cm (½") thick, ginger and garlic (5–6 cloves), finely chopped, 5 or 6 spring onions, sliced, and 2 or 3 whole chillies.

Soak all the vegetables in 5 cups of water, and 3 tbsp of salt, in a basin overnight, weighted loosely by a plate with a brick on top. Rinse this mix, tasting for salt, and pack fairly tight into a glass jar left in a warm place for three to seven days (taste for sourness). You can add a glass lid and refrigerate when sour enough. Serve as an apéritif before most meals, or sprinkle in a clear soup just before serving.

Kim Chee—Variation Two

Chop brassica leaf and cucumber, and mix with lesser amounts of flavourings; garlic, chilli, onion, peppers, ginger and 4–6% salt. Press the mass down in large earthenware jars; some shrimp or oyster can be added (up to 5% of the bulk). This matures in five to eighteen days 10–18°C (50–64°F) in cellars or buried in the earth, and can be further kept for many weeks refrigerated in closed plastic containers. Every Korean eats about 150–250 gm (5¼–9 oz) per day, so it is a very common relish.

Gundruk (also known as pickled greens) Nepal

Variation One

Several brassicas (leaves of radish, cauliflower, cabbage, mustard, rape) are gathered, washed and sun-dried to greenish-yellow. Then the wilted leaf is crushed and pounded to shreds in a mortar and stamped down tight in a jar lined with banana leaf, and the top pressed down tight. Stand for ten days to ferment.

This sour ferment is then sun-dried for some days and later used in soups and curries. It is stored in a cloth bundle and hung in the sun to dry from time to time. It is also possible to store sun-dried leaf ferment in an earthenware jar, tightly wadded with 15–20% salt to preserve it, and to use it as a salty addition to meals and soups. The Chinese sell small 600 mL (10 fl oz) pots tightly wadded with fermented, dried, and salted brassica leaf, almost like a wad of tobacco or silage. I eat it in very small pieces as it is very salty.

Gundruk—Variation Two

Leaves of radish, cauliflower, mustard, etc. are washed and pounded with a wooden mallet or put through a washing mangle and squeezed dry. A dry earth vessel (sun or fire warmed) is first layered with straw, then with the pounded leaves. Each layer hammered tight like a tobacco wad, until the whole is tight-filled. This is placed in the sun for a week—note that it froths. The jar is kept warm by a fire or the sun.

About a week later, the ferment smells of yeast and is taken out for sun-drying. It is left out day and night (for night dew) over two to three days, then kept in a clean, dry, covered pot. If everything is not "right", fruit fly worms get in and it is spoilt. However, it occurs to me that even 15% by weight of salt, plus a new compacting of the gundruk in the new pot would prevent worms, and make for a long, safe storage.

This leaf ferment closely resembles the kawal of North Africa.

Note—Overeating of raw or fermented radish can cause stomach pain. Any overconsumption of lactic acid will do this; it may also produce thirst. Sufferers recover, but as with many ferments or foods, these are intended as a regular food in modest quantity. (Oranges have the same effect if too many are eaten at once!)

Mustard Leaf Pickle (*Brassica juncea)*

The broad-leaved mustard is wilted dry, salted, and pressed under a heavy stone in a jar. The liquid is drained off, and the process repeated (salt and press). It ferments in five to ten days at 5–30°C (41–86°F). Repeat as before, then wash, add spices (garlic, ginger, cumin), pasteurise at 80°C (176°F) for fifteen minutes, cool, package or pack to seal. Keeps six months at room temperature. Used as a side dish at many meals. It can be dried, rammed down in pots and well-sealed to keep. To keep in this way, salt is retained.

Cacti (also known as Nopales, Nopalitos) Mexico

Young plates of prickly pear cacti are gathered in spring, using tongs to collect them. The spines are scraped off with a sharp knife, and the edges trimmed if necessary. Cut cleaned plates in strips and store in refrigerator, under water, for up to two weeks. Cook by stir-frying, roasting, steaming or boiling in water for fifteen minutes to reduce oxalic acids. Cooked strips are added to cold bean salads, leaf salads, or casseroles. They are also pickled in vinegar or oil-spice achars, and used pickled.

Drying Cowpea and Pumpkin Leaves in Africa

Leaves and whole young tips are sun-wilted and dried for a day, and that night wadded tightly into a large pot with a little water, then steamed for an hour, often turned, until soft and yellow, and then again sun-dried to crisp.

To store, the dried leaf is rolled into football-size parcels and hung from the roof in cotton bags or bundles bound with cord over large leaves. (Tredgold, 1986).

Kawal (also known as *Cassia* Leaf) Sudan

Leaves of the leguminous plant *Cassia obtusifolia* are pounded to a paste and packed in an earthenware jar, which (covered with sorghum leaves) is buried in a cool shady place. Contents are hand-mixed every three days, and at about two weeks, the fermented paste is moulded into small balls and sun-dried. This high-protein preserved ferment is added to spicy stews containing okra (which can also be dried) and onion, and can be kept for long periods. All the essential amino acids are present in kawal (*New Scientist*, 8 August 1985). Kawal is made in the Sudan, and obviously has promise as a food in emergency situations.

It is during the fermentation and sun-drying that the concentrations of some of the amino acids increase. A particular value of kawal is that it contains sulphurous amino acids, which are normally obtained from either fish or meat. Poisonous glycosides, often found in *Cassia* leaves, are destroyed in the ferment.

The traditional cheap way of making kawal is well suited to tropical Africa. The ferment demonstrates the value of traditional food preparation techniques.

Campbell-Platt (1987) notes that the kawal ferment breaks down toxic glycosides in the leaf, and that both bacteria and yeasts are present. The leaves are collected at flowering and fruiting times. The paste is stirred every three days over twelve to fifteen days. The container (30 litres [8 gals]) is buried in a cool and shady place. Protein content is about 25% of dry weight.

Ochse (1931) states the leaves of these leguminous tree leaves are edible. These could be tried for kawals:

- *Albizzia procera*—young leaves.
- *Bauhinia malabarica, B. tomentosa*—young leaves.
- *Cassia hirsuta*—cultivated for leaves, steamed.
- *Cassia laevigata*—as above, and unripe seeds cooked.
- *Cassia occidentalis*—leaves steamed, unripe pods eaten raw.
- *Cassia tora*—leaves steamed, seeds roasted or cooked.
- *Sesbania sesban*—young pods and flower buds, leaflets eaten.
- *Sesbania grandiflora*—white-flowered varieties are cultivated for their edible flower and young leaves. They yield large quantities of flowers, and can be steamed or stir-fried.
- *Erythrina variegata*—young leaves steamed.
- *Leucaena glauca*—buds, young leaves, and young pods steamed or baked in coconut. Ripe seeds are roasted and pounded, or germinated and eaten.

Leaves Eaten in Senegal

Aubert (1986) lists the following leaves commonly eaten in Senegal: peanuts, amaranth, *Cassia tora*, *Corchorus olitorius*, pumpkin or squash leaves, fig leaves, cassava leaf, *Moringa* leaf, and cowpea leaf. Dry powdered baobab leaf is added to couscous to raise the protein level to about 13%. Leaf protein (low in methionine) supplements cereal protein (low in lysine), as do eggs or *Spirulina*. Young leaves are particularly digestible and rich in amino acids.

Dry and powdered leaf can be beneficially added to cereal ferments, as they are in Asia, for preparing starter cultures or increasing protein available in porridges of grains. These leaves are part of the "secret" ingredients in yeast cultures.

Bulla

The stems of the Abyssinian banana (*Ensete ventricosum*) are cut into small pieces and a large container filled and covered with *Ensete* leaves, weighted, and left at 18–25°C (64–77°F) for two to four months. The mass is drained, and the floury ferment used as a fresh dough or dried as a flour.

The dough or flour is used in flat breads, alone or with cereal flours (Campbell-Platt, 1987).

Shiso (*Perilla frutescens*) — Japan

Also known as Beefsteak Leaf

There are green and red-leaved varieties, used as teas and a spicy, mildly mint flavoured addition to rice koji fish ferments. The salted, preserved leaves are used to spice tofu and tempeh. The young shoots are eaten with raw fish, and the seeds salt-preserved for flavouring pickles, miso, and ferments.

Samphire Pickle

Samphire is the fat, crisp, salty succulent that grows in fields inundated by the sea, or rims estuaries and such places. It has long been pickled as a savoury with cold meats and green salads.

Wash and drain 2 cups of samphire well and pack loosely in a jar with spices: a pinch of grated horseradish, 14 gm (½ oz) nasturtium seed, and 14 gm (½ oz) peppercorns. Boil together: 1 cup vinegar, 1 cup cider with a pinch of salt, and pour over samphire. Place in a warm oven for 1 hour, then cover and cool. Store cold.

Bracken-Fern Ferment — Japan

The young wild shoots of bracken fern (*woggerlee* in Tasmanian aboriginal language) are gathered in spring, boiled a few minutes; cooled, and immersed in nuka-bran with some salt. These are later dried and stored for preservation.

Ti (*Cordyline fruticosa*)

The leaves are greatly in demand as food wrapping in earth-ovens. The plant is propagated by cuttings for garden hedges. Every Polynesian, Micronesian, and Hawaiian garden has these useful plants growing. The roots, baked, become soft and sweet, and produce a sugary juice.

Beer is made from the baked root and sugars can be evaporated from the juice, or the cooked root dried and later grated for its sugars.

7.2 ONIONS AND GARLIC

Choose good-keeping varieties (Spanish brown) and dry tops in the sun before storing. Either plait tops into a string about 0.5 m (1.6') long or place in mesh bags. Hang in a warm, airy place.

Pickled Onions

Peel in water, 4.5 litres (9½ pints) small onions, 2–5 cm (¾"-2") diameter, and soak overnight in brine: 454 gm (1 lb) salt to 1.2 litres (2½ pints) water.

Next day boil 2.3 litres (5 pints) vinegar, with 2 cups of sugar and a *bouquet garni* of 4 chillies, black pepper and cloves in a muslin bag. Remove spices after a few minutes, throw in onions, boil briefly, and pack in well sealed jars. For these, definitely use malt vinegar.

Lahsunko Achar (garlic condiment) — Nepal

Mix 10 peeled cloves of garlic, the juice of 4 lemons and 4 tsp of salt, and leave in a warm place for four days.

Fry a few chillies with 1 tsp of fenugreek seeds in 1 cup oil. Take out the chilli and seeds and grind them well. Combine oil, powdered spices, and garlic solution. Matures after four days. (Majupuria, 1986).

7.3 STEMS AND SUGARS

Bamboo

Young stems, 4–6 nodes long, are peeled until the white shoot emerges; this is then sliced crossways, nodes rejected as tough, and cautiously tasted. Sweet shoots are eaten fresh, or parboiled and frozen. Bitter shoots are boiled in two changes of water for twenty minutes in total. They can be frozen or salted, pressed to remove excess water, and dehydrated. They are used as for any vegetable, particularly in stir-fry foods, for their crisp texture (Creasy, 1982).

Bamboo Shoots — India

Bamboo shoots, peeled and sliced, are allowed to ferment in some water for several days before frying briefly until golden, and then added to curries or made into a pickle. Bamboo is also eaten fresh if tender, and canned in a light salt solution. As many bamboos have a hydrocyanic acid content, it is wise to get advice as to edible species, and to soak suspect species in changes of water for three to five days before cooking or pickling.

Bamboo Shoots — Japan

Bamboo shoots, in the husk, are boiled in water, to which a small amount of rice bran or bran is added, until they are tender. They are then cooled, husked, sliced, and any white sediment washed away. They are used in stir-fries, soups, or simmered in soy, sugar, and mirin with sliced seaweed for a pickle (eaten hot or kept cool for a few days). Bamboo slices are also part of nuka box pickles.

For a pickle, fried bamboo is immersed in oil and spices in an airtight jar, and later eaten as a lactic acid pickle.

Sugar

Jaggery (also known as gur or solid raw sugar) — India

Variation One

The top 2–3 joints and leaf (containing glucose) are cut from fresh canes, and canes are passed through three roller crushers (with over 50–70% sap extracted). The crushed canes are used as fuels to heat wide pans 1.22 m–1.52 m (4'–5') across, 22.8 cm–25.4 cm (9"–10") deep and the juice vigorously boiled to 115–117°C (240–242°F), continuously stirring, for fifty-one minutes. Lime is used to clear the juice before heating and straining into the pans to boil. Foam is controlled with crushed castor oil seed extracted in water, sprinkled on frequently. Today the hot syrup is simply ladled into buckets (plastic) which become moulds as it cools. Jaggery can be best kept by sealing into earthenware jars (mud or dough seals). Some gur is cooled in the pots in which it is sealed.

Crushed cane and their tops are fed to cattle but can also be processed into many products. They can be used as mulch or composted, or used as litter for poultry or livestock. Field trash is also composted, used as fuel or litter.

As one might expect, jaggery is a much more valuable nutritive sugar than refined sugar. Per 100 gm it has: 6–20 gm of glucose and fructose, 1,500–2,800 mg of mineral salts (versus 30–50 mg in white sugar), 600–1,000 mg of potassium, 60–130 mg magnesium, 40–110 mg calcium, 14–100 mg of phosphorus, and 4–40 mg of iron. Refined sugar lacks almost all of these minerals (Aubert, 1986).

Jaggery — India, Papua New Guinea

Version Two

Use only plump, fresh-cut, stems with the leafy tip and roots removed. Weigh this cane carefully. A good three roller crusher may be essential for efficient juice removal, but less efficient hand juicers are common in India. Pounding the cane helps for small mills. The juice is boiled in open pans two thirds filled (it can foam over); a 1.25 m (4') pan holds 75 kg (165 lb) of juice, and juice is removed hourly. Steel bottoms are usually 5 mm (⅕") or more thick, and liquids are measured by volume.

As boiling starts, scum skimming can be constant; the scum filtered and any juice returned to the pot. The juice as it is slowly heated hisses, and before this stops, clarifiers are added to remove such impurities as gums, fibres, waxes, clay, colouring matter, and albumin. The albumins carry these materials to the top as froth and are obtained by repeatedly washing and squeezing albuminous plants in water until a mucilaginous liquid is produced. Plants used to clarify are: peanut and castor oil seeds, okra, comfrey, *hibiscus* species, or any plant yielding mucilages that will entangle the impurities and form a scum.

After this stage, boiling over is a risk and must be watched, stirring maintained to prevent burning, and cold water sprinkled on to stop the scum mixing back into the juice. Vegetable albumin is added until the scum froths white, then the boiling rate is increased until a drop of sugar syrup snaps (I am told about fifty-one minutes). The temperature is then 115–117°C (240–242°F); aqueous extract of castor seed is always ready to sprinkle on in this vigorous boiling stage. Then the pan is taken off the fire, or the fire cut off the pan and the mass cools as stirring (always constant) continues, and begins to solidify.

It is then poured into moulds (plastic buckets) to set. Lime or clean wood ash can be added (1 tbsp of lime for6 litres [10.5 pints]) after clarification, to reduce acidity. About 2 kg (4½ lb) of green plants clarify 45 litres (10 gals) of juice. At home, in the last minutes of boiling, banana or sweet potato chips can be dipped in to cook (in a sieve) as can dough balls. Some sugar can be set over nuts and dried fruit, or grated coconut stirred in. Practise makes perfect.

Sugar Stems

An aquatic *Panicum* grass (*P. burgu*) grows to 3 m (9.8') in the Sudan, along the Niger, and is used for sugar in Timbuktu. Cut in April, when the river falls, the stems are trimmed of leaves and dried in the compounds or under shelter. When needed, they are cut into small pieces about 1 cm (½") long, crushed or pounded, so that water soaks into the tissues and extracts the sugar, then these fragments are put in baskets made from doum palm, and cold or warm water poured or dripped through into large pots, as a thick syrup. The liquor is called koundou-hari, the habitual drink of the inhabitants. It ferments quickly and needs to be bottled to prevent it turning to vinegar (it "turns" to vinegar in three to six days). The juice is thickish, caramel in colour, and is most appreciated by the locals.

Reduced by boiling, a sort of acrid honey is produced (katou or katoo), and is made into toffees and sweetmeats. Lime may improve acidity.

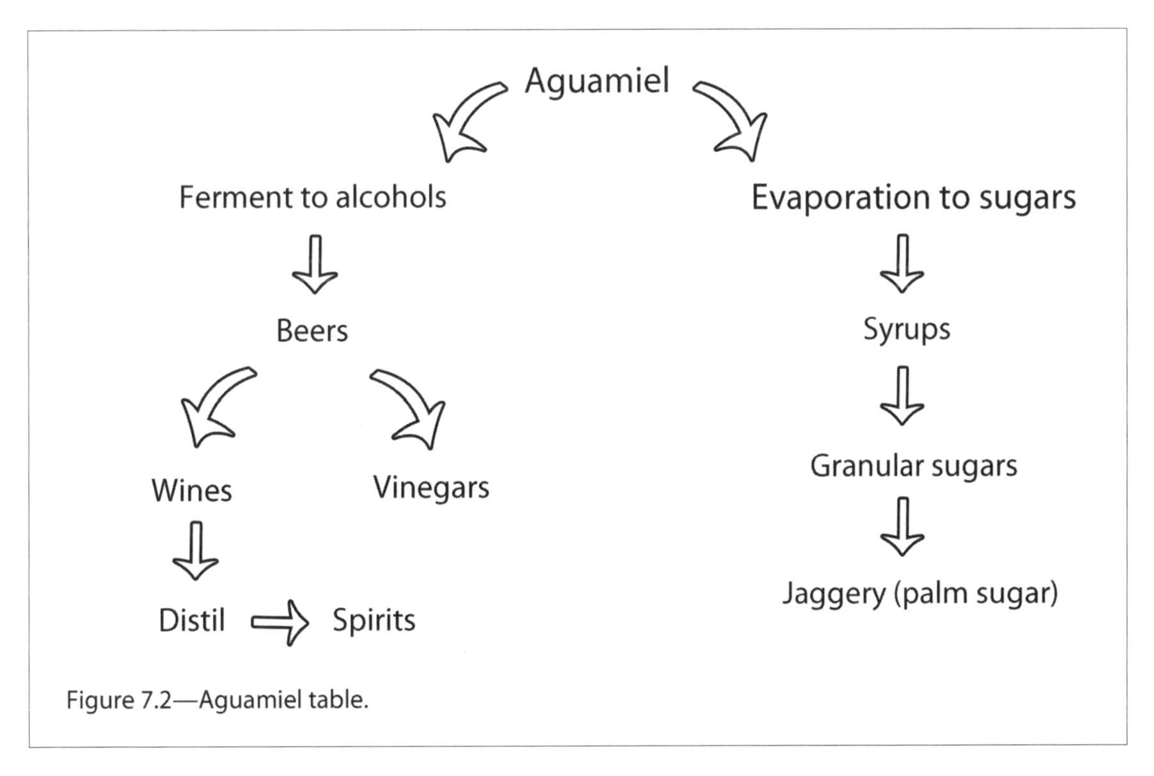

Figure 7.2—Aguamiel table.

Freeze-Concentrating Liquors

Several tree species, including sugar maple (*Acer saccharum*), cider gum (*Eucalyptus johnsoni*), and many of the cold area deciduous trees (walnut, birch) are tapped for sugary saps from early spring. Although today such sugars are reduced by heat, indigenous peoples concentrate them by freezing in tight-woven and gum-sealed shallow baskets 8–12 cm (3"–4¾") deep and 40 cm (16") wide. Exposed to frost each night, the thin ice on top was thrown away each day, and the sugars so concentrated to 30–50% of their initial volume for storage. Salts, sugar, and alcohols all have lower freezing points than water alone, and are left as concentrates after successive partial freezing. A physicist friend (John Reid) treats poor-quality wine this way, and creates a fortified sweet wine by the judicious addition of fruit syrups! As students, we usually drank cheap wines suited to this treatment!

Seawater, syrup, and even liquid ferments such as soy sauce and semi-solids such as seaweeds, agar, fish, and meats can be freeze-dried to remove water, and sun or wind dried by day, or sheets of ice removed. Freezing, unlike heating, does not destroy vitamins or create invert and caramelised sugars, so that this method is preferable to (and more economical than) fire heating in frosty climates or seasons.

Freeze-drying in fact mummifies whole animals, such as seals strayed and lost in the Antarctic ice cap, where they are preserved for hundreds or thousands of years by desiccation. Many museums now use freeze-drying by liquid nitrogen to preserve animals. Another physicist (James Lovelock of Gaia fame) states that he uses a flask of liquid nitrogen in his kitchen to make parsley brittle, and thus easy to powder, and he cannot imagine a kitchen without liquid nitrogen!

Larch Bark—Winter

The soft wood and inner bark of larches is easily scraped off, mixed with water, and fermented into a dough to be mixed with rye meal and buried below snow for a day. Fermentation commences, and the dough can be cooked as a camp bread or dumplings, the sweet wood pulp acting as a sugar for the yeast in rye. (Kirby, M.E., 1973).

Willow Underbark

The underbark of Arctic willows (and the starchy roots of some desert eucalypts) can be tasted for flavour. If not too bitter, or too tough, gather a good bundle, scrape off the outer bark, wash, and then scrape the young underbark (do not cut it off). Mix with a little sugar and seal oil to freeze, and use to thicken stews, or as an ice-cream. Especially good willow groves can be planted by selection of "sweet" varieties. (Olanna, 1989).

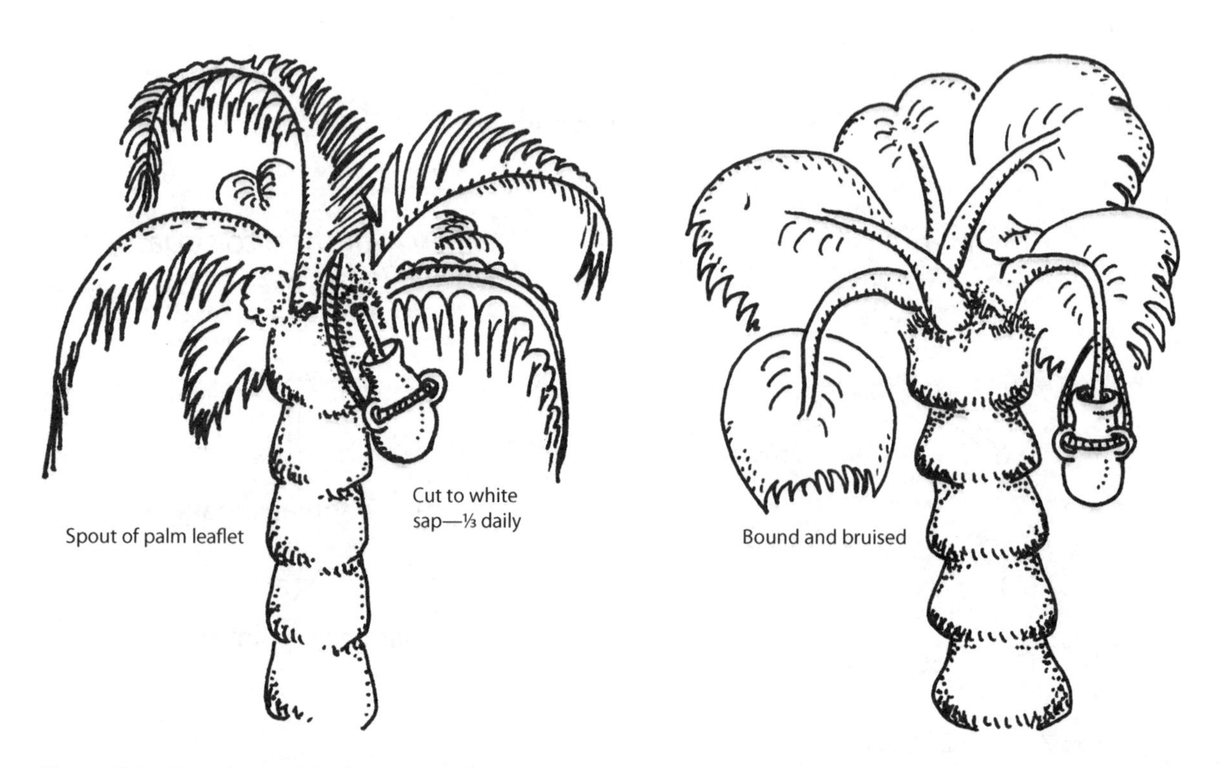

Figure 7.3—Tapping palms for aguamiels.

Palms and Saps

The flowering stalks of agaves, palms, and cacti yield sugary saps (aguamiels) if crushed or cut, as does the sap of many tree species (sugar maple, sugar eucalyptus). Agave flower stems may be earth-oven cooked to free the sugary pulps. The coconut palm flower stalk will yield eight litres (17 pints) per day of sweet sap, which naturally ferments to 8% alcohol in hours. Typical aguamiel processes are shown in Figure 7.2.

In Africa, India and Oceania a great variety of palms are tapped for sugars. These are the basic techniques:

- The new inflorescence is cut off, and the cut stem beaten to release sap daily. There may be an acid wash (vinegar or acetic acid) to renew sap flow, and cutting back and beating the stem as needed.
- The softer crown just below the growing tip of the palm is carefully cut back, a spout of palm leaf fixed at the base, and a jug tied on below to catch the sweet sap. See Figure 7.3.

Some Native Teas

Hot teas, made by steeping fresh or dried leaf or bark material in hot water, are drunk in many areas. A few regular teas are:

- Lemon Grass tea—Young shoot bases; also added to "hot and sour" Thai soups. (*Cymbopogon citratus*)
- Labrador tea—Larch shoots, young stems and needles, and young pine needles. These are a good source of vitamin C. (*Larix laricina* or *Pinus* spp.) Found in the Arctic and north areas.
- Cape Barren tea—Tasmania. Young stems and buds of *Correa speciosa* are used for tea.
- Rooibos tea—South Africa. A popular tea. (*Combretum apiculatum*).
- Yerba Maté tea—The tea of the gauchos of Argentina, Brazil, from *Ilex paraguayensis*.
- Herbal tea—Made from chamomile, lemon balm, clover flowers, alfalfa (lucerne) shoots, and so on (fresh or dried). This is widely consumed.
- Arabic tea—Brewed from Khat (*Catha edulis*) in Yemen.

Wild Date Palm Sugar — India

Palms are tapped when they are 0.5 to 1 m (1½'–3') high (seven to ten years). The leaf bases below the crown are cut back until the soft white core is exposed, and a palm leaf drip channel fitted to drip the sugary sap into a lime-lined pot. The white core area divided into three, is cut again every three days, yielding 70–130 kg (155–285 lb) per season. Ten percent of this weight is crystallised as palm jaggery (gur). The lime prevents ferment for a day or two.

Palm sap has B vitamins and vitamin C, also thiamine and niacin. The vitamin C is stable over five days. Most sap is consumed as toddy (eight to ten hours spontaneous ferment). Toddy is a nutritious drink and a source of thiamine, riboflavin, and niacin. Gur is higher in amino acids than cane sugar. Unfermented sap is used for gur, as fresh as possible, as ferment reduces the sugar content.

Date palms are variously tapped by running a trough around the base of the terminal buds for *Phoenix dactylis*; near the roots for *Phoenix reclinata*; just below the crown on one side only for *Phoenix sylvestris*; and coconut palms (*Cocos* spp) have the flowers cut off and the flower stem bruised, bound, and cut back daily.

Ayyangar *et alia* (1977) notes that liming takes place as soon as the fresh juice is weighed (or the volume measured), and the pH brought to 8.0–8.5, neutralising acidic fractions and forming mainly insoluble calcium phosphate deposits (part of the "mud").

Muds are washed to extract sugar, are dried for fertiliser, and the washings returned to vats. If sugar is needed to be decolourised, activated charcoal or bone char filters are used. Clarified sugar juice is 80% water, reduced to 40% water in pans or evaporators. In closed vessels, pressure is reduced over a series of four to five evaporators and the syrups are indirectly steam-heated.

Sugar crystallisation can be initiated by the addition of "seed" grains of sugar as nuclei. The "strike" is the whole contents of a pan when mixed crystals and syrups are poured out for cooling. Molasses is the heavy syrup separated by centrifuge from the crystals, the latter being also lightly washed and sun-dried.

Molasses is used in breads, and for yeast, bacterial culture, and alcohol production.

Coconut Syrup — Palau, Micronesia

Aguamiel tapped from the coconut inflorescence is reduced by slow boiling over coals until a molasses-like consistency is obtained, or more rarely it reduces to a crystallised form. This process needs close attention to prevent burning. The thick brown syrup is used with coconut cream and starches to make sweets, or to sweeten cassava and taro porridges. Traditionally it was men who took on this work, to spare their women for hard work.

Ferment Retardants

Palm sugars ferment naturally in hours, but those intended for reduced syrups or thick sugars (jaggery) may be kept from fermenting by smearing lime in the gathering pots, or ash pastes, and by adding the bark of *Manilkara*, a sapote (*M. hexandra*) to the liquor. This bark is 10% tannin, and other tannin barks may help. Note that "ferment inhibitors" may also be used in silage for higher sugar content of the forage.

Bark from various species of *Shorea* or small pieces of jak-fruit wood (*Artocarpus*) are used to delay fermentation of palm sap in South East Asia. Chalk is also added to the collecting pots or bamboos as well as the lime paste used to line the pots.

Agaves

Several species of agaves are used for aguamiels, the flower stalks cut and tapped for liquor, which is fermented eight to thirty days to pulque. Some recently discovered agaves were also trimmed, pit-covered, cooked and eaten. Slow pit-cooking over hot stones, banana leaves, (earth-covered) converts starches to caramels, invert sugars and inulin.

Agave angustifolia and *A. tequiliana* leaf extracts are used to promote bacterial growth in composts, sewage, and holding ponds. As a food, the leaf bases and stems are cooked in earth-ovens, then fermented into a mild beer, later distilled to mescal, tequila, or bacanora. Roasted cores yield a sweet, calorie-rich food.

7.4 HERBS

Most culinary herbs, especially seasonal herbs like basil, are spread on trays and dried at 140°C (284°F), and later air-dried to crisp, and powdered or flaked via a sieve for airtight storage. Almost all oily culinary herbs retain flavour if rapidly dried.

CHAPTER EIGHT

MARINE AND FRESHWATER PRODUCTS

• FISH • MOLLUSCS • ALGAE

8.1 INTRODUCTION

Seafoods generally have been grossly overexploited in the recent past by overefficient modern boats serving mass markets, usually for overfed people. There are increasing numbers of freshwater or indoor fish farms, some carefully conserved re-introductions (giant clams, trochus, turtle), and a few total bans in order to allow stocks to recover (crocodile, shark, whale, bluefin tuna, dugong).

Carefully conserved, the sea is a rich source of food. Overfished or polluted, it is as barren as overexploited land. Increasingly, we need to **locally** conserve stocks, and to disallow large and very wasteful boats and netting which destroy the ecology of seabeds, crush or overheat fish masses, glut local markets, supply pig or chicken food, or kill non-target species (dolphins in drift nets).

The modern culprits are marina developments, dredged mangrove swamps, large beachside hotels spilling wastes to the sea, and eroded lands yielding silt and agricultural sprays to kill seagrasses and corals. Like the land, the sea is destroyed for short-term profit and this means food deficits in the long-term. Only by conserving mangroves, river forests, shorelines, and shallows can we preserve seafoods. Worldwide, mangroves are being destroyed for golf courses, shrimp farms, and canal "developments". All these are essentially economic vandalism.

Nutritional Value of Fish

Fish are sources of iodine and fluorine. Oily fish contain high levels of histidine, and shark, halibut, and tuna livers contain very high levels of vitamins A and D.

Roe, 20–25% protein is rich in nucleic acids.

Fish oils (oily pelagic fish, seabird and whale oils) are rich in vitamin A, particularly the livers. Some livers (polar bear) have too much vitamin A and can cause illness or death in people eating them (carotenosis, which resembles rheumatic fever, but also causes painful lumps near bones). Do not eat too many seabirds at one sitting! Sled dogs also have "poisonous" livers.

Fish bones contain calcium, e.g. whole sardines, also the beards of mussels dissolved in vinegar, or pearls in vinegar.

Oyster flesh is rich in zinc (100 mg per 100 gm) if zinc is normally present in local waters.

8.2 FERMENTED FISH

Introduction

Tanikawa (1985) gives a wide range of fish processing methods. Without this book, much of the fish ferment material would be less precise. The book is on loan to me from my marine friend Atsuko, whom I thank. There can be few better references on fish processing than this book, and although much of the content is available elsewhere, the compendium is unique in its nuka and koji (rice-fish) ferments. I have used a nuka box for years, but only for vegetables until I read these salt-bran fish recipes. Such material truly enlarges our potential for good ferment and good food.

Recently caught fish undergo a rapid change as enzymes within the body are released. These enzymes are usually located within the livers and guts of fish, which is the reason many fish are gutted and cleaned as they are caught. Also, the almost neutral pH of fish muscle is attractive to microbes which waste no time in attacking and reducing the fish to a strongly-flavoured and dangerous substance.

Many countries, particular in South East Asia and China, have developed ways to preserve large catches of fish through drying or fermenting techniques. Oriental fish sauce is a popular example of an important fermented fish product.

Most drying and fermenting techniques use salt to control the preservation process. Salt inhibits enzyme and microbe attack, so the faster salt is added to freshly-caught fish, the better. The high levels of salt found in most fish preserves means they are safe from such toxin-producing organisms as *Clostridium botulinum*, which causes botulism.

Unfortunately, high salt levels in such fermented products as fish sauces mean they cannot be eaten in large amounts, so that as protein extenders they are limited. However, salted and fermented fish remain an important flavour enhancer to peoples living on bland rice diets.

Ferment certainly prolongs the useful life of fish products, from hours to days or months. Most ferment also reduces the fresh product weight, sometimes considerably, so that transport is easier.

In fish ferment, the following factors reduce spoilage or putrefaction:

- The use of fresh, wholesome, well-cleaned and washed fish fillets, larger fish, or viscera. Only small fish 6 cm (2½") or less are used whole, ungutted, and rinsed.
- Thorough salting.
- Pre-drying for small whole fish and shrimp.
- Anaerobic ferment containers or conditions.
- Low pH (high acidity) produced by lactic acid bacteria, acid fruit juices, or vinegar.
- Clean utensils and containers.
- Sterilisation of packages and vacuum packing or good sealing.

Compared with other processes to preserve fish in hot, humid or tropical locations, or in the event of an unexpectedly large catch of fish for the fresh-fish market, ferment is certainly one of the least expensive fish preservation methods.

Basically, there are four types of products from preservation-fermentation processes: (See Figure 8.1)

- Dried or dried/fermented fillets or whole fish (including dried squid, and bonito flakes).
- Fish sauces, sometimes processed with soy sauce, sold as liquids or expressed from pastes and miso-fish preserves.
- Moist whole fish or fillets fermented in grain bases or as a part of the production of sauces.
- Fish pastes from pounded or blended fish (usually small fish), shrimp, fish entrails etc.

Wood, concrete, food-quality plastic, and glass are all suitable bulk or home food containers for fish ferments. For sale, foil or plastic pouches, often vacuum-sealed are now used. Once corked bottles served for most sauces, and most home ferment was in earthenware jars, or wooden tubs, or barrels.

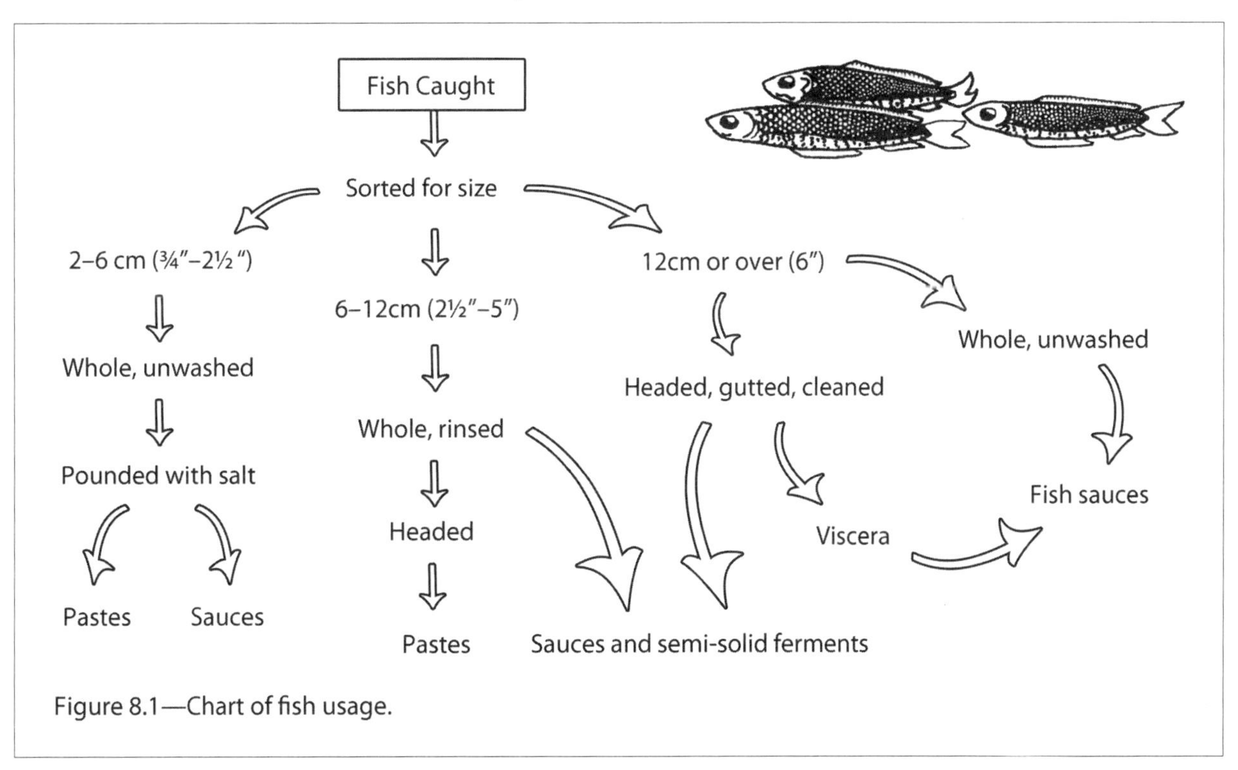

Figure 8.1—Chart of fish usage.

Fermentation of fish can be of small whole fish, whole shrimps, roe, milt, washed visceral mass (split before washing), gut and stomach meats (split if necessary) squid meats, octopus, oyster, or shellfish, and heads or trimmings.

Fish can be fermented in salt, in boiled rice, in bran, in sake lees, in vinegar, or even in compost heaps, as is done in Sweden. In salt ferments, fish meats and visceral mass become liquid after a year of ferment. Visceral masses of abalone, sea cucumber, roe and milt of sea urchins, roe of crabs and so on are all suitable ferment bases for popular Asian foods. For fish meat hydrolysis (liquification) to sauces and pastes, it is essential that enzymes are provided by fish or crab livers, visceral mass, etc. This is also true for squid meats.

Fermented Sushi Fish — Japan

A variety of fermented fish products is sold with rice (sometimes also fermented) in Japan. Some methods are as follows:

- Clean fish bodies (mullet, carp, bream, salmon etc.) are weighted down under 20–30% salt; salt is especially added to body cavity. Leave for one to two months. Then wash and drain the fish to de-salt (ten to eighteen hours). To 100% fish weight add 45% of this in rice, and 20% as koji. (See section 2.3 for description of koji starter)

  Alternate layers of rice and de-salted fish, filling the bodies with rice and koji. Weight the mass. The ferment is at first lactic, then later goes to yeasts. The fish is ready to eat after ten days, but tastes better after two months.
- Alternatively, fish fillets are brined for ten days, then soaked and de-salted for eighteen hours. They are drained, soaked for fifteen minutes in vinegar, and added to boiled rice to ferment (no koji) for ten days. The fish is sold with the rice.
- Smaller split fish are salted for twenty-four hours (454 gm [1 lb] salt per 100 small fish). The backbones are removed and 100 mL (3.4 fl oz) vinegar (with 8.3% acetic acid) is sprinkled on the fish. These are let drain for one hour and then alternated with a layer of boiled rice, a layer of bamboo or lemon grass leaves, a layer of vinegared fish, and so on until all are used. 100 small fish need 400 mL (13½ fl oz) vinegar, which is sprinkled on the bamboo leaves.

  The whole is pressed under weights and sold from ten days onwards.

  Note—Press weights are 10–20 kg (22–44 lb) of stones or iron over a wooden lid.
- Carp fillets are salted, and pressed at 20% salt for one month. The backbone can remain. Then they are de-salted and drained. Now, de-salted bodies are simply alternated with boiled rice layers, the top layer covered with bamboo leaves and parchment, and a lid fitted, on which a very heavy weight 150 kg (330 lb) is placed. The fish and rice are eaten after three months.
- De-salted salmon or other fish is fermented this way at times with fine-cut cabbage, radish (daikon) or carrot with the fish. Here 30% of the weight of the fish is rice. Koji is optional. Weighted at 20 kg (44 lb) for thirty days the ferment is best quality.
- Split fish are packed in barrels with 25–30% salt by weight and pressed ten to fourteen days with 20 kg (44 lb) weight. Excess expressed water is drained off. The fish are then drained, soaked in fresh water for ten to eighteen hours, and packed in a barrel in layers of cooled boiled rice. The fish layers are sprinkled with vinegar (or fish can have a fifteen minute vinegar soak) and cured for ten days, then sold. Variations are to sprinkle in layers of rice koji, or fine-cut vegetables, between the fish/rice layers. Bamboo leaves are commonly used as a layer above the fish and vegetable layers. After layering, the wooden lid is heavily weighted to 150 kg (330 lb). Curing can take three to four months but often best quality is said to be at thirty days. Squid are preserved in the same way, having been cleaned, boiled briefly in salt water and the epidermis scrubbed off, then salted, washed, and layered as for fish with rice, vegetables, and vinegar.

  Fish roe is also first pressed with salt, soaked, drained, and layered with koji at 50% weight of the roe or fish. For mackerel, tuna and bonito the fillets are salt-cured for two to three days at 20% salt, drained and sprinkled with koji. The whole is weighted over a bamboo leaf layer. Thirty to fifty days are allowed for thorough curing.

Fish Ferment in Koji

- Small round gutted fish are salted (20% of salt) for one day and the liquor poured away. Then the fish are stacked in a container with 50–60% by weight of koji and sold in glass or wooden containers after thirty to fifty days. Fillets of bream are similarly fermented, as are pieces of large fish such as bonito, mackerel, or tuna meats.
- Dried fish like herrings are soaked one day in rice wash-water and cut to 6–7 cm (2¼"–2¾"). Then salt and koji (for koji; use half the amount of the salt) is sprinkled over the top, vegetables and fish laid on, with vegetables twice the bulk of fish. Thus, layer up the barrel. Salt should weigh 15–20% of combined fish and vegetables. Weight to 10 kg (22 lb) for a month and then eat.
- Fresh, split, cleaned whole small fish bodies are laid up in a barrel with 25–30% salt, and pressed for two months. Excess liquor can be removed or added to other ferments; the fish are washed to de-salt, and drained after sixty days.

  About 50% by weight of fresh cool boiled rice and koji are layered with the fish, and the whole pressed and fermented six to sixty days before sale or consumption. Fish plus rice can be stored in coolrooms after a second ferment.

  Larger fish can be split, bones filleted, and heads and gills removed for this process.
- Use larger fish fillets or smaller whole fish. First, salt under weight for ten days (as above). Remove water, re-salt, and press for four to six weeks. Then de-salt in cool fresh water for ten to eighteen hours, (longer time applying to larger fish). Drain de-salted fish.

  Soak in used vinegar (of 5.3% acetic acid content at 400 mL [13½ fl oz] per 100 fish pieces) for fifteen minutes, then layer in cooled boiled rice (30–50% by weight), or a mix of cool rice 30% plus koji 10% by weight; some sliced vegetables can also be layered in. The barrel or crock is then weighted down to ferment. Traditionally, bamboo leaves are laid between strata of fish, and on top, before pressing. The fish are ready in ten days but can remain thirty days in the barrel, longer in cool storage.

Nuka (Bran) Fish Ferment

- Salt cure fish or fish roe with salt as usual (30% salt for a few days, pour off liquor and drain). Sprinkle sliced radish with salt and drain and wilt in the sun for two to three days.

  Now immerse the fish and radish in the nuka box when the weight of the fish is 30% of the damp bran, and press lid with a stone to exclude air, 20–30 kg (44–66 lb). Taste every week to decide on maturity.

  Store refrigerated in sealed containers.
- The liquid liberated from salted fish (adjusted to 30–35% salt) should be saved. Cleaned fish fillets are layered with 30–40% by weight of cooked rice or wheat bran in a barrel, and this liberated liquid is poured on, then the barrel is lidded and weighted.

  Long ferment (one to four years) produces excellent fish weight at 30–40 kg (66–88 lb). Liberated salt liquor can be boiled, skimmed, cooled, and brought to 35% salt for storage in barrels or jars.
- Sardines, headed, cleaned and washed are salted (at 30%) for one to three days, washed in fresh water, dried one day in the sun, and laid up with 40% of rice bran and 20% of koji, each layer sprinkled with ⅔ cup of sake.

  This is all weighted at 20 kg (44 lb) for five days, then saturated salt solution (liberated liquid plus some fresh salt) is poured in. The whole is weighted at 35 kg (77 lb) for four months.

Sake Lee Ferment

Fish or squid meats, stacked at 20% salt for ten days, are washed thoroughly and layered with an equal amount of sweet sake lees for ten to thirty days. Though fish is somewhat soft, it is delicious.

Abalone meat in 10% salt for one day is used, as are other large shellfish, rinsed in light brine after salting and with equal weight of lees for twenty days.

Colombo Cure — Sri Lanka

Mackerel, seer (*Scomberomorus* spp.) and large sardines, fresh landed, are gutted, gilled, and well washed in sea water. 3 parts by weight of fish are mixed with 1 part of salt in large tanks. To increase acidity, acid fruit pulp from the fruits of *Garcinia cambogia* or tamarind are added. The fish are submerged, and weighted with stones on mats. Here they remain for two to three months. The liquor is a "blood pickle".

For long storage and maturation, the fish are held in large wooden barrels in the blood pickle for up to a year, the flesh remaining firm and flaky.

Juices from the pickle are used as a fish sauce, and can be reduced, sterilised, and bottled after the fish are sold. This sauce resembles all fish sauces.

Fessih — Egypt

In the Middle East, fish is first split and salted, then buried in a wrap of paper or leaves in hot sand or hot silty mud to "mature". The fish is eaten from three days onwards.

Soaked and de-salted, these fish are often fried in sesame oil and served with bread, tahina, and parsley.

In Japan and China the meats, roe, cleaned visceral mass, livers, and cleaned fresh wastes of many fish and shellfish species are fermented and salt-cured to yield a great variety of local and commercial products.

Basic ferment is in the following media, or mixes of these media:

- In newly boiled and cooled rice (salmon meats mainly)—Sushi.
- In steamed and cooled rice inoculated with *Aspergillus oryzae*. For mackerel, herring, oily fish, fish roe, and other ferments and sauces—Koji.
- In the standard household bran tubs or nuka; rice or other bran can be used. For herring and puffer fish (fugu)—Nuka-zuke.
- In the lees or ferment liquor of sake (rice wine) after pressing or distillation; sake itself is added to many ferments—Kasu.
- In spiced vinegar pickle. For sardine, mackerel, octopus, or squid—Soused fish.
- In ferment to a liquid sauce with meats of fish and oysters. Ferment usually for longer than a year. Shottsoru or in China—Chiangs.

As well, liquors from salt-curing processes are used to "top up" pickles and new ferments, as they are rich in the yeasts and bacteria that enhance fermentation. Rice wash-water is used to "top up" drier ferments.

Warnings:

- Puffer fish (fugu) and all members of the toadfish family are poisonous, although these are eaten in Japan. Much of the poison is decanted by repeated changes of the salt liquor. Even so, we do not recommend that any poisonous fish be used in sauces, pastes, or ferments at domestic levels, or where no expert analyses are made.
- Careless primary preparation in hot weather can result in *Clostridium botulinum* (poisoning by botulism) culture in fish or ferment. Although such cases are very rare, good processing hygiene must be observed. For this reason, the fast chilling of fresh fish, cool (iced) wash-water, and in warm weather 20% salt, is recommended. Very clean tubs, utensils and bottles are also necessary. Never use stale fish for this reason, or fatty fish or oils, unless strictly to recipe. Bad flavours develop from fish fats, so fats are removed and washed off viscera and meats.

The process of natural ferment proceeds from a yeast-dominant fluid to a bacteria-dominant fluid; for example, with salt at 20–30% by weight, yeasts occur at 10^3 to 10^4 cells per mL of fluid, bacteria at 10^4 per mL. After six or so days of ferment, bacteria occur at 10^8 per mL of fluid.

Typical yeasts are:

- *Rhodotorula mucilaginosa*
- *Cryptococcus*
- *Rhodotorula minuta*
- *Sporobolomyces*

At low salt levels (10% or so) *Candida debaryomyces* can occur, or be added, at 10^4 cells per mL. At the later ferment stages, *Staphylococcus* bacteria dominate, with *Lactobacillus* species.

In bran (nuka) ferment where bran is 30–70% of fish weight, and where vegetables such as cabbage, daikon radish or carrots are fermented in bran, yeasts are valued for their good flavour. Bacteria here are restricted by the alcohols and esters produced by yeasts such as:

- True yeasts of some three to five species of *Zygosaccharomyces, Saccharomyces,* and *Torula*.
- Film yeasts of three to four species of *Hansenula,* and *Mycoderma* spp.
- *Bacilli* such as *Lactobacillus cucurmeris, L. fermentati, L. lindneri,* and *L. helormensis*. Ideally the *bacilli* should be restrained by the yeast alcohols and esters. It is the yeasts that give a good flavour to fish, radish or rice.

In salt the bacteria *Lactobacillus plantarum, L. pentoaceticus*, and the yeast *Streptococcus faecium* have all been identified.

After twenty days of ferment, the mono-amino acids produced by bacteria increase. Fish and meats are preserved by salt, yeast alcohols, and lactic acid from bacteria.

Pressing or Pickling in a Barrel or Crock

- Cleaned concrete tanks and marine stainless steel tanks can also be used.
- For fish roe ferment: 10 kg (22 lb) weights are used (usually stones).
- For salt ferment: 20 kg (44 lb) weights are used.
- For rice-fish ferment: 25–35 kg (55–80 lb) weights are used.

Note—that for all oily fish (mackerel, herring, or fat mullet), antioxidants of citrus fruit oils can be used, especially lemon terpenes, to prevent spoilage or rancidity if this becomes a problem. Such fish are often brined, air-dried, and smoked to preserve them, or hard salted and air-dried as board-like "stockfish" used in soups, or softened before cooking or fermenting.

Dried fish of any kind are cleaned, split, washed well, and dried on mats in an airy screened place until reduced to 20–30% of their weight, bringing them indoors before evening. Efficient drying (sometimes under plastic tents to prevent rain or insect spoilage, or in solar driers) usually takes two to three days of good weather. An auxiliary heater and drying shed may save part-dried fish in rainy or humid conditions.

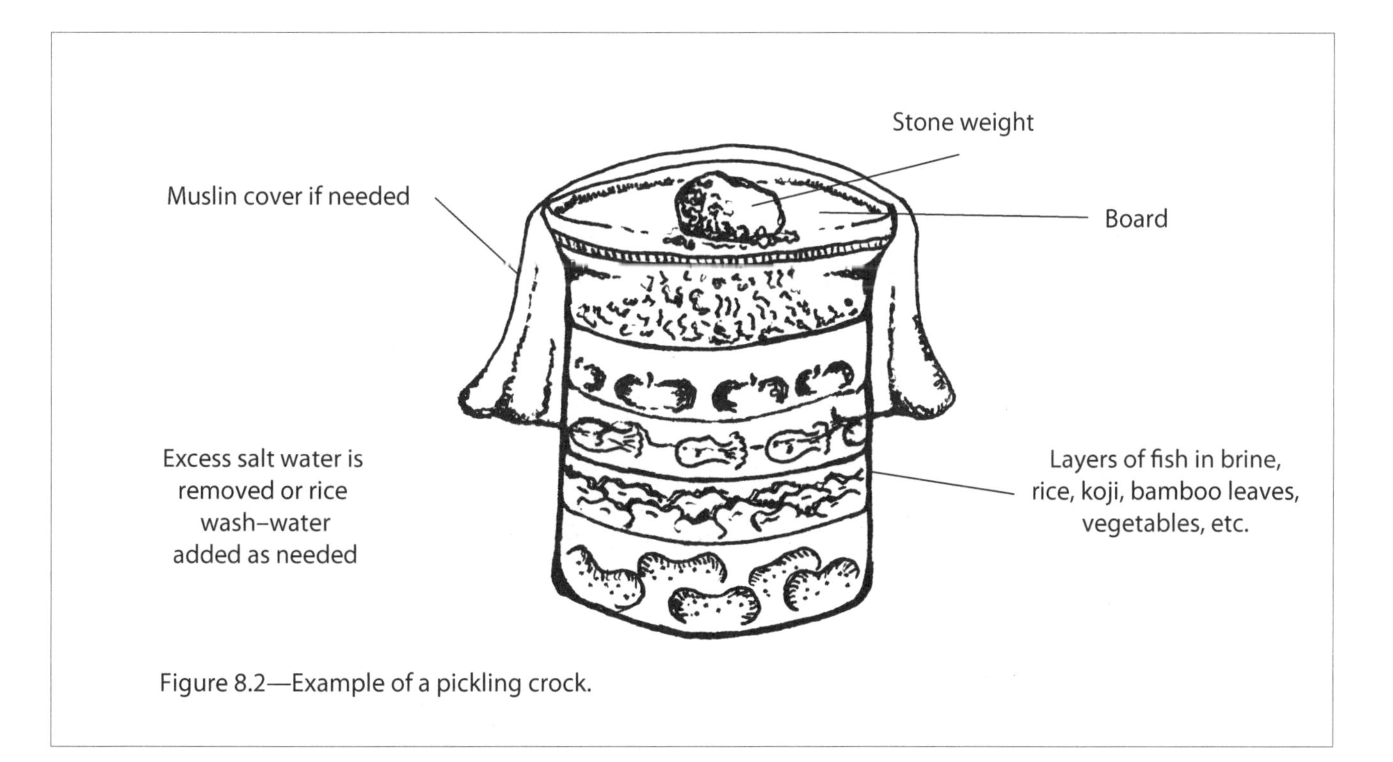

Figure 8.2—Example of a pickling crock.

Fish bones can be soy-dipped and oven-dried to a crisp, powdered, and used as a "salt" for calcium phosphate.

Oyster shells and mollusc shells can be dissolved in vinegar for calcium, as can pearls, or burnt and broken shells used to add lime to gardens.

Shark and ray fins can be air-dried, and the cartilage used in soups or casseroles. Such cartilage blocks the growth of solid cancer.

Fish Sauces

Fish fermented as sauces and pastes are critically important in many tropical areas where refrigeration is absent or costly. The result, as a salt-stabilised or sterilised liquid, is a high-protein product with good mineral content and some iodine. It can be transported and eaten with cereals inland, in areas of leached soils, or where rice or millet is a staple.

A word of warning—if fish is excessively fermented, ammonia is released, and amino acids are lost (hence a loss of protein). If the fish begins to smell like ammonia; then the ferment has gone too far.

The following is a basic fish sauce recipe. Any such sauce can be made of just cubed fish meat, just oysters, or just viscera, or mixes of ingredients. Due to the long processing time, it is wise to make a large quantity or barrel at any one time, as such sauce stores indefinitely in bottles. The addition of sake, sake lees, or koji, sweetens the sauce. Soy lees flavours as do any herbs added before filtering. The final addition of dilute alcohol in the bottles prevents any further ferment, (thus $C0_2$ gas formation in bottles). Some boiled and frozen ground krill adds lysine to vegetable or soy ferment (freeze, boil, grind and add to ferment at 2–5%). Pelagic shrimp of any kind add a red colour to ferment; this includes freshwater shrimp.

Viscera

All fats and gall bladder are removed, and the viscera split, cut up to 3 cm (1") lengths, thoroughly washed in fresh water and rinsed in a light brine. Any floating fats or wastes are skimmed off. Viscera includes liver, roe, milt, stomachs, "gizzards", pyloric caecae, intestines, but not gills or kidneys, blood or fat.

Livers

The livers of fish, squid, crayfish, crabs, and so on (gall bladder removed) are added at 2–4% by weight to squid and shellfish, and fish meats fermented to sauces to provide the enzymes that cause proteolysis (the combining of proteins with water or brine). This process occurs in 15–20% salt over one hundred days, enriching the liquid with fish proteins and the amino acids of ferment.

Visceral masses and roe, including the mantles of clams and shellfish are taken from sea cucumbers, abalone, scallops, squid, sea urchins, and fish either in the normal processing operations or for special product processing.

Wastes

Use fresh wash-water on gardens (this includes diluted salt water) or in orchards, pastures or crops. Bury any heads, fins, fats, and blood washings near fruit trees or below seedlings of tomato crop, using small pits, well covered and mulched over.

If there is a large bulk of wastes, convert it to silage (see Agricultural section).

Salt

Salt is used in all curing processes, from 15% by weight in winter or cool periods to 20% by weight in summer or hot periods. Sugar, sake, sake lees, herbs, and vegetables can be added to ferment and pickles. The pH can be adjusted with alkali salts or vinegar.

Liquid Fermented Fish Sauce—Variation One

Any number of fish, fish fillets, visceral masses, roes, molluscs etc. can be used for fish sauce. Here is one method, using small sand fish:

The small fish with head, fins and viscera removed, are thoroughly washed in fresh water, and drained. About 20–30% of the total weight of the drained fish is weighed out as salt, and almost 20–30% of the volume of the fish is koji, or *Aspergillus oryzae* culture, as used for miso. Layers of fish, salt, and koji are packed into a barrel and weighted at 15–20 kg (33–44 lb) for a year, in a cool place. At twelve months, weights are removed and the mixture stirred occasionally for six months.

Then the liquid is pressed out through a muslin filter, and aged in a container for up to seven years. The sauce is then again filtered and bottled in 2 cup lots. Curiously, fish sauce is most used to flavour steamed white-fleshed fish, or over rice dishes with fish.

Alternatively, the fish filtrate of the sauce can be added to soy sauce lees or koji to further ferment, then filtered as fish soy sauce.

Liquid Fermented Fish Sauce—Variation Two

Fresh whole shrimp, anchovies, or sardines and fresh trimmings of larger fish plus heads, backbones, and the meaty parts of the washed intestinal meats are layered in a concrete or wooden trough or earthenware jar with coarse sea salt, layer for layer, and well pressed down. The top is cloth-covered and capped, the contents weighted.

Stirred every week or so and well turned over, the mass ferments for six to twelve months. It is then strained and the liquor bottled. The solid residues (bagoon) are often dried and ground for animal feed. Both yeasts and lactic acid bacteria are involved over time. Salt is heavy, and the condiment is used sparingly over rice or in fresh fish dishes as a flavouring. Roughly 3 kg (6½ lb) fish to 1 kg (2¼ lb) salt is the bulk used. Fish or fish wastes and salt in this preparation can be added from time to time.

Bottled liquors can be sterilised by boiling at 100°C (212°F) for thirty minutes, but the sauce is often transported and used unsterilised.

Liquid Fermented Fish Sauce—Variation Three Japan

This is a bottled liquid derived from the ferment of many species of fish meats, fish viscera, oysters, or clams. The general process (using only very fresh fish) is as follows:

Thoroughly wash and clean viscera or shellfish, remove heads and fins from all fish, then fillet and cut up large fish. Wash in cool fresh water, (adding ice to water in hot periods) skim off fats or wastes, then drain well over a mesh or sieve. For viscera, be sure to remove the gall bladder from the livers and to cut up and wash all viscera.

Add 17–20% by weight of salt to the drained material and mix well with fish or shellfish.

In a clean barrel, lay down layers of fish and some koji (10% of bulk) and cover with a wooden lid weighted with a 20–30 kg (40–60 lb) stone. Cover the barrel with muslin.

Stir well three times per day for one week, once per day for one to two weeks, then once a week thereafter for six months. After a six month ferment, remove the stone weight and stir occasionally for six more months. Soybean lees can be added at the end to give a good flavour.

After one year filter the whole ferment through cheesecloth, boil this liquid to sterilise 100°C (212°F) for thirty minutes and bottle in 75–300 mL (3–10 fl oz) bottles. A dash of sake or vodka can be added to each bottle to prevent further ferment. Keep in bottles at least forty days. Some sauces are sold after seven years ageing in the bottle!

A fast method uses 4–5% livers added to meats and viscera in only 4–5% salt for four days, stirred frequently, then 17% more salt is added. The sauce is sterilised at 100°C (212°F) for thirty minutes, pH adjusted to 6.2–6.6, then strained and bottled as above. Note that the addition of fish livers causes rapid protein hydrolysis.

Nuoc Nam — Vietnam, Thailand, Cambodia

Small fish (sardines, anchovies) and shrimp are processed much as for bagoon. 6 parts fish to 4 parts salt are kneaded in earthenware pots, sealed, and buried or kept cool for three to four months. Then the clear floating sauce is carefully decanted from the top of the fish.

Alternatively, the whole (unwashed) fish are piled in large vats with layers of salt, and the pickle fluids are slowly drained over three days, first allowing three days to ferment (six days in all).

The remaining fish paste is trampled down flat and covered with fronds of leaves. Baskets or trays are tightly fitted into the vat and weighted or wedged down. The previously drained liquid is then returned to cover the fish to 10 cm (4") deep, and these fish left to mature for three to twelve months.

The juice then run off is top quality fish sauce. If salt is now again added and the mass again weighted, and fresh brine poured over, a lower quality sauce is produced in a few months. Sterilised, the fish sauce is bottled and the fish remains are used as agricultural fertiliser.

Molasses, brown lump sugar, fruit pulp or fruit acids, vinegar, and lactic acid may also be added (if sources are available), to give a good colour and assist hydrolysis. These are added early in the ferment, with the first salt. The added sugars may be first caramelised or part-caramelised and roasted rice or corn added for colour and flavour.

Fish, crabs, or squid livers and viscera, well cleaned and rinsed, are often added for their enzymes, also pineapple juice or the latex from papaya or figs are used to add other enzymes.

Budu — Malaysia

In Malaysia, 1 part salt to 5 parts of fish is combined with anchovies, palm sugar, and tamarind pulp. This is buried for six months in a sealed pot, and produces a dark, sweet smelling sauce. It can be bottled for up to two years before sale.

Shiokara — Japan

Strips of fish flesh and washed viscera (gall bladders removed) are placed in barrels (1 part salt to 4 parts of fish material), stirred well for several days, and then sealed and fermented for from a few weeks to a year. When fermentation ceases, malted rice or barley is added, to 30% by volume, and two to three days more ferment is allowed. Sauces and pastes are made of the liquids and semi-solids so produced.

Palau Sauce — Micronesia

Various reef or white-fleshed fish (**not** tuna) are made up as a soup with pork or chicken stock, onion, garlic, ginger, and salt. A large quantity of soup is made, some is eaten fresh, some strained and then reduced on a slow fire to a brownish, molasses consistency. This keeps well and is used on cassava, taro, and rice as a sauce (Robert Bishop, *pers. comm.*)

Fermented Visceral Masses

Viscera of large fish are split, washed, and all fats and contents removed. Carefully remove gall bladders and discard. Soak well in fresh water and rub fats off and discard. Cut masses up and add 20–30% of salt by weight. Stir once or twice a day for a week and then seal. Excess water can be poured off at first. Sugar can be added to redden the ferment, or koji or sake lees added to sweeten.

When fermented for six to twelve months, a little alcohol is added to the bottles to prevent carbon dioxide generation (and burst bottles). If the ferment is sugar-sweetened, a new ferment is continued.

Garum Sauce

The Roman and Greek garum sauce was a rich dark essence sold in bottles. It was made from cleaned fish intestines, livers, blood, small fish, and trimmings of fish pickled in layers of sea salt, and cured for weeks in the hot sun. In Vietnam, nuoc nam is the equivalent. Shrimp is salted, dried, pounded, fermented and made into cakes. All fish sauces enhance fish dishes or are used on rice. The enzymes from the livers and intestines are needed to hydrolyse the protein in fish scraps and offal, so that the sauce contains as much protein as the waste.

The livers of crabs and prawns are equally valuable. Anchovy essence, oyster sauce, and fish sauces generally, are made in much the same way; (more detailed recipes are given below) and may take months of ferment to mature.

Fish Pastes

Balao-Balao Philippines

Boiled rice is mixed with whole shrimps or fish pieces and about 30% of salt, then given several days to ferment when the shells of the shrimp are sufficiently softened for these to be eaten. The ferment is primarily a lactic acid one, and after four days pH is 4.0.

Bagoon Paste Philippines

This is a staple food in many places, and the expressed liquor makes a favourite sauce (patis). Sardines (*Sardinella, Decapterus,* and *Stolephorus*) and small shrimp are washed in clean water. About 3 parts fish to 1 part salt is added and packed into jars or tanks, covered closely with cloth for five days, then sealed. The containers are exposed to the sun for a week, then packed into smaller sealed jars or pots to mature over the next three to twelve months. Sealed, the paste keeps for many years.

Kamboko Cakes (short-term preservation)

Mince fresh or frozen fish (under two months frozen) with bone-free fillets. Wash the mince free of blood and fats in fresh water. The fish meat can be centrifuged or pressed-dry by hydraulic or weighted press, then blended with gluten powder, salt (3%), starch or a little sugar (5–10%) or paste from soaked soybeans. When the fish paste feels adhesive (if too wet, add cornstarch), mould onto bamboo sticks or mound on flat bamboo pallets. These moulded shapes can be steamed, grilled, broiled, boiled, or deep-fried. They then have a refrigerator life of four to six days but are best used fresh. Various spices, soy sauce, chilli, or hot sauce can be added, or a dip of soy sauce, mirin, and chilli made.

Minced squid or fish is excellent in dosa batters, as drop-cakes or steamed in cake cups in the oven, and a little chilli helps, plus soy and mirin dip. This is also a good way to cook minced abalone, oyster meats, mussel meats, or shellfish which do not mould to kamaboko pastes, unless mixed with 60% of fresh fish mince.

However, very fresh-caught fish can be mixed as surimi (uncooked kamaboko) at sea, which keeps well for up to three months if quickly frozen. It is cooked before fully thawing.

Fermented Fish Paste (gyomiso)—Variation One Japan

Mix 75% fish meats with 15% salt, then add 10% wheat bran and *Aspergillus oryzae*. This mix ferments in two weeks at 25–30°C (77–86°F). To stop enzyme action, cook, and the pH should fall from 6.0 to 4.5–5.0.

On cooking, the water content is reduced to 40% and the paste bottled and sold.

Similarly, cooked and frozen krill or whole shrimp is added to soy sauce ferments to increase flavour and amino acids or oils.

Fermented Fish Paste—Variation Two

Crushed or minced fish fillets (75%) are mixed with wheat bran (10%), salt (15%) and koji, and kept at 25–30°C (77–86°F) for two weeks. Watch the pH, and as it falls from 6.0–6.5 to 4.0–4.5 (or at pH 5.5) boil the mixture to stop the ferment. Reduce the water content (by cooking) to 40%, blend and bottle. The paste should yield 45–50% of the raw fish weight before boiling is stopped, so weigh the fresh mince and ingredients before cooking.

Dried Fish Paste India

A fish paste made from catfish in Mysore (India) is a blend of minced fillets mixed with either tapioca or sorghum flour (50:50), the flour mix being partly gelatinised in hot water. The mix is spiced with salt, chillies, and tamarind, and then extruded and dried. It is cheap to manufacture and has good keeping qualities. Experimentation should give equivalents for other tropical fish, and other flours or tapiocas. (Srivastava, 1985.)

Anchovy Paste — United Kingdom

This recipe is dated 1886. I have modified it a little. Boil an equal weight of anchovies and water, skim regularly, and simmer for two hours. Strain and filter through a coarse cloth or fine sieve, and add 2 gm (pinch) of salt per 454 gm (1 lb) of anchovies. For every tbsp of paste, add a tbsp of butter. Melt together and boil a few minutes, then stir in a little turmeric and chilli powder. When cooling, add the juice of a lemon per 454 gm (1 lb). Put into pots, seal with paraffin wax, lid tightly and refrigerate.

Ngapi — Burma

Small anchovy (*Anchoviella comersonii*) and shrimp (planktonic shrimp, adding a desirable red colour) are partly sun-dried for two days, then mixed at 3 parts fish to 1 of salt.

Half the salt is added at the first pounding until a paste forms, which is then tightly packed in wooden tubs. Here it remains a week before it is again pounded with the remainder of the salt. This final paste is spread on mats in the sun to dry for three to five hours. It is repacked in tubs for a month to mature, and is finally pounded and packed in sachets or bottles. Stored in sealed pots, the paste is said to keep for two years. Content is 43% moisture, 20% protein, 1% ammonia, 2% fat, and 22% salt.

Nam Pla — Thailand

In glass, wood, or earthenware, press down alternate layers of fish and 20% salt by weight, using a wooden plug and a weight. Add a little brine to cover the fish. In large vats, fitted with a tap at the base, the liquor is drawn off after three to four months. A second brine is added, and a second draining after three to four months. These liquors are stored in glass or earthenware, sealed, to mature.

To the basic liquor, heated, roasted rice or wheat flour can be added, and for red pastes *Monascus purpureus* culture added. The flesh of the fish, shrimp, or squid can also be cooked to reduce water, and thickened as a paste with roasted flour.

Such sauces, with lime juice and chopped chilli, or fresh-grated horseradish, are used as a condiment or side dish (nam pla).

Sardine Paste — Korea

Small sardines are mixed with 20% salt (by weight), and allowed to ferment and age. The solid paste keeps a year or more. The paste is periodically kneaded every few weeks, then matured in a sealed (and sometimes buried) earthen jar.

8.3 PARCHED-BOILED FISH

Tsukuda-Ni — Japan

This is the general term for fish, squid, shellfish meats, fillets, or chunks of fish, and shellfish boiled in soy sauce, sugar, and gluten of wheat for sixty to ninety minutes. Such products store well for sixty to ninety days, or until moulds grow on them. The meats are usually carried as lunch items with pickles and rice. Dried and de-salted fish pieces can be used.

Tsukuda-Ni of Fish Fillets

Fresh fillets of fish are scalded, bones and skins removed, and are then roast dried and cut into small cubes of 1.5–2 cm (½"–¾"). Set aside. Combine 3 litres (6½ pints) tamari soy sauce, 1.1 kg (2½ lb) sugar, 1 kg (2¼ lb) wheat gluten, 10 gm (⅓ oz) agar, and 150 gm (5¼ oz) ginger. Add the fish cubes, bring to a boil, and simmer, stirring occasionally, until the solution is absorbed into the fish. Keep cool, use within three weeks.

Tsukuda-Ni of Fish Flakes—Denbu (fish paste)

Flake, soak and press dry 4 kg (9 lb) steamed or boiled fish then crush to a fibrous mass with a wooden spatula. Combine 4.5 litres (9½ pints) soy sauce, 3.75 kg (8¼ lb) sugar, and 14.5 litres (31 pints) water. Bring to a boil, and add fish; simmer for twenty to thirty minutes. Take off heat and add 2 litres (4¼ pints) sweet sake. Sometimes miso is also mixed in (three times the quantity of fish meat).

The paste keeps well and is used in other dishes or as a rice flavouring.

8.4 SOUSED AND VINEGARED FISH

Clean and wash, removing dark meat along backbone of the fish, or clean shellfish. There are various levels of processing for food, for example:

- Dice and souse in lime juice, onion, and tomato (ceviche). Eat over a few days and keep cold.
- Slice thinly raw and eat in soy sauce dip with horseradish (sushi). Eat as fresh as possible.

Rollmops

Soak 6 medium fish (herring or mullet) for two to three hours in a brine of 56 gm (2 oz) salt to 2 cups water. Boil together 2 cups white wine or cider vinegar, 3 bay leaves, peppercorns, pickling spices, 1 large sliced onion, and 2–3 pickled gherkins. Let this cool. Drain and dry fish, wrap each fillet around slices of onion and cucumber and pack well in a glass jar. Pour on vinegar and tuck in a spare onion. Serve after four days with rye bread and sour cream. Keep refrigerated.

Sardines — Scandinavia and Russia

Use 10–12 cm (4"–4¾") fish.

Wash 2.2 kg (5 lb) of fresh fish, pulling out gills and intestines but leaving the belly whole. Rinse and drain. Measure 4.5 kg (10 lb) of salt and sprinkle some on the fish (reserving the rest). Prepare a solution of 3 parts 5% vinegar to 1 part water and cover the fish with this solution. Leave for twelve hours then drain. Now pack tight in criss-cross layers, covering layers with a mixture of the remaining salt, 2.2 kg (5 lb) sugar, and spices (pepper, bay, cloves, hops and ginger). Weight down with wood or stone. Onion rings and horseradish can also be layered in. Cure for two weeks, then start using; will keep for months in a cool, dark, or refrigerated place.

Soused Fish

This is a short-term preserve (seven to ten days in a sealed jar).

Fillet 4 medium fish, removing heads and fins. Season with salt and pepper. Roll the fillets and pack them into a shallow oven tray or dish. Scatter some peppercorns, a dash of soy sauce, and a few mustard seeds (or a tbsp of mustard) in the dish, and pour on a mix of ½ cup vinegar and ½ cup water. Cover dish and bake thirty minutes.

This can either be served as is, or packed in a jar and covered with vinegar (stores cool for up to a week). Sardines or small fish cooked like this can be eaten bones and all. To keep longer, a layer of oil can be added to the top of the vinegar.

Carp

Carp and musky fish are not generally prized in the West, but are frequently eaten in China and South Eastern Asian countries. Skin carp fillets and cut into 3 cm (1") cubes, wash thoroughly, and soak for one hour in 4.5 litres (9½ pints) water to which 1 cup of salt is added. Place in a saturated salt brine for twelve hours. Rinse, then pack down in a crock with spices and 2 parts vinegar to 1 part water (with some sugar added). Slowly simmer until fish is softish, then place into jars with a little onion and a slice of lemon. Seal immediately. Keep refrigerated and eat over twelve weeks. Bones are softened by the long vinegar marinade.

Escabeche — Spain—Originally Arabic

Thoroughly wash and cube 4.5 kg (10 lb) fish, oyster, or squid meats, then hold in saturated salt brine for half an hour. Drain and dry on paper.

Brown 2 cloves garlic with 2 large onions (sliced to rings) in 1 cup oil. Cool and set aside. In 1 cup of oil, mix spices (chilli, cumin, bay leaves and black pepper) with 4 cups of vinegar and simmer for fifteen minutes. Let cool, then combine with garlic and onion mixture. Pack fish in warmed jars and pour on the vinegar and spice mix. Matures in a day, but keeps three to four weeks in a refrigerator.

Ceviche — Mexico

Combine 1 kg (2¼ lb) cleaned, cubed, raw fish or salted salmon, 1 cup lime juice, sliced onions, cubed tomato, crushed red sweet peppers, pinch of chilli, chopped seaweed (optional), and salt and cayenne pepper to taste. Stand overnight. Eat within five days or refrigerate for a longer period.

Lomi-Lomi Salmon — Hawaii

"Hawaiian salt" is the reddish salt from sea-spray pools in red soil, mixed with chilli and roasted ground kukui nuts (*Aleurites*). It is used on all raw fish dishes—Hawaiian limes are ideal for this dish. Hawaiians eat small crabs, limpet meats and washed seaweeds raw with sliced salt salmon, poi, earth-oven cooked pork and vegetables; a veritable feast, much appreciated by all!

Swedish Pickled Herring (Inlaggning av kryddsill)

This recipe serves for small fish or strip fillets of larger fish. Fish are first rinsed, cleaned, and paper-dried.

Combine 6 cups water, 2 cups vinegar, 454 gm (1 lb) sugar and spices (black pepper, chilli, dill, bay leaves, garlic, ginger and cloves). Boil and cool (to dissolve sugar). Pour over the fish and leave for ten hours. It can be kept refrigerated for six weeks and can be consumed from day one! Keep vinegar pickles in glass if possible. Vinegars of 5% acetic acid are used. Shake jars when full to remove air. Usually, slices of onion are pickled with the fish.

Below 10°C (50°F) pickled fish will keep for months.

Gravlax

Split and remove the backbone of a large 4.5–9 kg (10–20 lb) salmon. Rub the following into the flesh sides, one by one; 2 cups olive oil, 1 cup salt, 1 cup brown sugar, pepper, mace, and chopped dill.

The fish pieces are tied tightly together, flesh to flesh, and pressed under a board, (or in a large dish) for forty-eight hours. It is served as thinly sliced fillets on rye bread, or as sashimi (raw sliced fish, Japanese-style) with soy and horseradish (wasabi) dip. This keeps a few days refrigerated, but is mature in two days.

8.5 SALTING AND DRYING FISH

Salting Stockfish

Enormous quantities of cod, whitefish, and schoolfish are split and salted at sea in coarse rock salt by the Portuguese, Norwegians, Argentinians, and British. In Portugal, the cod are taken from the brine and sun-dried to make bacalhau, and in Norway they are wind and sun-dried in the cold air.

Dried stockfish are a staple food of many coastal nations. The process is simple, but care must be taken in removing all blood from the split fish; in keeping fish well brined and weighted; and in drying to a 35–40% moisture content (down to 17% if fish are to be kept in tropical climates). Dried, salted, fish have a life of months or years if well prepared.

Drying, salting, and smoking are the traditional way to keep fish. "Slack salting" uses 10–20% of salt to raw fish waste (preserves for seven to twenty days). "Hard salting" uses more than 30% of salt to raw fish waste (preserves indefinitely).

Salting for very cold wind-drying can be brief; ten minutes in brine before hanging in the wind under a roof, for thin fish or thin fillets. More serious hard salting would be to lay up split fish under dry salt in a barrel, rinse them, hang to dry, and at the "tacky" stage hang in a cool smoke from dry hardwood chips or sawdust in a smoke room or tower. For long-term storage, fish can be dried over a full week. Some fish are dried only a few days, with a high moisture content, and these can be as pleasurable to eat dried (not cooked) as a good cheese. Fish with low moisture content can be as hard as a board and need a forty-eight hour soak before cooking.

An essential in preserving salted fish or red meat is to remove blood by soaking in fresh water or a light brine, and then draining well or pressing. Change the water several times if blood is still present. If fatty fish are salted, grated lemon skin added to the salt may help in preventing oxidation (lemon oil is released).

Soaked in fresh water overnight, skinned, boned, and finely shredded (like the bonito flakes of Japan), dried fish are baked with stock, fresh herbs, potatoes, or made into fish cakes with rice, bread, or mashed potato and fresh herbs, then fried crisp.

Fish for salting have heads and thin belly meats removed; these can be used for soups or as fish stocks. The dark meats of the backbone will need to be scrubbed out. The cleaned fish are poached in brine (⅝ cup salt to 4 cups water giving, 80% salt); the "water" has an equal quantity of 5% vinegar added. Leave in this pickle for about five days, drain, dry, and re-pack tightly with a sprinkle of dry salt per layer. Boil and skim original brine and store, use ⅝ cup salt to 2 quarts water. Leave the pickled fish for no longer than two weeks, rinse and consume during this period, or freeze in sealed bags.

Salted Fatty Fishes

This recipe is for such fish as herring, mullet, and mackerel. Head whole fish and mix with 15–17% salt, 5–7% sugar, and spices. Weight in a barrel at 5–15°C (40–60°F) for several months. More strong brine can be added at about a year. Fish are then washed and filleted, cut into pieces, and packed into jars with 5–12% vinegar added. Salt is at 12–14%, and pH 5.6–6.0 when mature.

Salt-Dried Products

Fast-spoiling fish (sardines, mackerel, whiptail, herring) are washed and split and put in salt (about 15% of fish weight). They are dry-salted or brined for six to twelve hours, and are then dried skin-down on bamboo mats. They are turned only when nearly dry. Slender fish, such as whiptails, are dip-salted in heavy brine for as little as fifteen minutes, and will then air or wind-dry under cover in cool winds to form a tasty yellow flesh, rather like a good cheddar.

Sardines are also wind-dried for five to six days, before bundling for sale, but are first salted a full four to five hours in Japan, where the same brine is constantly re-used (some salt added) to give a special odour to split fish (Jack mackerel) before these too are sun-dried.

Throughout the world (Portugal, Norway, Japan) cod is dry-salted or brined, headed and gutted, and well scrubbed inside and put in a heavy brine solution (20–30% salt), pressed down for two days, then sun and wind-dried. For dry salt, about 200 kg (440 lb) salt is used per 1,000 fish for ten days. They are then washed in a light brine and dried, and again washed in fresh water before final drying.

Such fish are water soaked, skinned, filleted, and painted with a soy-sugar solution before grilling or broiling in Japan. Elsewhere the soaked and freshened cod flesh is made into dishes with potato, or soaked in milk, thickened and served with vegetables.

In the New Kingdom of Egypt, large brick-lined pits were made in which salted and dry fish were stored in fine (grass) ashes. For larger fish, the body cavity, mouths, and gills were ash-filled, and some were cloth wrapped. Smaller fish were salted, dried, and stored whole in the ash.

Pressed Salted Sardines

Brazilian fishermen catching sardines, first clean them, then give them seven days in saturated brine. They are then pressed out in a mass to remove extra oil and water, using heavy stones over a bag of fish. The block of sardines is then stored in a plastic bag, and pressed free of air as much as possible. The mass remains wholesome and saleable for at least six weeks. (*Int. Agric. Dev.* September-October 1986).

Macassar — Indonesia

Small fish (anchovies) are headed, washed, and placed in pots with an equal weight of salt. After about a four day ferment, a rice-mould ferment ang-kak (inoculant *Monascus purpureus*—but koji would serve here) is added. Ragi (see chapter 2) and spices are then added, and in the next few days the mixture reddens. It is then packed into glass bottles for sale.

Macassar is 66% moisture, 16% protein, 1% fat, and 17% ash.

The rice-fish combined ferment is also made in Japan and the Philippines, where ang-kak or koji (see section 2.3) is also used, and both brackish water fish (Chanos chanos) and freshwater predatory fish like the snake-heads (*Ophiocephalus* spp) are used. Here, the product is called buro.

Note that acidity is added via the lactic acid ferments in the rice, and that the moisture level disfavours yeasts.

Cooking Salted Fish

For hard-salted cod, it is important to give it a full forty-eight hours of soaking. Cut 2 fillets from the "fat" central part, skin, de-bone, dip in butter and fry as fritters, or proceed as for brandade, below.

Brandade

Heat 1 kg (2¼ lb) de-boned, de-salted cod gently in an iron casserole. In separate pans heat: 1 cup of olive oil and 1 cup of 50:50 milk and cream.

Combine in a bowl, (unheated) a clove of garlic, salt, pepper, lemon juice, and parsley. Use a strong wooden spoon to pound some of the oil into the warm cod. Alternate the oil with the milk mixture, add some lemon juice with spices. Keep pounding this into the cod.

After the oil and creamy mix are finished, beat in lemon juice and spices. The result should be a creamy-white mix. Do not heat too much or the mix separates and will need a blender to recombine. Serve with honey on triangles of fried bread.

Bacalao — Argentina

Soak salted cod for twelve hours, changing water twice. Boil in fresh water with an onion and bay leaf. Drain when just cooked, skin and take out all bones. Place in a casserole and add 2 cups of cooked chick peas, 1 clove of garlic, 2 tbsp of lemon juice, olive oil, chilli, and olives. Cover with a crust of breadcrumbs and soft grated cheese and bake for twenty minutes in a medium oven. Sprinkle with parsley and serve.

Dried Fish Meats

Small whole fish or small fillets can be ground, cooked to a paste, and dried in cakes containing 55–70% protein, 2–15% oil or fats, and 10–20% of minerals as well as vitamins B12, B, A, D, K, and E. Drying should achieve less than 10% moisture content, fats and salt low; meal is long stored. Vitamin C can be added to cool cake before drying. A screw press is useful in reducing initial moisture and oils. Such dried meats are added to rice dishes or starchy foods.

Fodder meal includes bones and fins and is steamed and pressed hydraulically to remove moisture and oil and these cakes are dried for transport. The meal is added to animal feedstuffs.

Dried Herring or Mullet

The whole cleaned fish are first threaded on a cord (through the gills and out of the mouth) for two days sun-drying, then split and the backbone and head removed (for fertiliser). The fillets are hung over a rope to dry for two to three weeks. The milt is also washed and dried and the roe salted. Note that heads, viscera, backbones, and fins can make fish sauce or stock feed-cake. After a month of drying, meats should be below 20% of moisture, based on the fresh weight.

Small Fish Drying

Small fish (anchovies, sardines) to 8 cm (3") are whole-dried on mats, in which they are rolled up each night and placed out each day. They are dry at 20–30% of their original weight.

Boiled Dry Fish

Sardines and anchovies, about 10 cm (4") long, are well washed to remove scales, slime and sand. These are laid out on steaming trays and stacked in boiling salt water, 1 kg (2¼ lb) salt to 20 litres (42 pints) water until the fish float on the boiling saline (about fifteen minutes). They are then drained and dried for two to three days on straw mats or mesh, and packed in paper bags or cardboard for storage.

Freeze Drying

In cold Arctic winters, fish put out to freeze at night lose water when sun-dried to thaw by day. Pollack and cod are so dried over seventy days or so.

Dried Shark (mango maroke) Maori

Traditionally, shark was cleaned, split, and sun and wind-dried on rails or poles for four to five months. The roe were also dried. Small quantities were cooked with sweet potato, and steamed for thirty to forty minutes (mango maroke).

Whole dry shark, unskinned, was charred over a fire, cooled, and the skin scraped away. It was broken open lengthwise to eat; only a little is eaten at a time. Shark livers, gall bladder removed, were eaten after steaming in cabbage leaves.

Dried Shark Fin (also rays and skates)

Cut off dorsal, pectoral and caudal fins without including meat if possible. Wash fins well in sea water or diluted salt brine. Scrub with a stiff brush and bore or punch a hole in the fin tips. Thread fins by the holes on cord and hang in the air for two to three weeks to cure to 30–40% of their original weight. Dried fins are 1–2% of the total fresh weight of the shark. Hang from house rafters in a muslin bag to store.

Dried shark fins are fresh-soaked for four to five days to soften, then put in 90°C (194°F) water for twenty to thirty minutes to remove the skin. They are then brushed clean and cooked or dried.

Cartilage China

Cartilage from rays, shark, calves, and skates can be freed of flesh by boiling and cleaning off attached meat. The meats are then saved for batters, potato cakes, soups etc. The boiled cartilage is sun-dried and used in "shark fin" soups with chicken stocks and vegetables.

Cartilage is now recognised as a preventative of growth in solid cancers such as lung cancer, so shark fin soup has a preventative or stabilising effect. Cartilage in soup should be boiled in 2 cm (¾") squares until soft, then chicken and vegetables added.

Parch Boiling

Dried products are "parch boiled" by cooking in a relatively small amount of salt liquor (as in Boiled Dry Fish previous), and this is absorbed by the dried fish or squid. The process needs careful measurement and close watching to remove the fish rapidly, using a wide pan. Continuous stirring or timing is necessary until the solution is taken up. The meats are then dried in the air.

Float Boiling

Raw fish and mollusc meats are put into a boiling, shallow, wide pan and boiled rapidly for ten to twenty minutes, then on low heat for forty to sixty minutes. The fish is removed, the liquor again boiled rapidly, and more fish added as before. The remaining liquid has good flavour, and can be diluted with new solution, or finally cooled and bottled for table uses, or salted as a sauce. The meats are dried, and sometimes stored in oil.

Fushi Japan

Fushi is dried fish stick, used shaved into soups, etc. Bonito, albacore, tuna, mackerel, and sardine are used. The fish are cleaned and filleted, then cut into strips and placed on trays, skin down. Small side fillets are uncut, large fillets (fish of more than 3 kg [6½ lb]) are split; belly meat is another fillet used.

The trays fit a tank at 80–85°C (176–185°F) for fresh fish or 90–95°C (194–203°F) for stale fish and slowly the temperature is raised over twenty minutes to boiling. Fillets must remain straight and not split due to heat. Boiling continues forty-five to ninety minutes (longer for oily fish).

The trays are removed and cooled in a large tub of water. Any skin and small bones are then removed.

Boiled and skinned fillets are sun-dried to "tacky" and smoke-broiled over an open short chimney in stacks of six to seven trays, about 1.4 m (4'8") above the charcoal (smokeless) fire. Trays are rotated until all fillets are uniform yellow-brown in colour and dried. Any cracks are then filled with a paste made of some of the fillets. This process is repeated, the fire somewhat hotter and the time shorter for each smoking. Now all the fillets are air-dried in their trays or on mats (after ten or more smokings to remove water) for three to four days.

The fillets are carefully shaved with very sharp knives to reveal a brilliant red-brown surface and again air-dried for two days.

Boxed for thirteen to fourteen days, moulds of *Penicillium* and *Aspergillus* grow on the fish fillets. Those with moulds are taken out and first shade-dried then sun-dried. Thick mould is then brushed off and the fillets returned to the box together to grow the second mould. After four such processes, the surface of the fillets is light brown. The yield is 20% of fresh fish, the product is very expensive. Storage may have to be at 0°C (32°F) to inhibit insect borers.

Mackerel fillets, being small, are less processed (not boiled) and not shaved. Sardines are not boiled, simply broil-smoked to dry.

These very popular fish sticks are shaved, at home, into broths and soup dishes with miso and a few vegetable sticks.

Seasoned and Dried Fish

A paint of soy sauce, sweet rice wine (sake), and sugar is painted on split small fish or medium fish fillets. Alternatively, fish are marinated two to four hours in a solution of: 6 cups soy sauce, 45–60 gm (1½–2 oz) sugar, and ¾ cup sake. A little gelatine, agar, or glucose can be added to glaze the fish.

Thus seasoned, fish is sun-dried to 30–40%. Keep these in cool storage after drying.

Note—All dried fish and dried seaweeds need to be kept below 70% humidity in warm, dry conditions. Thus they can be kept indoors in muslin bags above the back of refrigerators or in warming cabinets above hot water pipes, or in special cabinet storage on hot water lines. White powders form at high humidity on squid, fish, and seaweeds; these may lose food value. However, some thick brown dry seaweeds are stored for three to four years in moist dark places to "mature" and soften.

Fish Pemmicans

Fish pemmicans were a staple of coastal and river people (Salish, Tlingit, Shoshone) in the USA. To prepare, pound salted and sun-dried fish flesh into granular form, and combine thoroughly with dried fruits and the oils of fish or seabirds poured on. Lard and suet are also suitable fats for long preservation.

Large skins (or plastic bags) of this pemmican are stored in leaf-lined tight baskets and can be buried for long storage. Burial should be in cold coarse sands, ash, or peats. However, do trials in a small way before risking a lot of fish preserved like this. Probably a slow ferment takes place after burial, as it does with the small whole seabirds the Inuit stuff into whole sealskins (blubber intact) and bury under snow in hut floors for midwinter parties.

Any of the spices used in jerky or pemmicans (hops, lemon oils, cloves, ginger, garlic, turmeric, and nutmeg) can be bacteriostatic. Dusts of lime, ash, or turmeric can also be used. Soaks or marinades of salts and sugars prevent insect larvae surviving, as does dehydration and careful burial or storage systems.

Eels

Dried Eels — Maori

Traditionally, skinned, headed, and filleted eels are dried and part-smoked. These could be kept for some months. If too dry to cook, they were soaked a few hours in water before grilling or pit-steaming. Also, elver eels were prepared as above, half cooked by grilling, and dried in the wind and sun, stored dry and airy, and watched for moulds. They were usually cooked by steaming.

As eels, like snails, can produce copious mucous from their skins, they are first rubbed vigorously in coarse salt (or large numbers placed in a salt bin). They are then gutted, and either cooked or hot-smoked to conserve. Even fresh eels, de-slimed in hot water, split and headed, are salted and dried twenty-four hours before frying skin-side up, then grilled until the skin is crisp.

Jellied Eel

Skinned eel is cut into 4–8 cm (1½"–3") pieces (boned), and cooked in salt water and lemon juice. At first, cold water is brought gradually to the boil and the eel simmered forty-five to sixty minutes until very tender. Add 1 tsp gelatin to 300 mL (10 fl oz) of the warm stock, but if the stock is reduced no gelatine is needed. The eel pieces are arranged in bowls, the stock added, and the eels set in a cool place. Keeps seven to ten days if refrigerated. Use a little less volume of water than of eels. Herbs can be added to the warm stock such as chopped parsley or a little mint.

Roe

Salting Mullet Roe

Choose only perfect roes in skins, roll in salt and soak in absorbent paper towels. As papers saturate, replace them, each time sprinkling more salt on roes. When the roes are hard and dry, they are stored and are eaten thinly sliced, with thin bread dipped in olive oil. Roes keep for months refrigerated, each wrapped in a plastic bag or waxed paper. Good dried roe dipped in wax travels well, and can be used on boat voyages.

Dried Mullet Roe

For dried mullet roe, 6 kg (13¼ lb) salt is used to 50 kg (110 lb) roe as a brine or dry salt. Then the roe is pressed under a board after ten to twelve hours of salting, drained, and sun-dried, being pressed each night for about ten days, and further dried for ten days in sun alone. This method was used in ancient (Pharonic) Egypt and is illustrated in tombs.

Roes can be dried after a three to four day fresh water soak, to remove blood and firm up, then placed out on bamboo mats in the sun until 20% of their original weight.

Sujiko (whole roe) — Japan

Wash and drain fresh roe. Then either dry-salt or brine soak them.

- Dry-Salted—12% by weight of dry salt is spread over layers of roe on a mat in frames so that the roe can drain and dry. The first drainage takes one week. Storage is long-term, in jars or boxes with 3% of salt newly added to replace lost drainage salt.
- Brined—Saturated salt brine is used. Whole roe are often stirred to get salt penetration. After a few days, the roe is packed in boxes of 3% dry salt for keeping. Drain for one to one and a half hours before storage.

Fish Roe Pickle — India

Wash roe (do not break skin), salt, and add 1 tsp turmeric to each piece of roe and drain well in a colander for thirty minutes. Fry roe in 1 cup olive oil on low heat until cooked to the centre (keep this oil for later use). Cool, then cut into 5 cm (2") lengths.

For 1 kg (2¼ lb) cooked roe, blend together into a vinegar paste the following ingredients: 1 cup of vinegar, 1 cup oil, 100 gm (3½ oz) mustard powder, 3 tsp turmeric, 1 tbsp peppercorns, 1 tbsp cumin seed, ground paste of 2 garlic cloves, and salt to taste. Now mix this together with the reserved cooking oil (2 cups of oil in total). Mix this combination carefully with the pre-cooked fish roe pieces. Insert into an airtight jar with a secure screw down lid. Mature at ten days, it will keep for months.

Taramasalata (salted roe)

Taramasalata is a famous Greek recipe, a spread for biscuits and breads. The chief ingredient is salted fish roe (cod, mullet, tuna, carp). Soak, wash and drain 110 gm (4 oz) salted roe. Soak 4 slices of white bread (two to four days old without the crust) in water and squeeze dry. Place roe, bread, 1 small grated onion and 1 tsp chopped dill in a blender, and slowly blend, adding 1 cup olive oil and ¼ cup lemon juice until pink and creamy. Chill. Keeps a week in the refrigerator. Optional additions: black pepper, pimiento, anchovy, cayenne, cream cheese, etc.

Salted Fish Roe in Koji — Japan

Whole roe, carefully removed, are used after deveining and washing. They are pressed for two days in 30% salt, then washed and drained, and layered in koji (30–40% by weight). After four days the barrel can be topped up with a liquor from rice wash-water, and the ferment continued for fifty days before storing or sale.

8.6 PRAWNS AND SHRIMP

Fermented Shrimp (kung-chom) — Thailand

Shrimp is minced or pounded with 7% salt and stored overnight. 2% cooked rice and 1% garlic (minced) is mixed next day with the shrimp, and the whole mass fermented seven days at 20–40°C (68–104°F), then aged seven to ten days at 30–45°C (86–113°F). This solid reddish paste stores for six months. Protein is 11–17% and it is used with rice dishes.

Dried Shrimp — Japan

Wash and boil for ten to twenty minutes in 360 gm (12½ oz) salt to 20 litres (42 pints) water, then dip out and sun-dry on mats or blinds. For large shrimp, heads can be removed once dried, but dried shrimp is best stored whole, or large prawns peeled and dried, the peeled meats boiled ten to twelve minutes before drying.

Drying Prawns — India

Fresh prawns are boiled in 6% brine for two minutes, shelled and immersed in saturated brine for fifteen to thirty minutes. They are then drained and dried to leathery and can be packed in tins under carbon dioxide. They keep well for months in tins or plastic-sealed sachets.

Shrimp and Fish Cakes — China

Blend the following to a coarse paste: 375 gm (13 oz) fish fillets, 375 gm (13 oz) raw shelled shrimp, 3 slices of crisped bacon, 1 cup of ground peanuts or fermented lentil pulp, 2 tbsp of cornstarch (cornflour), 1 tbsp of soy sauce, 1 tsp salt, 1 egg white, 1 tbsp of olive oil, 4 cloves of garlic, and 1 knob of ginger.

Heat 1 cup of oil in a pan and spoon in tbsp of the mix, fry crisp and brown. Makes 2 dozen.

Prawn and Fish Crackers (these fluff up when fried)

Clean, shell, and grind up meats of fish or prawns (500–750 gm [1–1½ lb]). Beat an egg with 2–3 tbsp of sugar, and stir thoroughly into the pounded meats. Add 1 kg (2¼ lb) of tapioca flour, or cornflour, and enough slightly salted water to mix a stiff dough, which is shaped into a roll, and steamed wrapped in cloth until stiff and well cooked. Cool overnight.

Next day, cut into thin slices and sun-dry discs. Store dry in a tight jar until cooking, briefly, a few at a time, in deep hot oil. (Davidson, 1978).

8.7 GASTROPOD SHELLFISH—LIMPETS AND ABALONE

These are best just placed on coals or hot stones, upside down, and eaten from the shell. Or they can be steamed fifteen to eighteen minutes in very little water and twisted out of their shells with a strong thorn, a bamboo sliver, or a heavy needle. As children we ate gastropods by the sea in just these ways, dipping our food in salt and vinegar, with fresh bread and butter.

Squid, octopus, abalone and limpet are best tenderised by patient tap-beating with wood until tender. Commercial abalone factories use a very large wooden mallet, a wood block, and one mighty hit; but at home we patiently tap until the meat is doubled in area. Abalone and octopus are more easily minced in a food grinder or processor then served as patties with an egg, ginger, garlic, crumbled dry nori, and breadcrumbs (⅓ of the bulk) to bind and dry. Just fry on each side until brown, about five to six minutes. Do not overcook!

Serve with a soy-mirin sauce, shallots, and steamed potato or taro.

Very heavy, large conch shells can be hung from a branch with a hook in the foot until the meat is free and the shell drops off, then the meats are cooked, or eaten raw as sashimi, as is done with giant clam meat or trochus in Palau. These are soused overnight in lime juice or coconut-sap vinegar, salt, ginger, garlic, mirin, soy sauce and lemon rind, and are eaten next day as a pickle or with onion and tomato salad and poi (fermented taro), or bread.

Dried Squid and Cuttlefish

Split squid on belly side and take out viscera, ink sac, and "pen" (skeleton). Split head and remove beak and eyes.

Soak bodies well in fresh water, scrub in saline or salt water to clean and remove skin. Then wash again in fresh water so that the squid are not hygroscopic (meaning they do not take up water from the air in storage). Either lay squid out flat on a wire mesh, or tie tentacles together to hang in pairs over a rope or wooden bar.

Dry in the sun for ten to sixteen hours, and bring in to a roofed area at night. Cover with straw or a cotton sheet. As squid dry, pull and twist to a desired shape then tie up to six squid together on the third day (when drying is complete in fine weather). Tie in bundles of ten using the tentacles, and pack in cotton bags to market, or to leave hanging from the house rafter for home use.

Skinned squid can be packed together in straw bags on the second day to generate a white powder on their bodies (aspartic and other acids). Several days later, they are finally dried on mats in the sun for storage.

Even large cuttlefish (all are sold with their shell) dry in seven to eight days. The eyes are always removed from the head. Final water content should be 18–22% of normal fresh weight.

Larger quantities of squid can be skinned by close-packed agitation in water at 50–55°C (122–130°F) for ten to twenty minutes. Enzyme action plus rubbing removes the skin.

Fermented Squid

100 parts of cleaned squid meats* cut in thin strips are mixed with 4–8 parts of squid livers and 15–20 parts of salt by weight (summer 20, winter 15). The whole is put in a barrel or large vessel, and for a week stirred two to three times a day. The vessel is then tight sealed for a year. Add 2–4% of the ink if black squid is desired, and if red squid is wanted 2–4% of the liver is added. White squid is made from skinned squid and only 2% of liver.

*When cleaning, remove livers and ink sacs undamaged, for weighing, and wash the meats thoroughly; remove and discard eyes, beaks, and horny sucker rings (rub with salt).

Squid Pickle

Clean squid or octopus are boiled in lightly salted water and the skins removed. The squid is then salted down (15–20% salt) for several days, then moderately de-salted. Make up layers of rice, koji, vegetables, squid, and cover with parchment and a 20 kg (44 lb) stone for twenty to thirty days.

Vinegar-Pickled—Octopus and Squid Meats

Octopus or squid are cleaned and the tentacles quartered, and these are boiled with the body in fresh water for thirty minutes. Cool rapidly in cold water in a colander and cut into small pieces.

To each 50 kg (110 lb) of meats add 454 gm (1 lb) strong vinegar, 454 gm (1 lb) salt, and 10–12 kg (22–26 lb) water. Keep cool to preserve, or refrigerate or freeze meats when pickled through.

Tsukuda-Ni of Squid

Dried squid is softened or boiled, and the skin removed. Roasted to dry out, the heated body can be rolled out to 3 times its length. This is now cut into strips and parch-boiled in 2.5 kg (5½ lb) sugar, 4 kg (9 lb) wheat gluten, 0.2 kg (½ lb) salt, and 2.5 litres (5¼ pints) water.

This is stirred continuously for twenty minutes, boiling. When the solution is absorbed, the fish is cooled for sale. Use in three weeks. (Note—that no vinegar or soy sauce is used.)

Octopus

Large octopus are best minced and made into patties with an egg, some dry breadcrumbs, and some oatmeal to dry the mince. Fry in oil.

Otherwise, as done in Greece, clean and cook the octopus in its own juices over a gentle heat for forty-five to sixty minutes, or until pink and tender. No water or salt is used to cook octopus.

Cool, chop into pieces (not minced), and mix with ½ cup of olive oil, 1 clove of garlic, ½ cup vinegar, salt and pepper to taste. Marinate twelve hours in a cool place and serve with lemon juice and parsley.

Dried Crayfish and Shellfish (koura maroke) — Maori

Crayfish are "drowned" in cold water until the tails are easily removed (the fronts are steamed twenty minutes and eaten cooked). The tails are shelled, and threaded on a wire to dry for two days, then pounded flat and hung until quite dry. They are stored in bags in a cool, airy place. Shellfish, scalded to open, are also so dried. They can be eaten as is, or cooked in sauces briefly after a soak.

Dried Abalone

The simplest method is to clean and wash the meat, salt for about five days, wash, boil for five minutes and dry for seven to ten days. I have sliced abalone and sun-dried it successfully raw. Dried abalone is grated to flavour soups and other dishes.

Almost all tough shellfish meats can be treated this way, e.g. mussels, short-necked and long-necked clams, etc. Where large quantities of boil-dried fish are processed, the same water can be used over and over again to process each batch. Such fish-rich water can also be used in soy sauce brines or soup stocks.

Fermented Mussels — Thailand

Mussels, shelled (90%) and salt (10%) are fermented at 21–45°C (70–113°F) for fourteen days. This is a large home industry in Thailand. They have a reported shelf-life of six months. (Saono, 1986).

Tsukuda-Ni of Shellfish Meats

Raw fresh shellfish is shucked, or steamed and shucked.

Make Tsukuda-ni solution of 2 litres (4¼ pints) soy sauce, 750 gm (26 oz) sugar, and 250 gm (9 oz) wheat gluten. This solution is first boiled down to 80% of volume, then for every 7 litres (15 pints) of this first solution add and boil 2 litres (4¼ pints) tamari soy sauce.

Now in this 8.5–9 litre (18–19 pints) solution, add and boil for thirty minutes, 5.5–6.5 kg (12–14 lb) shellfish meats, 350 gm (12 oz) sugar, 350 gm (12 oz) wheat gluten, and 150 gm (5¼ oz) ginger.

As the meat becomes black-brown, take it out and cool; this latter process is repeated as often as necessary to boil all meats (start at 2 litres [2¼ pints] of tamari soy sauce and repeat). Cool and use in three weeks.

Crabmeat Fish Balls

Cheap white fish fillets are pounded with drained crabmeat after soaking in seawater or 3% brine. The paste is formed into 3 cm (1") balls, or slabs of about 10 x 5 x 5 cm (4" x 2" x 2") which are thinly sliced and carefully deep-fried before adding to hot bowls of soup.

The balls can be used to fill peppers or placed on cucumber slices in a steamer; added to soups a few minutes before serving; or spiced with chilli, pepper, and herbs, all finely chopped. (Balls containing prawn meats are mixed with egg yolk.) Balls can be cooked in a dough made up of 2 eggs beaten with 3 tbsp of rice flour or cornflour. The author likes big balls made up of 150gm (5¼ oz) of meats, but feel free to be creative.

Cornflour is also used to cover balls before frying.

Clam Chowder

Steam-cook any shellfish in very little water, in a covered pot for five to ten minutes. Then remove from the shells, cut up if necessary, and save any cooking liquid. Wash cooked meats to remove any sand or grit.

In another pot, cook in water (using the same amount of bulk as the shellfish): sliced carrot, chopped onion, grated ginger, and small cut potatoes (sweet potatoes, or yams). Drain cooked vegetables, saving the cooking liquid.

Now make up a white sauce of: 2 tbsp butter, 2 tbsp flour, and 2 cups milk. Heat gently until thick.

When sauce is thick, cool it down. Add the clams and vegetables, and then gently re-heat with salt, pepper, chopped parsley, or scallions to taste (½ cup).

Simmer a few minutes, adding clam juice and vegetable water to thin. Serve with toast.

Crisp bacon, if available, can be crumbled into the chowder, and at times I think the bacon may be the only meat in canned "clam" chowders! A bisque or thick soup is made in the same way, but using a tomato base instead of white sauce.

Oyster Sauce

Combine 100 gm (3½ oz) oyster meats, 454 gm (1 lb) anchovies, 6 cups white wine, juice of 2 lemons, and some grated lemon peel. Boil ½ hour and strain. Add spices (nutmeg, cloves). Boil fifteen minutes, and add a few sliced shallots.

Bottle, seal well, keep cool. Oyster flesh can be added to chiang or miso ferments at 10% of ferment bulk, or layered in vinegared koji to make oyster pastes or sauces.

Oyster and Shellfish Powders

Cook cleaned shellfish briefly, cool, mince fine, mix with powdered biscuit (matzo or similar), fine-chopped lemon peel, and mace. Pound to a paste, roll into thin cakes, place on paper in a slow oven until crisp, then pound the cakes to powder and store in a dry place, well sealed.

Krill in Soy Sauce

Sea krill or zooplankton is boiled and frozen when fresh. These can be ground and added to soy sauce raw materials, adding lysine and good flavour to the sauce. The same would be true of miso or long bran ferments of fish and 2–5% ground krill could be tried in soy ferments or other long ferments.

Sea Cucumber

Sea cucumbers or "sea slugs" (*Stichopis* spp.) expel their intestines on capture, and these may be gathered and made into a type of "clam" chowder. They grow back, and so are a perennial harvest. Alternatively, the body meat (white meat) is extracted and fried, or dipped in batter and fried. Smoke-dried, the whole cucumber is stored in baskets and keeps for months. Soak well in fresh water before re-cooking in soups or as minced patties with breadcrumbs or root starches to bind.

Chopped sea cucumber meats can also be marinated in a soy-vinegar sauce overnight (chilled) and eaten with salads.

Sea cucumbers are gathered mainly by women, from the coral sands at low to mid-tide, and placed in a capacious tub towed from the belt, or in a towed net. Like clams, turtles, and indeed any shellfish, sea cucumber need careful protection for a sustained yield. They may need to be re-introduced into exhausted environments.

8.8 SALTED SEABIRDS

Dry Salt only

Fish and seabirds are dry-salted, layer by layer, in barrels, stoneware, or glass jars. Sometimes, dark sugar is added 1:3 to coarse salt. The birds or fish are then weighted with a wooden cover under stones to create their own juice, and a muslin cover tied over the barrel. They are best eaten in the first six months (soaked overnight) but will keep for two years and then need forty-eight to sixty hours soaking.

A Bass Strait recipe for salted seabirds or gamebirds: Combine potatoes (washed and whole), carrots (washed), cabbage (with the stalk only quartered), 3–4 seabirds (salted and soaked) or 4 goose quarters. Boil fifteen to twenty minutes until potatoes are done, and serve all on a large platter for 4–8 guests to divide up. Delicious! Can be served with mustard and vinegar pickles.

8.9 NON-EDIBLE USES OF FISH

Fish Livers — India

- In villages, livers are marinated and exposed to sun heat in pots to ferment and decompose; the released oils are strained off and collected for rust-proofing and lights (lamps).
- Livers are chopped and boiled or steam-extracted and the floating oils collected (low yields).

- Minced livers are steamed at 80–90°C (176–194°F) under pressure (2 kg [4½ lb] per square centimetre) and the oils drawn off from the float layers (often on-board boats using engine steam), and centrifuged ashore. Refined oil is blended to known vitamin A levels with vegetable oils, and encapsulated in gelatine. Steaming for forty-five minutes extracts all oils from the liver, with the water level carefully raised while oil is run off from the top. Any remaining moisture is removed by anhydrous sodium sulphate. This efficient method produces high-quality medicinal oil used as vitamin supplements.

8.10 ALGAE AND SEAWEED

Nutritionally, seaweeds are very low in calories and fats but high in minerals and vitamins, therefore ideal food adjuncts for western diets. They have been consumed by people for at least 10,000 years, and are eaten by cattle, deer, foxes, bears, and sheep. In deep storm-washed kelp drifts, kelp flies breed, and their larvae feed waterfowl (ducks and geese). Mullet and inshore fish swarm around them for food at spring tides.

Potash from kelp stems and iodine from the ash have also been used since the 12th century in Europe. Ancient "fields" of boulders were created off the western Irish coast as seaweed holdfasts, and are still harvested today, although their origins are ancient, even pre-historic.

There are about forty genera of seaweed eaten fresh, dried, salted, fermented, or in combination with other ferments. Seaweeds provide a wide range of minerals, and some have excellent vitamin content.

Novices should always choose well-illustrated books for identification, but only a very few seaweeds are poisonous. Beware of very fine green marine algal tufts (below coarse hair diameter) and of eating too much of the *Caulerpa* group. Christianson,1981; Arasaki, 1983; and Fortner, 1978 provide good illustrations for identity.

The use of agars and alginates in food has increased historically. They are used as a way of creating viscosity, to stiffen gels, make emulsions, thicken soups, to prevent curds breaking, and to clear beers and liquids. Many products (including cosmetics) now contain alginates. Sausage skins can also be made entirely of alginates, and alginate spun to threads is used in weaving wound dressings.

Notes on Specific Seaweeds

a. Chlorophyta (green seaweeds)

- *Caulerpa racemosa*. (Sea grapes)—Greenish yellow, 4 cm (1½"), in lava tidal pools. Eaten fresh, as it doesn't keep well, but see kombu and kim chee. Many *Caulerpa* resemble velvety green or purple buttons fastened to the rocks.
- *Codium edule, Codium reediae*—1.5–6 cm (½"–2½") soft green, velvety, spongy. Found on coral reefs. Fresh, or wilted in brine and eaten in salads. Doesn't keep.
- *Ulva* spp. (Sea lettuce)—Eaten all over the world as a garnish, salad, soup thickener, and snack. It is chopped very finely in salads (otherwise tough), and can also be dried, roasted, powdered, or deep-fried.
- *Enteromorpha* spp.—The long green "hair" found where the streams reach the sea. The weed is thoroughly rinsed, squeezed dry, salted with 1 tbsp of salt per cup of weed, then ripened (fermented) until dark in colour. It is used chilled or eaten toasted. It can also be fermented, dried, toasted, powdered, and added to salt as a seasoning or spice for fish, stews, and stocks.

b. Phaeophyta (brown seaweeds)

- *Dictyopteris australis, Dictyopteris plagiogramma*—Subtidal on corals, leaf-like fronds, 4–16 cm (1½"–6¼"), golden brown. Can be salted and kept cold indefinitely, and used as a spice on raw fish, or octopus.

c. Rhodophyta (red seaweeds)

- *Ahnfeltia concinna*—Found as intertidal bands on lava, Hawaii (limu 'aki 'aki). Golden yellow, stiff, tough 5–15 cm (2"–6"). Used to thicken stews, soups, or for jellies. Only tender tips are chopped fine and eaten raw. Stores indefinitely dried. Mild flavour.
- *Asparagopsis taxiformis*—In Hawaii, in seaweed gardens in fish ponds (limukohu) 2–16 cm (¾"–6¼"). Deep salmon-red, strong iodine flavour, and most popular. Intertidal lava benches, cleanest. Washed, lightly salted, and rolled into tight balls which keep indefinitely in sealed tins. Used sparingly as a spice with meat, or fish.

- *Gracilaria* spp.—12–50 cm (4¾"–20") on coral rubble. Light yellow-green. Japanese (ogo) chopped and salted; used as salad, pickle, candied, soup vegetable or dips.

General Seaweed Preparation

As soon as possible (within a few hours) after harvest, wash seaweed in lukewarm water (not hot, which softens weed, and not cold which does not remove animals and coral). Do not soak. Drain quickly.

Chop or slice into 2 cm (¾") pieces or squares, removing the tough parts. Then, do any of the following:

- Salt or combine with salads and eat raw, or combine with raw fish, limpets, abalone, lemon juice and vinegar.
- Add to soups, cooking for the last few minutes before serving.
- Dry in the sun, or half-open oven at 66°C (150°F), for long storages. Soak before use.
- Salt in layers and ferment for two hours to three days (depending on species). Store chilled or dry for storage, or sun-dry after ferment.
- Half-dry and press into cakes for later slicing.
- Blanch in steam for a few minutes and freeze for later use. Often species are blanched and dried for "quick cook" later.
- Ferment to alcohol pickle, using yeast and 2% sugars.
- Can with beetroot, cabbage, or tomatoes.
- Boil down for agar, jelly or gel.
- Fry with fish or as a wrapping.
- Lightly toast dry flakes before eating.
- Make tea (for kombu only).

Cultivated Seaweeds

Japan:

- Nori—*Porphyra*. Dried in thin sheets after washing and chopping.
- Kombu—*Laminaria japonica*. Dried and powdered for soup stocks.
- Wakame—*Undaria pinnatifida*. Dried and folded.
- Hiteogusa—*Enteromorpha* spp. Dried and powdered.

Europe (Ireland, Norway, Iceland):

- Dulse, dillisk—*Calmaria palmata*. Dried, soups, a "spice".
- Sea lettuce—*Ulva lactuca*. Dried or fresh chopped.
- *Enteromorpha*—as above.
- *Monostroma*. Dried as a spice.
- *Laurencia pinnatifida*.
- *Irdae edulis*.
- Sugar wrack—*Laminaria saccharina*. "Seaweed bread" is made from the alginate. Stalks eaten fresh.
- Murlins—*Alaria esculenta*. Eaten fresh, cooked in porridges and milks.
- Laver—*Porphyra laciniata*. Dried, fried, soups, snacks.
- Iceland moss—*Chondrus crispus* as carrageenan, also boiled with milks as a jelly (South America, Tasmania).
- Giant kelps—*Durvillea antarctica*, *D. potatorum*, *Sarcophycus potatorum*. Soaked and roasted in the Southern Hemisphere. Dried and powdered for further processing, or hung to mature in the home.

Seaweeds are used as spices, salted or dried, and except for fast-rotting seaweeds, many were traditionally dried, salted for long storage, or for fermentation. Favoured seaweeds are cultured, transplanted, conserved by careful harvest, and even cleaned of corals while growing. Old timers like their seaweeds fresh, on the beach after cleaning, or fried with fish, bacon, or eggs. Gourmets like special mixed fresh chopped algae salads, with raw limpet and octopus, sea snails, or raw fish.

Seaweed should be thoroughly cleaned, rubbed, and sloshed about as you pick it, but often the storm waves do this for you. There are no poisonous seaweeds except for a very fine green weed of temperate areas, and that you can avoid. But there are many rich-tasting seaweeds that taste of the sea itself, and these are truly marvellous as little dried pieces, often offered to you by strangers on an Irish coast, or in Hawaii, on lonely rocky headlands.

Almost all seaweeds freeze well, most dry well, or can be kept salted or "ripened" to taste, and refrigerated. Only *Caulerpa* and *Laurencia* cannot be preserved.

Drying Seaweed

Pre-drying and post-drying treatments:

- Wakame—(*Undaria*) is dipped in lye or limewater before air-drying.
- Several species—(*Fucus, Plevetia, Cytoseira, Sargassum, Durvillea, Hizikia, Analipus*) are all fresh water soaked two to ten days (taste to test for bitter principles) before drying to reduce astringency, and to leach out tannin-like substances.
- All large dark-brown algae are stored dark and dry for two to three years, after air-drying to soften. (*Laminaria* and like spp.).
- Mozuku—(*Nemacystus, Tinocladia, Endisme, Cladosiphon, Sphaerotrichia*) are all preserved in brine at high levels of salt, with soy sauce, vinegar, and sugar.
- Red algae—(*Gracilaria, Meristotheca, Grateloupia, Chondrus, Eucheuma*) are simply dried or pre-treated with ash or quicklime before drying.
- Fresh water algae—(*Nostoc*) are either simply dried or stored in concentrated salt-sugar solution.

None of these seaweeds are dried in direct sunlight, but rather in a cool oven or in shade.

Plants are first washed clean in fresh sea water, then warm fresh water, dried, and soaked in water and vinegar or limewater before sun-drying to bleach. Brittle-dry plants are pounded to powders and used as soup stocks, or as a condiment mixed with sesame oil, or made into teas.

Dried Seaweeds (Kombu)

Kombu is a style of *Laminaria* seaweed which has been dried to leathery sheets up to 2 m long. Usually a few inches are cut off and used in soup stocks with fish flakes, and fresh sliced vegetables. It is also simmered in mirin to make a dip for tempura.

Storm-washed or cut long fronds are hung on lines under roofed (no walls) areas, or laid out on beds of shingle pebbles in the sun, where they are turned daily. Smaller fronds dry in a few days. The dried product is "shaped", trimmed, bound or tied for market, as bundles, flat folds, folded "packets", or if too short, in straw bags.

They are used as soup stocks, teas, vinegared, or wrapped around cold dishes.

Basic Kombu Stock

20–50 gm (¾–1¾ oz) dried kombu is stood in water for thirty to sixty minutes, then slowly heated. Just before boiling, the kombu is removed. Now add bonito flakes (or dried fish, or shrimps), boil, then cool and strain. The process can be repeated for a "lesser" stock, using the strained seaweed and fish. The extracted kombu is now parch-cooked in soy sauce and sugar and is so preserved as tsukuda-ni, eaten with rice.

The stock, seasoned with soy or salt, is heated and poured over parboiled eel, snow peas, sliced chives and kombu (one slice) in a bowl, garnished with fine strips of citrus peel.

Be careful never to boil kombu, as it releases bitter or off-flavour tastes. The secret is to heat to near boiling, and to parboil any dry ingredients separately, or add them uncooked to bowls (e.g. parsley). Raw eggs can be broken into bowls, and hot rice, ladled over them in the same way, as for sukiyaki.

Light Cucumber Kombu Pickles

Take ten 100 gm (3½ oz) cucumbers, salt, and roll on a board to soften skins. Dip briefly in boiling water, plunge into cold water. Cut the kombu leaf, a piece 5 cm (2") long, fine-chop 3 de-seeded chillies. Arrange all in a deep jar, sprinkle in 3 tbsp of salt, press and weight. Stand for three days and eat, or refrigerate for longer keeping.

Dried Seaweeds (Laver)

Porphyra (sea lettuce) seaweeds are washed, trimmed, chopped, and spread as a thin continual sheet on straw mats in the sun. They are then cut into square sheets and stacked for sale. Cultivated, the spores are caught on rope mats, brush bundles, bamboo screens or branches bound with rope. These are hung mid-spring (September in the south, March in northern waters) and harvested until growth ceases in the warmer early spring water. The chopped, washed fronds are "sieved" out of suspension in fresh water by the drying screens, and matted down, like papermaking, the frames reversed to dry the underside.

Porphyra is unusual even among seaweeds for its high (dried) protein content (22 mg per 100 gm), high in calcium and phosphorous, with 44,500 IU of vitamin A and 250 mg of thiamine, good niacin and vitamin C content. It also contains iodine, so it is an invaluable food additive in iodine-deficient areas. As well, it dries well and stores in sealed bags (in the dark) for years.

Porphyra capensis in South Africa is gathered and air or oven-dried, then soaked again in any intensively-flavoured liquid ranging from acidic (vinegars, lemon juice) to spicy (soy sauce, chilli, garlic and ginger powders). Then the seaweeds are again dried, crushed into flakes, and bottled for addition to rice, soups, omelettes, steamed or raw vegetables, or eaten just as a stock.

Sheets of laver in Japan are called nori and are wrapped around vinegared rice, raw fish or pickles, and are cut into rounds, or crushed in soy and sugar and boiled briefly, toasted and eaten, or seasoned, re-dried, and eaten. They are sealed in cellophane or plastic bags for sale.

Laver Bread

Boil fresh laver ten to fifteen minutes until tender, then add water to make a thick jelly. Coat a spoonful of this jelly in oatmeal, and fry with bacon until crisp. Serve with potatoes and butter. It is better if coated in egg before the oatmeal dip, then fried crisp.

Hot English Laver

2 cups boiled laver are mixed with ¼ cup beef gravy or boiled with 2 beef cubes, then lemon juice and pepper added and a little butter.

Matsumo Ferment

In Japan, the seaweed matsumo (fir needle) *Analipus japonicus* is sun-dried and salted. Mushrooms washed in salt water are inter-layered in pots with salted matsumo to ferment.

Namasu — Japan

Cucumber, carrot, daikon radish, ginger, and kombu seaweed, are all briefly salted and pressed, then marinated in rice vinegar.

Ake — Hawaii

Raw liver is chopped with onions and seaweed. The chopped liver is washed in salt, squeezed well to remove blood, further squeezed, rolled in flour, then rinsed. Eaten raw.

Kim Chee

This salty ferment is part of all Korean meals. It contains Chinese cabbage, salt, garlic, ginger, chilli, vinegar or lactic acid, sugar, and chopped seaweed, if available. Variously made overnight, over weeks, or stored for years. *Gracilaria* seaweed can be used instead of cabbage, and is "ripened" a few days in the refrigerator. (See section 7.1).

Seaweed Salt

Wash and squeeze *Enteromorpha,* dry in the sun for three to four hours or to brittle in a warm 65°C (150°F) oven until it is dark in colour. Powder with nori, add salt and crumbed hulled sesame seeds. Sprinkle over fish, and rice. (Dry *Ulva* can be included, and a chilli).

Limu Oki-Oki Pickle — Hawaii

All seaweed in Hawaiian is called limu. Mix roasted, crushed kukui (*Aleurites*) nuts 1:2 with sea salt, then mix with 4 cups of fine-cut *Gracilaria* or *Grateloupia* seaweeds (cleaned and part dried). The mix is spooned into small glass jars and refrigerated and will keep for months as a pickle with bread and butter, cold meats, cooked pasta, and root vegetables grated or cooked. It is a type of Hawaiian caviar! (See also Hawaiian salt).

Pickled Ogo (*Gracilaria verrucosa* and spp.)

(Also known as ogo kim chee)

Combine and set overnight, 1 kg (2¼ lb) fresh ogo (pieces cut 2"–3" long), with ½ cup of sea salt. Drain next day. Add and mix (without cooking), 2 cloves of garlic, 1–2 medium onions (chopped), 1 tsp of chopped chilli, and ½ tsp of paprika.

Pack tight into jars. Use after a few days. Refrigerate. *Codium* and *Ulva* (5%), treated, as for ogo, can be added. The mix is kept refrigerated. Small bowls are served with all meals, soups, or snacks.

Hijiki (*Hizikia fusiforme*) in sauces

Dried hijiki is fried, then simmered with soy sauce and sugar. This "sauce" can be made in large quantities as it keeps very well, and is used on rice dishes and fish.

8.11 SEAWEED EXTRACTS

Green Algae Powder in Oils

Green algae (*Enteromorpha, Chaetomorpha, Monostroma, Ulva*) are quickly washed in fresh water, drained (squeezed) and quickly air-dried to crisp. They are carefully toasted, held over a gas flame for a few seconds (not burnt) then mixed 50:50 with raw sugar and added to sesame oil for a spice on raw foods or salads. This keeps well for long periods.

Agar

- The seaweed *Gelidium amansii* and other species are boiled in water to extract agar-agar. About five *Gelidium* spp., and species of the genera *Gracilaria, Ceramium, Ahnfeltia,* and *Acanthopeltis* are all used. These are dried on racks and sprinkled with fresh water to bleach them. The dried weed is transported, soaked in fresh water, and boiled in a tank to which vinegar or sulphuric acid is added to achieve pH 6.

  Boiling continues for twelve to fourteen hours in about 75 kg (165 lb) batches, using the same water each time. Then the hot "double" extract is strained through cloth (20 meshes per 2.5 cm [1"]) and the liquid runs into small wooden trays holding 14 litres (30 pints) each to solidify. The gel is cut into bars and subjected to freeze-thaw. This frees the agar strips of water and with it the agar is purified of salts and pigments, and dehydrated. This dehydration can be achieved by refrigeration freezing each night, and sun exposure daily. Quality varies very much based on species mix, pH, and care in preparation.

- Made from *Gelidium* spp. or any seaweeds with good gelatinous extracts. First extract in hot water, then reduce the extract to a gel before freezing. A gel sets at 35–50°C (95–122°F) but melts at 80–100°C (176–212°F). Agar is released from seaweeds by boiling alone, or by heating water to 80°C (176°F). Long boiling develops off-flavours, so seaweeds are removed before boiling to condense the agar liquor.

  Red seaweeds, heated in water to 80°C (176°F) or more lose their agar and produce a viscous liquid. Strained and frozen, it becomes dry and bleached, whitish, spongy, and this freezing process can be repeated to produce whitened agar.

- Choose fresh *Ahnfeltia* or high agar containing seaweed, and cover with water in a large pan. Boil slowly, stir well, simmer for thirty-five to forty minutes when seaweed should be mushy and broken up. The water should be like a thick, smooth paste. Strain, freeze in 1 cup quantities for soup stock, gelatines of cold meats, pickled fish and eggs.

  Tomatoes or fish fillet pieces can be later cooked in this gel. Pickles, spices (pepper, lemon juice, celery salt), and chopped and salted seaweed can be added, and the mix poured into a mould to set. Eaten cold with parsley or salads.

Alginates

Alginic acids are released from the structural cells of seaweeds, and need chemical dissolution with sulphuric or hydrochloric acid and then a warm wash of sodium carbonate. Thus seaweeds from which agar has been extracted can be further processed to alginates, which break down the cell walls of brown seaweeds. It has a very wide range of food, medicinal, industrial, and cosmetic uses. Used mainly as a thickener or stabiliser in foods.

Of alginic acid, Arasaki (1983) says it is extracted from such brown seaweeds as *Laminaria* spp., *Macrocystis* spp., *Nereocystis, Undaria, Ecklonia,* and *Eisenia*. He remarks on the competition for the use of seaweeds, and the fact that edible seaweeds now make alginate production uneconomic. In the future we will probably find that the use of seaweed for fertiliser will also become uneconomic, and that increasing demands for nutritious food products will inevitably reduce less direct uses for such unique assets as seaweeds.

Low levels of dilute sulphuric acid are first used to macerate brown seaweeds, followed by dilution with a warm alkaline solution (sodium carbonate), which precipitates alginic acid, which is dried to its alkaline salts (not soluble in water, but the sodium salts are able to hold ten to twenty times their volume in water). Alginine acids are at their highest content in algae (47%) in autumn and early winter, and lowest (10%) in spring and early summer.

Carrageenan

Carrageenan differs chemically from either of the above, and is extracted only from the red seaweeds (*Chondrus, Gigartina*). It forms weaker gels than agars, but is extracted in hot fluids (milk, water), and widely used in foods as a thickener and a stabiliser.

Carrageen is bleached by repeated seawater soak and sun-drying of *Chondrus* to a light yellow before the extraction of the carrageen.

Other than for bleaching, a correct preservation technique for *Chondrus* is to wash it in fresh water, dry rapidly in the shade, and store dark and dry.

Other Seaweed extracts

Glutamic acid (monosodium glutamate)—is the flavouring agent derived from dried kelp in Japan.

Inosinic acid—Responsible for the flavour of dried bonito flakes, and multiplies the flavour of glutamic acid, thus reducing the need for the other to one-fifth. Both are now produced from yeasts cultured in molasses, then enzymes are used to strip the cell walls from the yeasts to obtain the nucleic acids.

Spirulina (*Spirulina platensis*)

In Chad, this algae has been used for centuries as a high-protein additive to grain diets (at 70% high quality protein per dry weight). Algae is produced in shallow fresh water ponds (part shaded), at ambient temperature, with a constant trickle of water. The dried algae is of the same or better value than are eggs for amino acids. Culture of good varieties of fresh water algae in ponds or glasshouse ponds is of real benefit for nutrition. Calcium phosphate salts may be needed for good algae production. Beware, however, of collecting unidentified blue-green algae from waters polluted with sewage and agricultural wastes, as these may be extremely poisonous.

CHAPTER NINE

Meats • Birds • Insects

9.1 MEATS

"I didn't claw my way to the top of the food chain to eat vegetables!"

(Lapel button, USA)

Meat products can be divided into two groups: those which are salted and/or dried and those which are made into sausages and cured (fermented). The main fermentation organisms present in sausages are the lactic acid-producing bacteria (*Pediococcus* and *Lactobacillus* spp.), which lower pH. Meats can also be salt pickled and smoked. In this chapter we take a look at both Western meat processing techniques and traditional meat meals, which often include birds and insects.

Hanging or Ageing of Game Meats

Strong meats such as venison, hare, jackrabbit, quail, pheasant, partridge, boar, kangaroo and the like are hung (in whole or part) in a cool airstream, or are wrapped and buried. The small animals and birds are left ungutted, but haunches or parts of larger animals are buried as parts, dusted with bran or turmeric (wrapping can be paper and cloth, well-tied to exclude worms). Meats are normally aged three days in summer, ten days in winter, or in chill rooms.

A hare or quail, buried or hung from four to seven days is "ripe" when the feathers or belly fur is easily plucked; it is then skinned carefully, and the now "set" entrails removed. The contents of the chest cavity (liver [gall bladder removed], heart, lungs, and congealed blood) are cooked separately (outdoors if possible) and made into strong gravy. The body is jointed and cooked as for standard recipes.

When the meat is ripe, it may have a greenish tinge over the belly, but once cooked it is not only tender, edible, and delicious, but more easily digested. I have often prepared hare and quail in this way. Some other partial ageing or rotting of meats, with body heat plus natural bacteria and algae, is practised by various people.

In Switzerland, I ate and enjoyed venison aged in a running stream for ten to fourteen days then cooked as usual. Obviously, such water should not go to village supplies but can be used in irrigation of crops or fish ponds.

Butchering meats where refrigeration is unavailable:

- Beef or game is hung in cool, airy places up to a week before cooking.
- Dust the joints or cuts with turmeric, or bran, and hang in a dry cellar. Most meat offered for open sale in Nepal is turmeric-covered and this extends its life briefly. For longer keeping, meats are salted.

Unto — Basque area of Spain

Unto are pieces of aged salt pork, taken from the brine and allowed to spoil until very pungent. This is added to salted pork shoulder, turnip greens, potatoes, and part-cooked and drained dried white beans to make the winter soup caldo gallego (Galician broth).

Fermented Walrus (after Olanna, 1989)

After the spring hunt, walrus is cut into 50 x 50 cm (20" x 20") slabs, with meat attached, and stacked in a hole 1.5–2 m (5'–6½') deep, in permafrost. The slabs are placed "skin-side down", then skin-side up and so on. The hole (in permafrost) is then covered with canvas or bags (not plastic), filled over, marked with a cairn or a stick, and there the walrus remains until fall.

When winter comes, the fermented meats are dug out, the fur scraped off, and pieces axed off and eaten, with salt.

Note—that even in deep freezers, an axe may be the best tool to cut up larger pieces of frozen meat or any material for cooking (fermented walrus is not cooked).

Seals

Fresh blubber from seal or whale is cut up and gently heated over a low-heat fire, then the blubber removed and the oil used for cooking, frying, or for oil ferment of meat and intestines, also to make ice-cream in following text.

Mukluk

Inuit or Eskimo

In the Arctic winds, seal flippers are aged, wrapped in blubber, and are eaten raw or cooked. Scalded seal flippers, brined, were a favourite food of the Tasmanian aboriginal sealers, roasted on the open fire or boiled. Narwhal blubber exposed to Arctic winds on racks is invaded by a green mould, and tastes like a good Stilton cheese.

Arctic Ice-Cream

Fat from moose or caribou (reindeer) is stored in racks to freeze and dry. In a large bowl grate equal (small) quantities of dried fat and seal oil, beaten or blended to a smooth cream. When the mix is smooth and light, add mixed berries (fresh or dried) e.g. salmon berries, cranberries, crowberries, or blueberries. Add sugar to taste and freeze to store (Olanna, 1989).

Note—that such a mix, poured warm over dried meats, stores very well frozen or chilled as pemmican, and that the dried meats can be of dried, salted, and pounded salmon flesh.

Note also—that boiled, flaked, and then cloth-dried ling cod flesh can be added to the smooth mix of oils and fats for storage.

Seal Intestines (after Olanna, 1989)

Intestines are cleaned, as for sausage casing; the outer skin is scraped off and the intestines cut up, cleaned out, and the inner lining skin well washed.

They are then either:

- Hung in the cold air to dry in long strips, and when dry but flexible, bundled to keep, under seal oil, in containers.
- Cut up fresh and immersed in seal oil (out of the sun) for five days, and thereafter eaten raw (or fermented).
- Cut up smaller and boiled one and a half to two hours with fresh pieces of seal meat, and thickened for a stew.

Seal Flippers

The tail flippers of seals are prized for salting down (as for salted beef) in barrels (Tasmania), or are well wrapped in burlap and placed in an open-topped hole in permafrost (Alaska). The hole is covered over with skin or boards, and with rocks if bear or wolves are present. After four to six weeks, try to see if the fur comes off easily, then slip off the fur, poach a few minutes in hot water, and serve in slices.

Salted seal flippers, freshened one to two days in fresh water, are skinned and boiled for one and a half to two hours with vegetables. They keep, salted, for up to a year. Fresh flippers can be de-furred by scalding before salting.

Seal flippers are also fermented in shallow, grass-lined holes; covered with seal blubber until the fur slips off easily, then poached a few minutes and eaten (after Olanna, 1989).

Note—that well managed seal populations are quite sustainable for harvest, and that a few seals always drown in gill nets even inshore. At present, large pelagic gill nets kill far too many seals and are being banned. Northern grey seals, southern fur seals, and sea elephants are recovering from excessive historic harvests in large numbers. As well, the Pribilof Island herds, managed by Russia and the USA, are at full strength.

Note also—that marine pollutants may accumulate in seals, walrus, whales, and porpoises, (at the top of the food chain) and, like the caribou, radioactivity can accumulate to dangerous levels. Where native peoples harvest sea animals, periodic analysis is therefore essential to health.

Kiviar

An Inuit delicacy. About 800 little auks, fresh caught, are sewn into a sack made of the whole skin and blubber of a seal (the seal being skilfully "peeled" from the mouth end), and flipper holes sewn up. The heat of the birds causes a ferment, insulated by the sealskin and blubber, and the whole lot is buried in the permafrost below the igloo for later partying in winter. Then the seal and auks are cut up with axes, and eaten as is.

Jugged Hare — Ukraine

Shoot hare carefully with a .22 rifle and plug the hole in the head or chest with cotton wool, hang overnight to cool without skinning or gutting. Wrap in 3–4 thicknesses of paper and bury about 46 cm (18") deep, weighting a board over the covered pit to keep cats and dogs from resurrecting the hare. (Burying is equivalent to hanging in cold weather and can be used for quail, pigeon, pheasant, wild duck, and deer.)

After seven to ten days, uncover hare, unwrap, skin carefully, open abdomen and carefully remove the now solidified stomach and intestines; discard these. Carefully split the chest and remove heart, lungs, and clotted blood and keep in a small basin. Carefully remove the gall bladder from the liver. Joint hare and cut into pieces. Brown hare pieces in a little oil, together with 454 gm (1 lb) tender steak cut into thin slices, bacon in broad strips, 1 onion, and a muslin bag of *bouquet garni* (peppercorns, mace, sage, thyme).

Pour over just enough port wine to cover, and place a lid on the casserole, sealed on with a 2.5 cm (1") thick roll or strip of pie or biscuit dough. Stand the casserole in a baking dish of water and bake in a moderate oven for two hours.

Gravies

- Strain then thicken pan juices with cornflour and heat for gravy and/or
- While hare is baking, gently boil liver, heart, lungs, kidneys, with water to cover 2.5 cm (1") deep, and add chopped garlic and shredded ginger. This is best done outside! After one and a half hours, blend, thicken with cornflour and heat to serve.

Now remove all meats from the casserole with a slotted spoon and serve meat on a thick board in the centre of the table. Use two jugs, one for wine gravy, and one for blood gravy with the heart, etc. Serve on hot plates with squares of toasted bread, new potatoes with parsley, a few tender carrots and beans or cabbage.

Dried Meats

Jerky or Biltong

Jerky (USA) and biltong (Africa) refer to the dried lean meats of kangaroo, bison, beef, elk, turkey, etc.

- The most basic jerky in dry sunny areas is simply strips of meat 1–2.5 cm (½"–1") thick dipped or rolled in ashes and hung to dry under shade in the heat and wind.
- Meat strips are beaten out flat with a mallet which is occasionally dipped in a bowl containing crushed garlic, pepper, salt, turmeric. The thin strips are sun and wind-dried.
- Strips are dipped, or marinated for twelve hours in a solution of: 1 cup salt; ½ cup brown sugar; turmeric, or black pepper (ground); 2 crushed garlic cloves; and 2.2 litres (4½ pints) water.
- In wet, cloudy, or humid weather, strips are hung in any one of the following:
  - 1 m (3') above the coals of open fires
  - Over a cooking fire on a rack
  - In a cabinet with a warm fan
  - 1.5 m (5') above a "smokeless" charcoal fire in a wide chimney
  - In solar-heated cabinets
- Marinated and salted meats may be dried and smoked.

Note—that meats used for jerky are trimmed of all fats that could cause rancidity. All jerky is best kept hung either in bags, or as loose-packed strips in warmed or heated jars (let cool after sealing) or in a press-top tin. Jerky keeps forever.

It is eaten as is, cubed for stews, cooked with butters or oils as pastes, or made into pemmican with fruits and fats. Jerky is a popular snack in southern Africa and south western USA.

Dried Beef or Game

Cut 2 x 3.5 cm (¾" x 1⅓") strips of lean meat, trim off all fat, beat lightly with a wooden hammer dipped in turmeric, salt, and garlic. Hang strips in an airy place or in the sun. They should be dry in three to five days and can be stored in tall narrow jars. Usually eaten as is, but can be dipped, pounded, and cooked as a hash with potatoes. Best meats for jerky are the lean meats of kangaroo, deer, and reindeer. Some jerky is dipped in strong boiling brine before drying.

Whole skinned and split carcasses of rabbit briefly dipped in brine can be salt or ash-dipped and dried in cold dry or hot dry air. They are soaked before making into stews.

Dried Meats (chimukyu or nyama hukutu) Zimbabwe

Experts (the Shona of Zimbabwe) make a drying platform; a wooden structure with four legs, each 60 cm (24") high, and the roof of parallel sticks; placed over a fire or in the sun. The meat is turned and dried for two to three days, then one day in the sun to ensure storage. (See Figure 9.1).

Sometimes meat strips are hung from a rope in the main hut above the cooking fire. Such dried strips are preserved in clay pots for up to one year in Nepal.

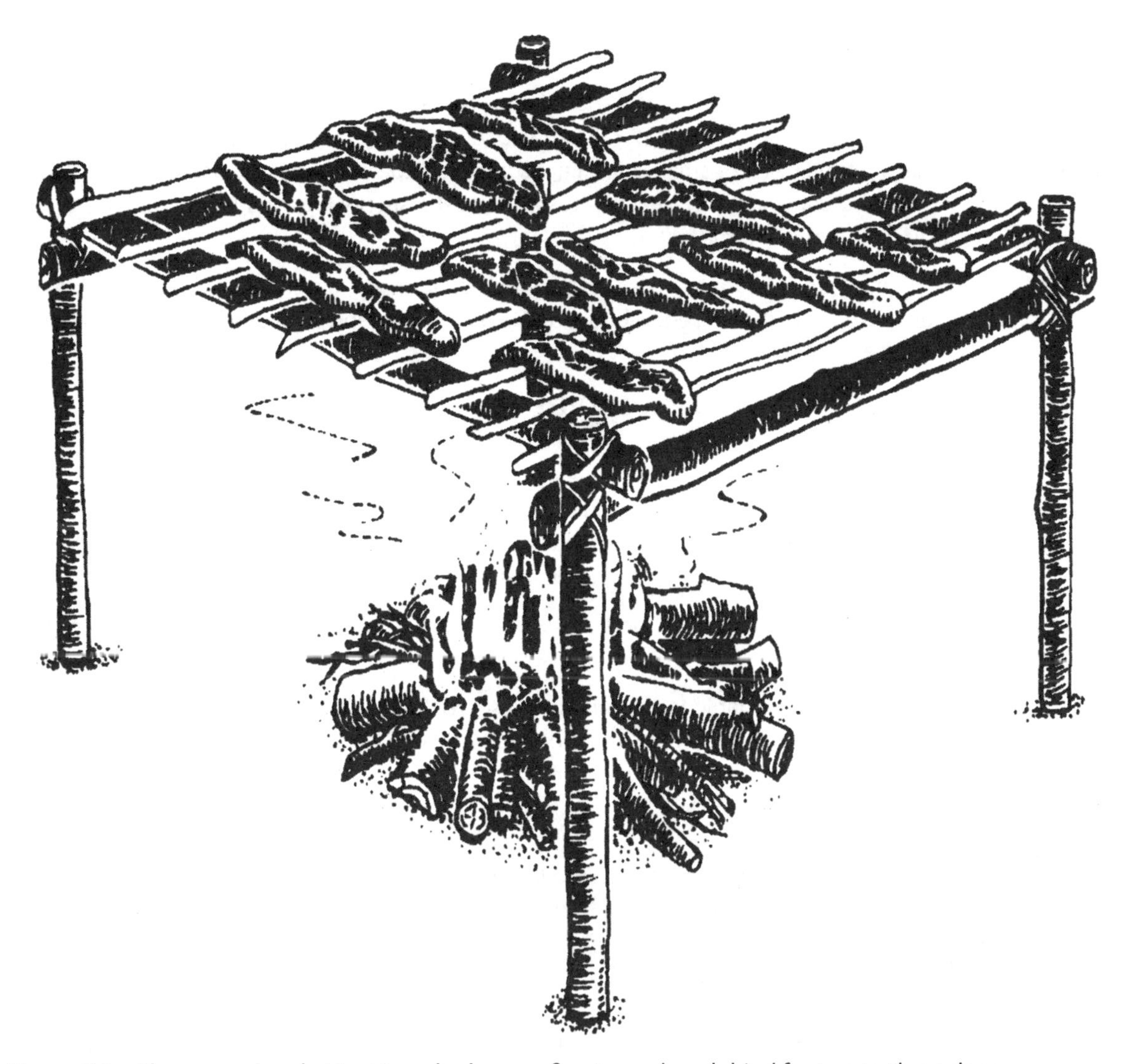

Figure 9.1—Shona meat rack. Meat is racked over a fire, turned, and dried for two to three days, then one day in the sun to ensure storage.

Game Jerky (Boer biltong) — South Africa

Game carcass like impala, yield 58% useful meat in live weight, compared with 44–50% of cattle live weight. Various game meats (ostrich, gnu, buck) are taken in early winter for jerky. About 3.5 kg (7¾ lb) of fresh meat reduces to 1 kg (2¼ lb) of biltong when dried. All meat is cut into 2 x 2 cm (¾" x ¾") strips.

A basic recipe is to marinade the meat (50 kg [110 lb]) in the fresh skin (which is laid over a trough in the ground) or in a tub with 1–2 kg (2¼–4½ lb) coarse salt and 500 mg sugar. Typical spices are coriander, anise, garlic, allspice, and herbs.

The meat, sugar, salt and spices are packed in a crock with larger or thicker strips at the bottom, and sprinkled with spices and vinegar on each layer.

The fresh meat lies in juices for six to twelve hours. If too salty, it can be rinsed in warm vinegar (1 part vinegar to 10 parts warm water). It is then hung to dry on wire hooks or racks. The strips can be dipped in ash, turmeric, pepper mixes to prevent insect attack. It dries in about three days, and is stored in cotton bags in a cool, dry place, or sliced up and stored in glass jars.

The marinated meats, some treated with chilli, are the basis for boervos (coiled sausage), which is grilled, fried, or casseroled to cook.

Dried Meat — Assam

Lean mutton is first sun-dried, then preserved in burnt lime for storage. (Assam has heavy rains and high humidity).

Pemmican — Native American

Jerky is pounded into shreds with a heavy mallet, and mixed with tallow (goose, pig, beef, or suet fats), dried berries or fruits, even vitamin C powder (1,000 mg per 500 gm) for essential food. Pound, mince, or combine, warm the mix, and stuff into jars, plastic bags, or small skins. Eat as is or add to soups, and stews. An energy food for trails. Ideally, such jerky can be dehydrated in cleaned stomachs over a smoky fire, or inside skins, also smoked. Pemmican stores a full year, or longer if refrigerated.

Optional additions to pemmican are oatmeals, dehydrated milk, soup mixes, onion salt, caraway, black pepper, turmeric, paprika, tomato paste, dehydrated vegetables, dried yeast powder and so on. Keep suet or fat at 30% or better.

Miscellaneous Meat Dishes and Preserves

Chasu — Philippines

Chasu keeps refrigerated for some months, and is usually sliced or cut into cubes to flavour rice and vegetable dishes. Traditionally, the centre of the pork loin is used, but you can use the whole loin. The meat should weigh between 0.5–2 kg (1–4½ lb), but only 5–7 cm (2"–2¾") in diameter.

Marinate the pork overnight in a mixture of 1 part soy sauce, 1 part honey, and spices (a little dry mustard, paprika, and salt).

When ready to cook, add a squeezed garlic to the mixture. Reserve the marinade as a basting liquid. Roast the pork at low temperature, 115°C (240°F) for twelve hours or more, basting occasionally. When cooled, wrap the meat in foil and refrigerate. If mould grows, simply wipe it off.

Fermented Pork — Thailand

Combine 80% pork meat, 12% thin-sliced pork skin (boiled to cook), 7% salted rice (steamed), 1% pounded garlic and mix well at room temperature. Then age at 25–45°C (77–113°F) for seven days. It keeps a month or two, is pH 4–5, and is used as a staple side-dish (23.1% protein) with rice.

Tapa (thin-sliced beef) — Philippines

Combine and mix well, 96% very thin beef slices 3.5 cm (1⅓") thick, 2–3% salt, 0.8% sugar, and spices (0.2% black pepper, 0.05% potassium nitrate). Incubate at 2–4°C (36–39°F) for two to three days, then at room temperature for twelve to twenty-four hours. Dry the beef. Keeps a month sealed in plastic trays.

Potted Beef Paste—Variation One

Rub lean beef well with salt and leave to drain for twelve hours, then repeat rubbing in salt and leave this just covered in water for three to four days. Dry the meat, rub well with black pepper, turmeric, garlic, and a pinch of nutmeg and cloves. Bake slowly in very little water in a covered pan for up to five to seven hours. Let cool. Now beat it to a paste in a strong mortar, adding spices to taste with a little melted butter and flour roux (see roux this chapter). Pack paste into pots and cover with melted or clarified butter, and cover pots with oiled paper. Keep refrigerated. Used as a spread.

Potted Beef Paste—Variation Two

Place 2.2 kg (5 lb) of lean trimmed beef and ½ cup of red wine in a covered pot in a steamer for four to five hours until tender (or steam in a pressure cooker for one and a half hours).

Then cut up the meat and blend or pound smooth with spices (salt, pepper, cloves), 4–5 anchovies (or anchovy paste), and 280 gm (10 oz) of clarified butter.

Pack into warm jars and cover with melted butter. Store cool.

Roux

Roux is melted butter or fat (low heat) thickened with a tbsp of flour and cooked until just brown, stirred frequently. It is used to thicken and brown gravies or meat juices.

Mutton—Variation One — Punjab

Cut 566 gm (1¼ lb) boned leg mutton into 2.5 cm (1") cubes and cook in a covered pan until its own juices are absorbed; measure out 375 gm (13¼ oz) of mustard oil; grind 250 gm (9 oz) onion, 250 gm (9 oz) ginger, and 60 gm (2 oz) garlic to a paste. If this is done in a blender it helps to use some of the mustard oil. Fry the paste brown with the rest of the oil, and add the meat to brown, adding 1 tsp chilli, ½ tsp each of cumin, cardamom, and cloves, 1 tsp salt, and ¼ tsp each of nutmeg and mace. After the pan is removed from the fire, add 1½ cups oil; pack the meat into sterilised bottles, covered tightly, and set aside for eight days to mature.

Mutton—Variation Two

Cut 566 gm (1¼ lb) chicken, pork or mutton into joints or 4 cm (1½") cubes and soak for ten hours in a basin with vinegar 1.25 cm (½") above the level of the meat. A paste of 50 gm (1¾ oz) each garlic and ginger is fried brown and the meat pieces added with 1 tbsp turmeric, 8 tbsp mustard oil, 25 gm (1 oz) chilli powder, 15 gm (½ oz) each of cardamom and cloves, ¼ tsp each of nutmeg and mace. Now add the vinegar and cook until it is all absorbed. Cool and screw down in airtight jars. This is said to last six months.

Moist, Delicious Hams and Turkeys

Wrap a salted ham or large stuffed turkey or any large piece of meat to be cooked whole in an old cotton bed sheet, well damped. Place in a pan and pour over a bottle of cheap claret. Place in a very hot oven 260°C (500°F) for one hour covered with a sheet of foil. Turn off the oven and go to bed. One hour before serving, unwrap the meat, paint it with a soy sauce-sugar mix, and brown at 180°C (356°F) for one hour.

This is the nearest we can get in a modern house to the juicy but thorough cooking of an earth oven. It eliminates "underdone" legs in the turkey, and leaves the breast meats juicy and succulent. I have cooked many large game animals and turkeys this way with great success. Be brave, try one, you will always use this method for large 6.8–18 kg (15–40 lb) meats. It would also work fine for 3–4 smaller birds wrapped together with other meats and vegetables. Of course, this type of cooking is very energy-efficient, as a large amount of food is cooked with minimal fuels.

You can also use a haybox oven for this, (see Figure 1.9).

A Great Pie (This recipe is just for fun) — Egypt

Marinate jointed game, chicken, or lamb meats (chunky or on the bone) with yoghurt, garlic and onion, and fill small specials such as whole pigeons or quail with various stuffings. The yoghurt marinade has pepper, ginger, garlic, cloves, coriander, and cardamom, mixed in. If meats are partly-browned or pre-cooked, so much the better.

Now mix 5 parts of fine wholemeal flour with 1 part oil (a small amount of sesame oil in olive oil is good). Knead well and divide into two slightly uneven balls. The smaller part is patted out flat and round to fit a special large round copper or steel plate with two handles that just fits your largest oven (or a brick oven).

Now the marinated meats, interleaved with onion slices, apple or apricot slices and dried fruit, and the crevices filled with savoury mince balls or whole uncooked eggs in the shell, are carefully built into a dome on the base of the round copper plate. Optional garnishes and stuffings include: pine nuts, grated cheese, spinach, stuffed cabbage leaves, sweet potato slices, and fried tempeh. A cup or so of cider can be poured over the meats.

The largest ball of pastry is rolled out round and draped over the dome of meat, then sealed to the lower pastry at the edges with a beaten egg "paint". The whole is put into the oven at 200°C (390°F) for twenty minutes and reduced to 120°C (248°F) to cook for two to four hours depending on how many pounds of meat are inside.

This pie remains on the oiled plate for transport and serving. Paint the crust with egg, and rosewater, if possible just before placing in the oven.

A Sea Pie (I had many such pies at sea!)

Layer up onions, fruit slices, bacon, potato slices, and jointed game birds or thick slices of game meats in a deep casserole, until 2.5 cm (1") from the top. Pour in 600 mL (20 fl oz) or so of good stock, wine, with soy sauce, grated ginger and garlic.

Cover with a thick 2 cm (¾") dough round to just fit the pot (made of fine flour, a little salt, baking powder, and water). Place in a 200°C (392°F) oven, then lower to 100°C (212°F) after half an hour and cook for two to three hours. Serve with slices of pie crust and green vegetables. Ideal for fresh lean game meats (jackrabbit, hare, wallaby, deer).

Bread and Onion Herb Stuffing for Game Birds—Poultry—Meats

For a medium-sized duck or chicken, prepare (chopped finely): 1 large onion; 6 slices wholemeal bread (stale or dry), blended or crumbled; 2 cm (¾") of ginger; 2 garlic cloves; 1 cup of "fine herbs", mainly sage and thyme, the rest parsley, chives, marjoram, coriander leaf, a little lemon grass, fennel, and mint.

Sauté the onion to a soft, golden stage, then mix all herbs, spices, onion and dry breadcrumbs in a large bowl, and grind on pepper and ½ tsp of salt.

Remove any fats from the body of the bird and place ⅔ of the stuffing in the cavity, the rest under the skin of the breast (loosened carefully from the neck end). For meats, beat out flat, lay on 1 cm (⅓") of stuffing, roll and skewer to casserole, or make a pocket in legs, ribs, or roasts and fill with stuffing and close with skewers.

Bake as usual for roasts.

Herb and Spices for Different Meats

Lamb—rosemary, apricots, turmeric.

Rabbit and hare—sage, quince, paprika, prunes, bacon.

Poultry—thyme, sage, onion, lemon slices.

Pork—grated apple, coriander, lemon.

Duck—grated rind and diced pulp of orange.

Stuffing can contain these herbs, fruits, and spices. Paint skins with soy sauce and brown sugar two to three times while roasting.

Breadcrumbs

Leftover stale but not mildewed bread can be kept ready to place in a "cool" oven at 65–93°C (150–200°F) until brittle-dry, then minced in a kitchen grinder, or briefly blended, and stored. Dry breadcrumbs keep for weeks or months in a tight-closed jar (left in a warm place). They are used in sausage, to coat egg-dipped fish or vegetable, or as a crust on baked milk or egg dishes and so on.

Salting and Brining Meats

Brine for Pork

Boil together 900 gm (2 lb) salt and 900 gm (2 lb) brown sugar in 4.5 litres (9½ pints) water. Cool and remove any scum. Wash pork and weight under the brine. (Kander, 1945).

Salt Pickling of Meats (legs and large pieces)

Dry-salt meat by rubbing in coarse salt and let pieces stand to drain for two days, turning often. Then rinse the meat in fresh water and pack it in a cask or drum. A salt solution for meat is made by mixing 680 gm (1½ lb) of salt to 4.5 litres (9½ pints) water; adding 226 gm (8 oz) sugar and a little potash; then boil and skim. Let this cool, and pour it over the beef, pork, lamb, or game pieces. Use a board and stones to weight the meat. It is ready in two to three months and keeps for a year, but can be best used over five to six months. Soak overnight before using.

Salt meats are usually boiled, but can be soaked long enough (forty-eight hours) to remove most salt, and then baked.

Traditionally pickling pieces are tongues, trotters, hams, legs, brisket of beef, flipper of seal, whale, etc.

Dry-Salted Meat

In a large wooden tub, mix equal parts of salt and brown sugar, add turmeric, ginger, pepper, and paprika. Cut meat into 10 cm (4") thick rounds and well rub in salts and spices. Weight down with plates and stones, and turn to rub daily for two weeks, then every few days. Ready at four weeks, can be briefly rinsed, dried, smoked, boiled, or baked if first well-rinsed, and de-salted for twenty-four hours.

Cooking Hams

- Boil the ham for an hour in a large pot. Take off the heat and stand the pot on five to six hessian bags; cover with ten to twelve more bags and let it "cook" like this all night. In the morning, uncover, let it cool, and while still warm skin the ham. Rub well with brown sugar and stick twenty to thirty cloves in the ham before letting it completely cool for slicing. I think this is the best ham. Very juicy. **Or**
- In the evening, cover the ham with a stiff paste of flour and water about 1.25–2.5 cm (½"–1") thick and place in a hot 260°C (500°F) oven on a tray. After half an hour, turn the heat down to 180°C (356°F) for one hour then turn off the heat. Keep the oven closed! Go to bed. In the morning, cool the ham, break off the crust and skin, and let it cool. Rub with brown sugar. Store cool and slice as needed. In a refrigerator it will keep a week or so.

Wiltshire Hams

These are cured in a paste made of 750 gm (1⅔ lb) of coarse salt, a litre (2 pints) of beer, 454 gm (1 lb) of treacle or molasses, 15 gm (½ oz) of ground pepper, and 30 gm (1 oz) of tumeric powder (all the above used for 5 kg [11 lb] of fresh pork).

The pickle ingredients are boiled together and left to cool, and the pork immersed to cure in a large crock, allowing two days for every 454 gm (1 lb) of meat, (i.e. 5 kg [11 lb] is left for twenty days). Meat is turned and the paste rubbed in every other day.

After two to three weeks, hams are dried, wiped off, rubbed down with bran, and enclosed tightly in a fine muslin bag, then smoked for twenty-four hours in cool smoke (hardwood smoke). Such hams keep at room temperature for up to a year. Once cut, they must be kept in a cool room.

Bacon

Bacon enters into many recipes, and even bacon "crumbs" (crisped bacon broken up) are used in salads as a condiment. Dry-pickled meats should only be attempted in cool to cold weather. Wet pickling is safer in warm humid weather, when there are less losses due to insects and spoilage in brines than in dry salting.

After salting, and sewing into muslin, most bacon hams or shoulders are cool-smoked to repel insect attack. Bacon taken fresh from pickle and sliced is called "green" bacon; bacon proper is hung in muslin, dehydrated, and smoked. There are various pickling techniques.

To Cook Bacon

To cook bacon from home-made hams: Cut slices a little thicker than usual and boil in a pan for five minutes or more, then discard the water and brown the bacon well before adding tomatoes, eggs, etc.

Dry-Salting Bacon Hams

Rub into hams some brown sugar (113 gm [4 oz] sugar to 9 kg [20 lb] meat) and coarse salt. Do this twice a day for three days, letting the hams drain, skin upwards. Then pack 1.25 cm (½") coarse salt in a barrel and lay on the hams. Cover hams with another layer of salt (1.25 cm [½"]), and so proceed until hams are salted.

Keep these for a month in a cool place, then take them out, wipe dry, pepper them well or rub turmeric over them, and smoke them in a cool smoke for three to four days. Tie hams closely in muslin and hang in a warm dry place. If you don't have kitchen room, wrap hams in strong paper and pack in dry ashes in a cool place.

Salt Beef

Use this recipe also for pork, mutton, wallaby, and so on. Boil and mix 4.5 litres (9½ pints) water, 680 gm (1½ lb) coarse salt, 226 gm (½ lb) brown sugar, and 1 cup of light ash. Skim and pour over meat which is cooled or cold. Meat so stored is good for a few months. It is best to dry-salt meats and let them drain overnight before pickling, to reduce blood content.

Salt Beef Paste

Put salted beef (soaked overnight if too salty) in a basin with a cloth tied over, and place the basin in a large pot to boil for seven to eight hours or until tender. Add at four to five hours, 113 gm (4 oz) clarified butter for each 1.8 kg (4 lb) beef. When cooked, beat beef and butter to a paste with 2–4 pickled anchovies from a can. Pack well into jars, pour on clarified butter and seal well. Store in a refrigerator.

Preserved Pork and Seabirds (poaka tahu) — Maori

Pork, beef, and seabird fats are rendered and kept for meat preserves (traditionally in leaf baskets in sand, now in pots). The meats are boiled in salted water, about half-cooked, then the skin removed (pork). All meats are slowly cooked in deep fat until well-cooked, and then left until warm, and placed into pots to preserve, well covered with fats. Pots are sprinkled with salt and sealed (e.g. with paraffin or hard fats). The longer it is kept the better.

Note on Fats

The "soft" fats, greases, and oils or waxes of seabirds, sea mammals, geese, ducks, game and free range animals are not known to increase the risk of heart attack, although the hard white, yellow, and saturated fats of grain-fed, penned and cage animals (and butter or cream) are implicated. Some fish oils may reduce cholesterol, especially pelagic school fish such as mackerel, herring, and sardine.

The same is true for some oils and fats of vegetable origin. Some are of little or no harm (olive oil, euterpe palm oil); others are less beneficial (coconut oil, oil palm [*Elaeis*] oils) if too much is used. **Any** rancid oils may be very dangerous to health, as are rancid meats of peanuts, coconut, and rancid nuts (walnut, pecan).

Briefly, well kept animals carry out numerous functions on any mixed farm, from pest control to gleaning, and normal exercise plus a good mixed diet ensures health if careful organic husbandry is the rule. We live on life forms, and to many of us it is no less a crime to kill a lettuce or tree than to kill a rabbit for food. In every case, to eat is to "take life", and we all eat.

Rendering Fats (beef and pork fats)

- Fat with skin of pork and birds, plus trimmings, can be put in a large heavy frypan or vat and just covered with cold water, then on medium-low heat rendered to clear fat which is poured off via a cheesecloth.

  Goose and pig skin are deep-fried until crisp, lifted out with sieves, and eaten as chicharrón (Mexico, pig) or gribenes (Germany, goose) cracklings. Slightly salted they are delicious even if not the perfect health food for every day. Rendered fats can be sprinkled with salt, cloth covered, and kept for months in a cool place, or used for soaps!

- Animal fats—Wash, cool, and if possible chill in iced water any clean animal fats, lard, suet and so on. Shred or fine-cut fats and add 50% by volume of water; add rennin, pepsin, or the chopped fresh stomach of a young pig or sheep which contain these enzymes (see also rennet). Heat under close control (best in a steam-jacketed vessel) at 45°C (114°F) for two to three hours, when the membranes should be digested and clear fat rises. Add a little salt, and draw off fats to cool, when the solid stearine separates from the oleopalmitic oils. Press fat solids later to further extract oils (placing fats in bags in a large press). Stearine is 40–50%, oleopalmitic oils 50–60% of the bulk.

  Fats are usually glycerides, solid at ambient temperatures. Oils are liquids at room temperatures; greases are softer fat. Note—that fig latex or rennin is also used to break down the fibrous binding of fat. See milks for sources of rennins.

- Extraction—Cleaned or fatty material is minced or shredded, and either directly heated at 40–60°C (104–140°F) or boiled in water. Fats are bailed out as they liquify, or let set solid and cut out.

  Neatsfoot Oil—Using bones, hooves, and hard parts of animals, it is rendered out by autoclave, under 60–90 lb steam pressure, as are other inedible fats used for soap and leather dressings, or used as for mineral oils as light-grade lubricant. Straining and skimmings are rich in phosphate and nitrogenous fertilisers, and can be added to fish silages for crops.

  Lard—Of all fats, the most useful is lard from pigs, and that dry-rendered from the kidney area is the best grade—called Leaf lard. Wet-processed kidney and back fat is called Neutral lard, and is second in value, and wet-processed trimmings from other areas is called Prime Steam lard and is used for soaps and cosmetics.

  Uses—Leaf and Neutral are used as shortening in bakery products, cooking and frying. Prime steam lard is used for ointments, pomades, and fine soaps.

  Tallows—From beef and mutton. As for lard, there are higher grades of white to pale yellow, and industrial or soap grades of yellow to brown colour. High edible grades are used in margarine, low grades in soap, and soft yellow grades as lubrication greases, soaps, candles etc.

  Bone and wool fats—Used in leather dressings, soap, and anti-corrosion paints. Wool fat (lanolin) is widely used as a base for skin creams.

  Fish oils—Have always been used to arrest corrosion in metals at sea.

  Guaiacum Resin—Protein extracted from the sawdust of *Guaiacum officinale*, a tropical evergreen tree (*lignum vitae*). It is non-poisonous and at 0.05% prevents rancidity in lard. Such substances are called stabilisers and enable fats and baked goods to be kept longer. The resin is extracted by boiling the sawdust or fine chips in salt or seawater and skimming off the floating resin.

Gelatines (or galantines)

Use calves feet and pigs feet, unsalted. If salted, soak for twenty-four hours in four changes of water, and change water if still too salty after twenty minutes of boiling. If unsalted, add a pinch of salt and boil, just covered with water, cook gently for three to five hours until falling apart. Skim and remove fat. Pour off reduced water through a sieve and let it sit overnight, using calves feet or trotters as cold meats if desired. Cut the set jelly into ½ to 1 cup pieces and freeze in individual plastic bags. These gelatines can be melted briefly and poured over stewed fruits (add honey if needed) to set to jellies.

Also, most are used as a quick soup stock with a knob of miso and fine-cut vegetables, or melted and poured over mixtures of hard-boiled eggs, cold sliced meats, ham, and sliced pickled vegetables as galantines or cold-jellied meats. Meats include large fish fillets (flaked), boiled eels, etc. Gelatine (warmed) can be painted on whole cooked (cold) birds to glaze them. Commercial gelatine is air-dried in thin sheets, or flaked, and sealed for storage.

Sausages

The sausage is one subject at which Europe probably excels in the fields of nutrition and ferment. There are probably two thousand regional varieties of sausages, and perhaps five basic ways to prepare or mix them, not counting varieties such as fish, cheese-based and vegetarian sausages. (Friends of mine who buy mass-produced sausages swear that **all** sausages are "vegetarian", but I suspect their gullibility).

Like cheeses, it is only possible to give general rules of good procedure, and to give a few recipes as examples. A very general content mix for good sausages (excluding any starter cultures) would be: 55–60% lean ground meats of pork, beef, game or poultry. Young animals are used for fresh sausage; older animals for dry fermented sausage such as salami; 20–30% fresh pork back fat, chilled and coarsely minced or hand chopped; 8–10% cereal binder, often dry bread soaked in marinade or broth; 5% (by bulk, not weight) of fresh herbs and spices; 2.5–5% of salt; a half cup of yoghurt, yoghurt whey, sour cream or dry starter can be added to mettwursts for long-keeping.

Sausage Types

There are perhaps five basic, common sausage types, although there are inevitable overlaps:

- Fresh sausages—of meat, fish, cheese/vegetables, uncooked and sold in bulk, as hamburger, or as strings of sausages. Kept cool, they are sold for frying, steaming, even used as forcemeats in baked meats. Examples: chorizo, plain pork or beef sausages, Oxford sausage, hamburgers, or fish sausage.
- Raw meats—carefully fermented and dried or smoke-desiccated for long storage. Bacteriastatic spices, alcohol marinades, and selected cultures create special flavours. They are long-keeping. Examples: mettwursts and salamis.
- Cooked sausages—simmered or steamed once filled, are kept well-chilled and re-heated by simmering to serve. These may also contain pre-cooked ingredients, together with other meats, and can be smoked before sale. Simmer at 72–75°C (161–167°F) for ten minutes to every 1 mm (25 mm equals 1") diameter of the sausage (normally about fifty to sixty minutes). Examples: hotdogs, wieners, saveloys, frankfurters.
- Sausages made from primarily pre-cooked ingredients—often from organ meats or blood, but also tongue, hearts, heads, tripe, brains, liver and so on. Cooked at 85°C (175°F), and cooled, they are then often smoked. They can also be combined with bacon and ham. Eaten cold or re-heated. Examples: liverwursts, blood (black) sausages, and so on.
- Pâtés and jellied meats—Basic sausage combinations baked in oven dishes, or set as galantines (gelatines), and eaten cold with salads. Pâtés are baked at 200°C (390°F) for from one to three hours (one and a half hours per kilogram [2¼ lb]). For galantines, thoroughly boiled pork rind, pork heads, pigs trotters, or calves feet are boiled one and a half to three hours with lean meats (venison, rabbit, shin) and set as a brawn in a cooling dish. If whole skinned tongues are used, they are pressed to cool in a mould. All are stored chilled, and eaten with salads or pickles.

See also Schaal, E. and Crowley, B. 1989. *The Great Aussie Sausage Book*.

General Notes on Sausages

Most meats for sausages are marinated, drained, and combined with chilled chopped pork fat (⅓ the bulk) before they are mixed with salt and herbs and passed through a mincer. Chilling makes the meats easier to chop. After packing meat and spices into a sausage casing, lactic acid bacteria develop in the meat. This lowers the pH value (which helps preserve the sausages) and contributes to the unique flavour.

Sausage need not be cased in skins or intestines, but can be steamed or boiled in pastries or cloth, large leaves or maize husks, bamboo joints, or simply made into rissoles or patties, grilled as hamburger, fried as meat balls, etc., or wrapped for cooking in earth-ovens in leaf material. "Sausage", in this sense, is ground meats (see kamboko) or meat-bean mixes.

Chopped or ground sausage meats for long storage are usually marinated overnight in pepper, wine, soy sauce, garlic, ginger, turmeric, mustard, salt, coriander, and a little vinegar (all adjusted to taste), spices approximately 4.5 gm (pinch) per kg (2 lb) of meats. Frequently cubes of coarse-chopped pork fat are added, and *Lactobacillus* cultures added as available, or whey sprinkled on.

Pork, beef, or turkey meats form the base material, pork fats the rest (60% meats, 30% pork fat, 10% marinade). All of the traditional marinades are bacteriastatic.

After casing and hanging for one to two days at 22–28°C (72–82°F) to desiccate, the sausages can also be smoked or further desiccated over a charcoal fire (1–1.5 m [3'–5'] above the fire). Sausages are stored over months at 10–15°C (50–60°F).

Lactic acid fermentation takes place with *Lactobacillus plantarum*, the pH falling from 5.5 to about 4.5. The skin and the fats suffice to seal the sausage sufficiently to prevent spoilage. Several bacteria and yeasts are found in semi-desiccated garlic sausages, where the fats are partially decomposed.

Although much of this semi-dry smoked sausage is consumed in three to four months, it will keep much longer chilled.

In recent times, meat ferments for sausage have been made using specific inoculants:

- *Pediococcus cerevisiae*, produces lactic acid.
- *Micrococcus* spp., to reduce nitrates and remove peroxides.
- *Lactobacillus plantarum.*
- *Lactobacillus brevis.*
- *Leuconostoc mesenteroides.*

The latter two are acid-producing, adding to stability and inhibiting action of other bacteria.

The use of starters of the above species suppresses many other spoilage bacteria, and are used mainly in dry or semi-dried sausages with a protein:moisture ratio of 1.5–3:1. Microbes are fermented at optimal temperatures before heating or drying sausages for long-keeping.

The ferments add to shelf-life, assist drying, and add to aromas and flavours. Fermentation is continued until pH drops below 5.3, to inhibit *staphylococci.* Incubation at 24–43°C (75–109°F) is continued for eighteen to eighty hours. Nitrites are used (not nitrates) to inhibit *Clostridium botulinum*, and starter cultures may also be used to rid bacons of residual nitrites, eliminating or reducing nitrosamines produced in frying bacon (Ward, 1989).

The mix of culture species, ferment temperatures and timing before drying and smoking sausage are not unlike the ferments of porridge meal (see ting in chapter 3).

Preparing Intestines for Sausage Casings (fresh intestines of pigs, calves)

Hose out contents where they can be washed into the garden. Separate fat, strip through fingers in warm water and soak well in water so that outer and inner lining strips easily. Threaded on a wire, they are easily turned inside out to strip. Sometimes they are drained and lime-dusted to eat, but for casings they are packed down in saturated salt water until used (fitted over the sausage filling nozzle of a mincer through which the sausage meats are ground). Every 8–10 cm (3"–4"), twist the filled casing to make a sausage, or tie larger sausages off with string.

Sausage Casings

Washed goat, sheep or pig intestines are packed in salt or brine. Natural (salted, or salted and dried) casings are soaked for an hour, cut into useful lengths of 1–4 m (3'–12'), rinsed through, and fed onto the mincer nozzle. The free end is tied, leaving a loop to hang the sausage or sausages to cure. Once a large sausage or string of smaller sausages (each twisted to separate) are made, tie off the other end. Sausages are then ready to mature, chill, simmer, or freeze.

Usual Sausage Fillings

Ground lean meats previously marinated twenty-four hours in wine, whey, yoghurt, vinegar, or soy sauce (all or any, to taste) are combined with 3–6% salt and spices (ginger, onion, garlic, herbs, turmeric, paprika, chilli, etc.). The drained meats are combined with the herbs and put through the mincer with enough soaked and squeezed bread or flour to bind all the other ingredients (trial ratios first). The mix is forced through the mincer into the casings. Make sure there are no air pockets. Hang in a cool place for two to three days. The sausages are dried (and a little smoked) over a clean (not smoky) fire or charcoal embers.

Some people like these lactic acid fermented sausages more after some weeks, and if well-dried and smoked, they keep for months. Once cut they should be fairly quickly eaten. Without marinades, whey, or some spices, they are cooked as fresh sausages. Fermented, marinated sausages are not always cooked.

A Good General Mettwurst Recipe

To equal quantities of lean beef and lean pork, use 25% by weight of skinless back fat, and to every 2 kg (4½ lb) add 3 heaped tbsp of salt. Before mincing, chill all meat and fat thoroughly. Mince lean meats fine and the fat coarsely.

Mix thoroughly into the minced meats: A clove or 2 of chopped garlic; a tbsp vodka or spirits (gin, brandy); 3 tbsp sugar or fructose; 1 tbsp peppers or ground pepper; 1 or 2 chillies (to taste); 1 tsp saltpetre; 1 tsp each mace, caraway seed, mustard seed, nutmeg, cinnamon, and/or paprika as available; or choose your own combination of spices to taste.

When thoroughly mixed, chill again, then fill casings, tie off end and hang to dry in cool air for three weeks, then smoke in cool smoke at 15–20°C (59–68°F) for two hours a day over ten to twelve days.

This sausage, stored below 18°C (64°F), keeps very well for eight to twelve months.

Simmered Sausage (Frankfurter)

Again use about 2 kg (4½ lb) of lean meats to 400 gm (14 oz) skinless fat, 2½ tbsp salt, 2 cloves garlic, 1 tsp marjoram, 1 tsp mustard seed, and 1 tsp pepper. Mince and mix well, then fill casings. Smoke in hot smoke 60–70°C (140–158°F) for an hour, simmer in water at 72–75°C (162–167°F) for thirty minutes, cool in cold water and store chilled.

Serve cold or re-heated.

Note—that pork fat can be replaced with skinned pigs' brains of equal weight (say 12% of total by weight).

Chorizo — Mexico

Beef, pork, pork liver, and some chilled pork fat are chopped or minced coarsely. The meats may first be marinated in red wine, garlic, ginger and so on. Spices such as peppercorns, paprika, chilli, pimento, coriander, 3–5% salt, and 100 mg (1 tsp) of sodium nitrate per kg (2 lb) are typical.

The mix of meats (usually just pork) and coarseness of the meats differs regionally.

When filled, sausage casings are first chilled two to three days at 3–6°C (37–43°F), then allowed to ferment at high humidity and temperature (30–38°C [86–100°F], 90–95% humidity) for two to six days.

They are gradually dried at 20–25°C (68–77°F), 50–60% humidity for weeks or months to mature. They, like salami and mettwursts, dry to 23–30% moisture.

Mexican chorizo is most often cooked fresh, uncased, and is quite spicy.

Longanisa (semi-fermented sausage) — Philippines

Combine lean pork mince 70%, pressed dry before mincing with 30% cubed fat from the pork back. Weigh and add salt 2% to all pork, sugar 2%, soy sauce 2%, vinegar 2%, port wine 2%, spices (black pepper and garlic, each 0.6%, for example 6 gm per 1,000 gm of mince meat), potassium nitrate 0.05%, for example 0.5 gm per 1,000 gm of mince meat, and phosphate blend 0.15%, for example 1.5 gm per 1,000 gm of mince meat.

Mix all ingredients. Fill sausage cases and cure at 2–4°C (36–39°F) for two to three days, then at room temperature 30–35°C (86–95°F) for twelve to twenty-four hours. Final pH is 5.2; keeps one to two weeks at 2–4°C (36–39°F) and one to two days at room temperature.

Salami—Variation One — Australia

Combine 95% beef, pork, or cubed pork back fat (fat is about 30%), with 3–4% salt and spices (pepper, sodium nitrate 12.5 mg per 100 gm).

Fill well-mixed ingredients into cases, and ferment at 20–23°C (68–73°F) for three to five days at 90% relative humidity. Then store at 15–20°C (59–68°F) for two to three weeks at 85% relative humidity. Placed in cool storage, keeps a month or more.

Salami—Variation Two — New Caledonia

Game meats, fish, and shark are cut up, as for a stew, and seasoned with salt and pepper. Oil is added only to dry meat (not shark). Vinegar (2 parts) and wine (1 part) are added to the pan until the meats float, then spices added (garlic, thyme, bay, cloves). It is left three to four days, then the meats are browned and served with the flour-thickened marinade, or (as in Tahiti where limes are used) eaten raw from the marinade if cut small.

Can be eaten (in Tahiti) in ten to twenty minutes from using a lime juice marinade.

Liverwurst

Using the liver of a pig, and half the head, together with the same weight of fatty pork belly as the liver, a bouillon is made with 3 level tbsp of salt, 3 or 4 carrots, a stick of celery, 8 spring onions, 2 onions, and a handful of chopped parsley. Keeping the liver aside, boil the bouillon for two hours, then remove 2 cups of the liquid and reserve this.

Remove flesh from the boiled head and cut up, cut the belly and skin into strips, sear the liver in boiling bouillon water for twenty to thirty seconds and cut this into strips. Add 3 pinches saltpetre, 2 tbsp black pepper, 2 tbsp marjoram and the 2 cups of bouillon to the meat mass, including the garlic and onions from the bouillon. Mince finely, using a 4.5 mm (3⁄16") disc, and fill casings. Now, cook these sausages by lowering them into boiling water, and adding cold water to bring temperature down to 80–83°C (176–182°F). Cook for ten minutes per mm (25 mm equals 1") of thickness (twenty to sixty minutes) at 85°C (185°F). Cool in cold water and dry out on boards. Cold smoke at one to two hours per day for three days, then store chilled (after Schaal and Crowley, 1989).

Black Pudding

To 1 litre (1¾ pints) of blood, use 1 kg (2 lb) of pork heart or fatty pork belly with skin, 700–1,000 gm (1½–2¼ lb) of skinless bacon, 3 heaped tbsp salt, 3 chopped onions, 2 tsp black pepper, marjoram, ½ glass of vodka, and 2 bay leaves.

Cook skins and bacon in water for thirty minutes, add heart (chopped) and cook thirty minutes more. Cut pork belly and bacon in strips and finely mince with cooked onions through a 3 mm (0.12") disc. Cube heart and fat to 10 mm (⅓") size, blanch fat in boiling water (in a sieve). Warm the blood and add to this mass, with herbs and spices. Mix well. Pour into casings and cook as for liverwurst. This black sausage is also sometimes smoked, but stored cool and usually re-heated in slices for breakfast (with eggs). Note—that pickled beef tongue may replace bacon and heart in this sausage, but procedures are the same, using 1 litre (1¾ pints) of blood to 2 kg (4½ lb) of meats.

Fat Preserved Meats and Sausages

Thoroughly cook or fry the meats or sausages. Pack into crocks and pour over enough lard to cover meat to 2.5 cm (1"). To use, take out and warm up, strain warm lard and return to jar. Serve meats just heated, with vegetables.

Duck Rillettes (Paste)

Bone duck or goose breasts, cut to 2 cm (¾") cubes, but cut pork fat into 1 cm (½") cubes (add at 150 gm [5¼ oz] pork to 1 kg [2¼ lb] of bird meat). Place meats in pot and add ½ tsp salt, ground pepper, 2 sage leaves, 1 tsp thyme leaves, and a glass of port or white wine. All ingredients are brought to simmer, skimmed after fifteen minutes, and slowly cooked (stirring at times) for one and a half hours covered, and thirty minutes uncovered. Cool until just warm and mash meats, then re-heat and mix well. Pour into earthenware, level the tops, pour on 100 gm (3½ oz) melted butter, so that a layer about 4 mm (⅕") thick is formed. Ready in three weeks. When opened, use within three days.

Pork Rillettes (Paste)

Proceed as above, but arrange meats to back fat at 1:2, add spice with juniper berries. This time pour off a little fat before pounding, pack into jars, then pour fat back on top. Keeps three weeks or longer in cool storage. Eat once open.

When cooking meats for long periods, place a diffuser (toaster grill or mesh) under the pot to reduce hotplate heater flame intensity.

Offal (off fall) Meats

Livers, kidneys, heart, lungs, gizzards, sweetbreads (pancreas and thymus), tripe (dressed stomach of oxen and sheep), feet, ears or heads, brains, chitterlings (washed intestines), fins, flippers, maws (stomachs), desert oysters (testicles), penises, and udders, as well as shins, trotters (feet), and blood juices are all valuable protein foods. As Davidson (1979) sagely remarks "carnivorous animals do not thrive on muscle meats alone." Many offal meats contain concentrated minerals, enzymes, and vitamins.

Blood—rich in iron, is often cooked with salt, grain meals, and fats in "black sausage".

A traditional herb is mint. Blood sausages are boiled or steamed, dried, and slices fried with eggs, tomatoes, etc.

Blood is collected from stunned or shot animals as follows:

- Animal hung over bowls or vats: cattle, deer, chickens.
- Animal held on table over vats: pigs, sheep, goats, rabbits.
- Blood left in chest until cold: game birds, hares, small deer, "hung" game generally.

Blood is strained, and whole blood added to cereal (breadcrumbs or meal), herbs, pepper, salt, and garlic. This is cooked "as is" for a breakfast, or further mixed with forcemeat as "black pudding" or sausage, which in turn is boiled thirty minutes before cooling for storage.

Slices are fried with eggs, bacon, or added to poached eggs for farm breakfasts. In France fresh-killed chickens are simply hung over pans of the cereal mix, which is later cooked with a little olive oil. Quantities depend on the quality of blood, but the cereal mix can be added until the mix is just moist.

Traditionally beef intestine is used for black pudding casings. My grandmother always added chopped pennyroyal to the herbs (about 5% of the total bulk), which can include coriander, thyme, sage, tarragon, garlic, and a little ginger. Blood from the chest cavity of game is used in gravies.

Livers—with the gall bladder or bile removed, should always be eaten "just cooked", and are often consumed fresh by hunters. They form a large part of the offal, and are shared carefully in desert tribes among menstruating women. Liver helps iron, anaemia, and is higher in vitamins and available iron than muscle meat. It is also rich in enzymes. The bile (with 6% salt) is kept to ferment *Parkia* seeds in Africa (as it emulsifies fat and water).

Testicles—in deserts, as well as small ground ear-bones, and pumpkin seeds, are a source of the rare zinc in alkaline areas. The penis of game is also eaten, roasted. Testicles go to the adult males but would benefit everybody, if there were enough of them!

Livers, kidneys, and pancreas contain nucleic acids.

Brains—are a source of phospholipids. They are usually egg-dipped, fried and served with a mushroom sauce, sprinkled with lemon juice and parsley. In camp, they are cooked (roasted) in the skull which is then split, and given to children and the aged.

Muscle—has potash, phosphates, peptides, nucleatides, creatine, vitamin B12, and other vitamins, but is not preferred by hunters to offal! Lean muscle (neck, shin, hocks) is reasonably fat-free and is casseroled for one to two hours.

Adrenals—are rich in vitamin C; they can be added, fine-chopped, to soups just before serving.

Cartilage—(shark fin, the cartilage of young animals) has anti-carcinogenic properties, for solid cancers. It is cooked for a long time in soup stock, and a few vegetables added late in the cooking for soups. Mostly, cartilage is used as gelatine. The soups used are gelatinous stocks of chickens' feet or pigs' trotters.

Whole Small Meats

The Ovambo people of southern Africa take the whole cleaned bodies of frogs, waterfowl, rodents and small birds and dust them with ash and salt. Then they are spread on thatch or mats to dry two to three days (Becker, 1975). The same peoples also roast insects, pound them with peanut butter and salt, and use this sauce on porridge. Caterpillars from the Mopane tree are cooked with beans and spinaches.

9.2 BIRDS

Preserving Birds

Birds and Meats (Le confit)

Fats from geese and pigs replace vegetable oils in cool, mountainous areas. In France, Confit™ is a trade brand goose or pork preserve. In autumn, when most small livestock are killed, all fats are rendered from geese, pork and poultry and saved in great storage jars to last all year.

Meat is cut into handy pieces (geese are quartered, chicken halved, pork in pieces), then these pieces are thoroughly rubbed in sea salt (coarse crystals) on the skin and flesh sides. About 113 gm (4 oz) of salt is used per goose (3.5–7 kg [7¾–15½ lb]). All internal fats and trimmings are rendered down.

Fresh meats are salted for three to six days, washed and put to cook very slowly under fat for two to three hours, then the meat pieces are placed in a deep earthenware jar.

After the fats have cooled, they are strained over the meat pieces until these are completely covered; then the top of the jar is tied over with paper. In a cool, dry place, geese and pork cooked like this will last for months. Periodically a jar is lightly warmed and melted and the meat pieces taken out for warming up, and heated again, gently, in the fat clinging to them. Such meats are also common in Russia.

Confit is served with potatoes, or on beds of lentils, split peas, haricot beans, rice or purées of lentils. Also, the plucked thick neck skin of the goose is turned inside out and stuffed with meat scraps, bacon, herbs, garlic and pork mince all mixed with 2–3 eggs. This sausage (tied at both ends) is cooked with the confit, and preserved in fat. It is served cold, in slices, or briefly re-heated whole in a casserole.

Preserved Goose—Seabirds—Pigeon—Duck in Bird Fats

From very fat birds, cut off fat and attached skin squares, remove entrail fats and chop up, sprinkle on salt and stand overnight. Wash, drain, put in a deep kettle with a cup of cold water and cook gently for one to two and a half hours, with a lid on. Strain fats, let cool, cover, and store in a jar for preserving bird meats as below.

Joint fowl, sprinkle salt over meats, and let stand. Soak in salt, garlic, turmeric, and sugar, then boil in fat until brown and tender. Stack meats in a crock, and pour over clear, hot fat until well covered. Cool and cover with a plate and stone. This will keep for months.

To serve, remove meat and some fat, heat, drain off fat and pour back into the crock. Serve meat.

Salted and oil-cooked geese and ducks can be smoked and dried to keep. Cut thinly to serve.

Salted Seabirds (*Puffinus tenuirostris*) Tasmania

Up to a million unfledged or just fledged seabirds (mutton birds, moonbirds) are taken in Tasmania yearly, on commercial islands under careful conservation rules (the species has increased historically). The birds are taken from their fledging burrows, killed, and the oil or wax contained in the crop expressed for medicinal and water-proofing uses (krill oils and waxes). They are plucked, and the feathers rolled to express the oils and washed for down sleeping bags or doonas. Dipped briefly in boiling water after the necks, legs, and tail are removed, they are fine-plucked by hand, rubbing with a wet sack. Birds are then split down the back, cleaned, then wings carefully disjointed with a sharp knife close to the breast. Birds are cooled on mesh trays, and when cool, are salted.

Salt is rubbed thoroughly in the wing sockets, and on the cut sides of the birds. They are then carefully and tightly packed in wooden barrels, with generous handfuls of coarse salt between the layers of birds; some pickles using 30% brown sugar. The heads of the full barrels are tightly fitted, and the birds taken to homes and markets. Most birds are used over the next nine to ten months.

The fresh bird market today uses fast chilling and freezing methods.

Fats (about 2 cm [¾"] thick on the breasts) are rendered for egg and meat preservation (as for geese), crop oils are boiled and strained to remove krill shells, and the dense feathers mixed with some duck feathers for bedding. Fats are preferred for fish frying, and the oils are used for chest complaints by the islanders.

Geese

In Scotland and Scandinavia geese are often salted (dry or in brine) for one to two weeks, then can be smoked to keep (Solan geese). They are soaked, boiled until tender, and served cold, or sliced cold with horseradish sauce.

Pheasants

On pheasants aged in feather, the *New Larousse Gastronomique* has this comment: "The ideal moment (to prepare the bird for eating) is when the pheasant begins to decompose. Then its aroma develops in an oily essence which requires little fermentation to reach perfection."

Small Birds

For crop-birds like weaver birds and sparrows, skin and boil whole (less feet and head) for five to six minutes. Preserve in a glass jar under salted white vinegar (pour on hot and seal well). Will keep for a year, eat bones and all. Alternatively, deep-fry in fats and store as for geese. Fresh, they cook well (split), firstly seared or browned in a little oil, then add 1 cup of cider, cover, and cook on a low heat twenty to twenty-five minutes. Serve on toast.

The weaver birds (*Quelea* spp.) of the grain fields of Africa can strip millet and sorghum fields. Traditionally taken when the grains are at the "milky" stage, these birds may add considerably to the protein of tribal peoples. Singed, roasted and added to normal porridge-grain meals, they represent a large (and largely unharvested) protein source. Harder (shattered) grains are ground-foraged by pigeons and doves, who nest in and damage thatch. These too can donate some protein to poor diets via mist nets or cage traps.

Feathers (to clean)

Goose and duck feathers, if clean, are simply sun-dried in a hessian (burlap) sack on a clothes line, but if dirty they are rinsed in a weak alkaline solution (sodium carbonate, chloride of lime, or weak lye), washed in fresh water, and dried as before. Washed and dried feathers are beaten to be used in pillows, as stuffing, or insulation in walls.

Feathers for fertiliser are hammer-milled with oat or rice husks and used in feather meal, or mixed with bone meal.

Seabird or oil feathers are put through a set of strong rollers to squeeze out the oil in the feather shafts, then washed in limewater or lye as above. A cup of ammonia per 40 litres (10 gal) of water helps with oily feathers.

9.3 SMALL VERTEBRATES

Rodents Generally

There are probably 60 or more species of "rats and mice" eaten by people as normal food. Rodents form a large part of the biomass in deserts (as in cities), and are always on sale in rice-growing areas like Hong Kong, together with edible dogs, snakes, and birds.

Note—that rice rats commonly take 50–100% of the paddy crop in South East Asia. Guinea-pigs are the domestic meats of Ecuador and Peru, and enter into their songs and legends.

How to Cook a Guinea Pig

Heat two 2.5 cm (1") thick steel plates, about 30 cm (12") round, over a brisk fire; one of these plates has a handle centre top. Tap the guinea pig on the head and throw it on one plate, lift the other plate and slam it down hard on the guinea pig. The fur singes or flares off, and the animal is cooked in a few minutes as a roundish, flat "pancake", and is so sold in markets. Fastfood indeed, excellent for nibbling on. This method is applicable to many small furred vertebrates.

9.4 REPTILES AND AMPHIBIANS

Dozens of species of snakes, lizards, tortoise, frogs, and the large aquatic axolotls (salamanders) of Mexico are gathered or farmed for human food. Reptiles are roasted, frogs and salamanders stir-fried or briefly boiled. The removal of too many of these valuable insectivores from crops allows insect pests to proliferate, so that farming these animals, rather than overgathering, is encouraged. This is not the case with grainivorous rodents and birds, which are very well able to proliferate despite our best efforts to capture and kill them.

9.5 INVERTEBRATES

In Africa, because I eat grubs, termites, invertebrates and unusual foods, the Tswana believe I will eat anything. When strangers come to our camp and want to know my totem, they ask: "What is it this man does not eat?" i.e. What is his totem? (One does not eat one's totem food, animal or plant.) This innocent, polite question reduces my friends to extravagant laughter, even tears. However, one day Louis Nkomo (a Zulu) solved the problem: "This man," he said, "will not eat the honey-bear (the ratel)." Even greater laughter, as the honey-bear will eat **anything**. Hence my name in Africa is Rra Takadu, and among the Sonora (Pima Indians) Iwo Jutum. In both cultures I am "honey-bear".

A very short and restricted list of the foods derived from invertebrates is as follows:

- Palm weevils (Africa)—are eaten fresh, or salted and dried after roasting.
- Coffee moth larvae (*Zeusero coffea*)—are roasted with rice and salt. Delicious!
- Termite alates (*Macrotermes* spp.)—are fried in a little oil, salted and stored as a seasonal staple in Africa. Alates are the winged stage of the termites. (See Figure 9.2).
- Locusts, grasshoppers, cicadas, crickets—are all high in protein, fats, and aresuperior energy foods to meats. Most are taken as adults, but the larval or pupal forms of cicadas can be dug from soils below shrubberies in the tropics and sub-tropics.

As food, locusts are 61% crude protein, 8% fats, 5% moisture, 10% fibre, 1.63% silica, and nitrogen 9.9% (largely as chitin), together with calcium, phosphates, and potash (0.6–1%).

Grasshoppers

Normal preparation consists of the removal of wings, wing covers, legs, and perhaps heads. The bodies are often boiled, lightly fried or roasted in pans, then sun-dried for keeping. Once dried and salt-sprinkled, many can be stored whole or as powders for addition to other foods.

Fresh insects are skewered and roasted, or simply fried and eaten. The larger insects are often likened to shrimp; many are added to omelettes and porridges in fresh and dried forms. The original (not shop-bought) North African couscous is made as follows:

Couscous

There are, of course, yeasts inside and on the bodies of grasshoppers from grain fields, and grasshoppers are the base of the wheat-cakes cooked as couscous, or at least are such a base in the countries of origin (in North Africa). Cooked wheat is cracked and pounded with uncooked grasshoppers and a little salt. It is moistened and packed into buried jars as a "nauseous ferment" (probably a normal ferment).

After six to eight days, the mass is formed into cakes 1–1.5 cm (⅓"–½") thick and sun-dried. Couscous is quickly cooked in water, in stews or alone to be eaten with sauces or hot meals.

One would expect that somewhere in grasshoppers, caterpillars, etc., are all the necessary yeasts and enzymes to digest plant leaf and grain. For couscous, some of the grasshoppers used are *Locusta bruchos, Attacus ophimacus* and like species. This recipe can be used for a variety of insects and grain crops.

Termites

Termites are widely eaten in the tropics (Africa, Australia) as a valued food. The large yields are from alates (winged forms) attracted to lights on wet or rainy nights in autumn, or at the onset of monsoon rains.

To harvest these effectively, preparations are made at large termite mounds before rains are expected. Surface soils are removed and most emergent corridors are plugged, but some of those likely to be used are left unplugged and a basket-like weave is made 0.5–1 m (1½'–3') above a trench dug in the side of the mound.

A bucket or pot is sunk in the lower end and a fire lit at the end of the trench. The catcher sits in a hut, sheltered from rain, and when the alates emerge they fly up, hit the basket, fall into the trench, shed their wings, and crawl towards the fire, so falling into the bucket. As the bucket fills, it is emptied into storage pots, and the termites later roasted and stored. (See Figure 9.2).

Often, a gas lamp is hung low over large shallow baths and when the alates fly they come to the lamps and drop into the buckets. At mounds, inverted pots can be "drummed" by several people, the drumming imitating the rain on the pots, so artificially initiating the nuptial flights.

Normally termites are eaten "as is" or browned in a large pan and sun-dried for storage. They can be added as pastes, powder, or whole to many dishes; used as a "relish" for porridges; or included in rice ferments.

In the first rains, winged termites fly as thickly as snow to lamps and lights. One can gather handfuls, eating some raw and later as fried-dried food; very rich and satisfying! When dining in Africa, I much prefer these to tough beef or mealie-meal, and subsist on them for weeks in season.

Maggots and Worms

Both African pygmies and Thais breed and eat maggots (fly larvae). Such maggots are "cleaned" in flours or brans for consumption. The British also breed maggots for fish bait. Fishermen keep their bait "lively" in their mouths (at least along the Serpentine while I was there), and the fish alive in a basket in the water.

Wet pig or poultry manure is deliberately exposed to flies, and then spread to dry with fly larvae and pupae also dried. Pulverised, this is added at 15% of bulk to stock feed, for protein enhancement and vitamins, or the spread manure raked out for free-range chickens.

Worms (in drums of fresh manure) and termites (in mounds of damp paper), or pill bugs (in piles of old wood and straw, damped down) are bred for chicken protein foods, as are cockroaches and (for turkeys) grass grubs dug up in pastures. Some pastures, rich in grubs, are better farmed for turkey than sheep. Shallow 6 cm (2⅓") ploughing in strips brings up the grubs for turkey poults. These cutworms are the larvae of pasture beetles.

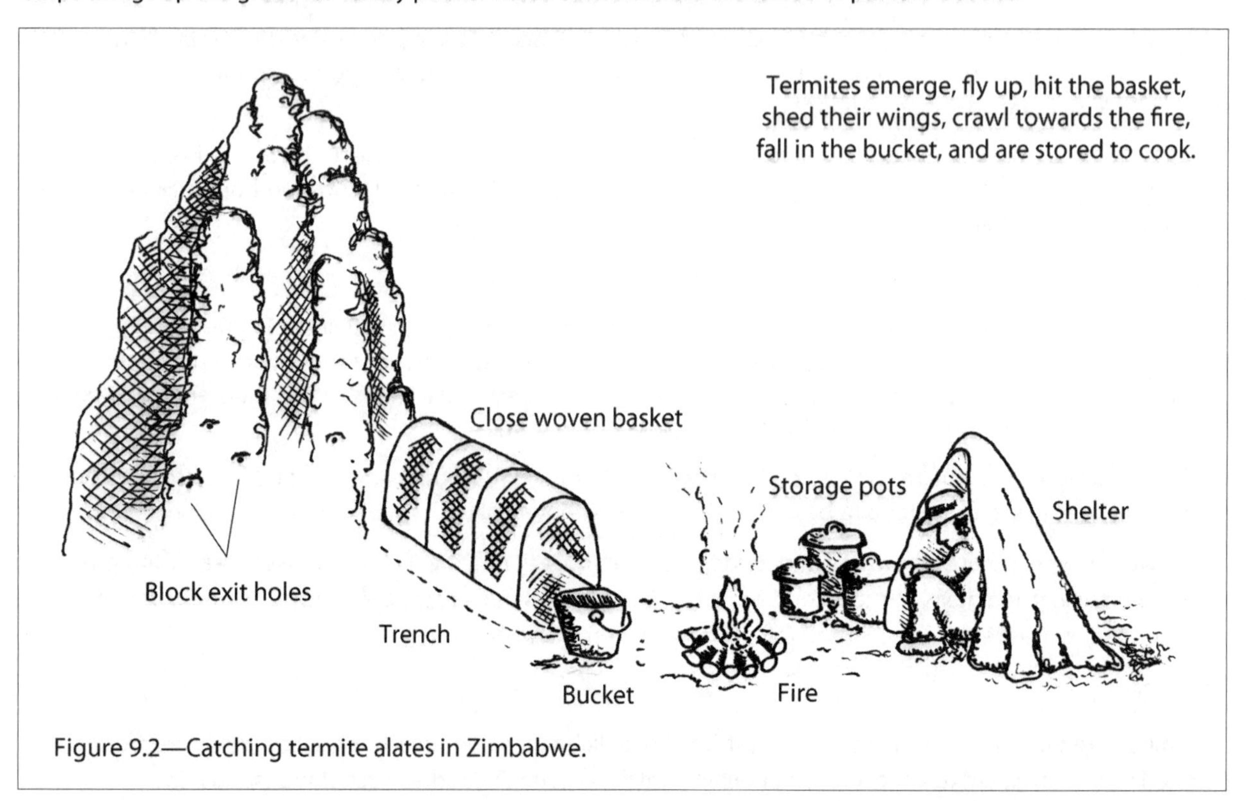

Figure 9.2—Catching termite alates in Zimbabwe.

Earthworms are collected and eaten by some peoples. Very large species (in Ciskei) exist in the sub-tropics, and in some cool areas such as Tasmania and Gippsland (Victoria, Australia) in forest soils. I have no recipes, but have seen small carnivores strip out soil carefully before consuming.

Caterpillars and Beetles

The Shona of Zimbabwe prepare grasshoppers, flying termites, and caterpillars in the same way. First these are boiled, then dried, and often eaten with sadza (stiff porridge); the insects add protein. Zinc and vitamin E are added by pumpkin seed or melon seeds, pounded and added to porridge. In 1954, the southern peoples named 2 crickets, 4 beetles, 8 locusts (grasshoppers) and 9 caterpillars as food, together with fifteen varieties of "mice". This ability to utilise small animals is vital to health.

Caterpillars

Termite mounds in Southern Africa support large trees like Mopane and Marula which in turn provide fruits, berries and "worms" (caterpillars of a moth). These are gathered by gloved hands (they have a terminal spike), the gut contents squeezed out, and are boiled or fried and dried by the ton. They are traded in village markets far from their origins. A cupful once cost about $1 locally, and they were a relatively expensive snack food. They are added to porridges as protein and are a comforting snack for we travellers.

On certain poisonous shrubs, the larvae of chrysomelid beetles concentrate poisons. When they pupate below these bushes, they are dug out and used as arrow poisons by the Khoisan people of the Kalahari, who take good care not to let that poison enter their bloodstream.

Grasshoppers

The grasshopper or locust is gathered for food in many countries, as in Africa and Mexico. In Botswana, the old folk tell me that a bin (or measure) of grasshoppers is exchanged for 4 measures of grain, and that as grasshoppers are most plentiful in drought years, they often substitute for grain. In recent years, the FAO (Food and Agricultural Organisation) and others spray the grasshoppers and remove these foods from people, who also mistrust eating those they do catch. So remote bureaucrats remove good protein food from hungry people at drought time.

In Mexico, grasshoppers are held in narrow-necked bottles to expel the bitter brown defensive exudate (which possibly contains plant poisons). Then the grasshopper jars are emptied into boiling water, and the parboiled hoppers rinsed several times in cold water. Several foods are made from the washed grasshoppers, including a "shrimp" omelette, but most are toasted in oil and eaten with lime juice, chilli, and salt. In Africa, the grasshoppers are stirred until near crisp in a little oil, salted, sun-dried, and stored as food.

Ideally grasshoppers or their relatives are held in large plastic drums overnight, which lessens the gut content. In order to make a meat sauce (or tortilla sauce), they are pounded with salt, chilli, garlic and tomato paste, fried as cakes, and dried.

(Some data from Kent Whealy's 1988 Summer Edition of *"The Seed Exchange"*, which everybody should join. For enquiries please go to www.seedsavers.org).

Insects

Smooth larvae of hawk moths and sphinx moths are dropped into deep fat in Mexico and eaten crisp, puffed, and salted like chicharróns (pork rinds).

Aquatic Insects

Coryxiids (water boatmen) are sieved from shallow lakes in Mexico when they are carrying egg masses. Dried (by the ton) they are added to pancakes, stews, etc., for protein and fats. Such beneficial production may well be susceptible to farming systems, as for brine shrimps and shield shrimps.

Ants

Whole green ants (*Atta* spp.) are eaten by many peoples, or crushed and added to cool drinks for their lemon-like flavour. (Eggs of many ant species are eaten by people including myself!)

Similarly, the "repletes" or honey-storage ants of Australia and Africa are dug and eaten as a special treat by native peoples, and indeed are so important as food they form part of the totem groups of tribes.

Grubs

The larger larvae of longicorn beetles, swift moths (*Hepialidae*), and ghost moths (*Cossidae*) are prized foods in deserts, where the larvae enjoy a protected, humidified life in tree trunks and roots. They are "fished for" in emergent ground tunnels, or split out of stem burrows, and eaten raw or lightly fried, tasting like roasted walnuts and almonds. I once ate twenty "bardi" grubs and was definitely overfed. About six to ten are a good meal as each grub is as large as a thumb, or a smallish sausage. One can remove soil below large trees to reveal emergent holes, lower a reed, knotted at the end, and smoothly pull out any grub that "bites". Hold the head to eat grubs, or they bite your tongue!

Snails (*Helix aspersa*)—"petit gris"

About six species of large snails are eaten by many discerning peoples. Bran-fed, and de-slimed in strong brine, thoroughly boiled, shelled and scrubbed, then grilled in garlic butter, they form a farmers' snack the world over. Snail farms are doing well in Australia and Europe. *Achatina* snails are eaten in the Seychelles, and large whorled Chinese water snails (pu pu) in Hawaii, where they are the most expensive aquatic crop.

The common garden snail is considered to be the best tasting snail (allow 2 dozen per person). Snails may be stored alive in cold winters or hot dry seasons, when they seal their shells. Storage is in wooden boxes with a mesh insert, in open sheds, for some weeks (check on hibernation, and process if the snails start moving).

When active, snails for food are "starved" a few days, in a box with 6 cm (2⅓") of bran. They are then transferred to a box containing a layer of coarse salt, which kills and de-slimes them. They are then well scrubbed under clean running water before simmering in a little salt, garlic, and water for up to six hours. The flesh is then removed from the shells, which are placed open-up, and a little garlic butter placed in.

To make the garlic butter: 6 cloves of garlic and 3 tbsp of chopped parsley, chives, and coriander are rubbed into 50–100 gm (1¾–3½ oz) of butter. The buttered snails are then grilled to re-heat, and served with fresh bread and butter (after Hudson, Lee. 1964).

Escargot — France

Fresh live snails are held in barrels or boxes of bran to purge, then covered with salt dampened with vinegar to make them express slime. They are then very well scrubbed and washed, and plunged into boiling salted water for five minutes to kill them, prior to removing them from their shells with a large needle, then again thoroughly washed.

The clean snails are cooked in salted white wine, herbs, and a chopped onion and carrot for four to six hours or so, strained out, returned to their shells and a dab of garlic butter placed in each shell. They are then grilled before serving.

Cockroaches

Australian war prisoners remained alive in prison camps in South East Asia by eating cockroaches with their rice, fermenting the rice with koji. They also allowed maggots to eat their wounds clean (and to secrete an anti-bacterial substance in the wounds). All cockroaches are edible, and (with rats) represent the larger biomass production of cities like New York and Sydney. They may be preferable to cannibalism in hard times.

9.6 EGGS

Egg yolks have a fair amount of vitamin A, thiamine, and riboflavin (free-range eggs have more folates and vitamin B12).

Eggs contain 30 mg of calcium, and protein of the albumen. A 60 gm (2 oz) egg has 6 gm (⅓ oz) of fat.

Preserving Eggs

- Gently heat 1 part beeswax to 2 parts olive oil, cool, and rub on unwashed eggs, using hands, and stand the eggs large end up in a cool pit, laying on sand. Stored in cold sand (below a shady tree and a metre [3'] deep). Eggs will keep for cakes for about a year.

- Mix water glass (sodium silicate) with water (1:10) to a 10% solution, and cover the fresh, unwashed, strong shelled eggs. Fill a barrel or tub and store in cool place. Keeps eggs for a year. Add eggs as needed.
- Break into patty pans and freeze, transfer to freezer bags to keep frozen.

Eggs

Storage and Preservation

Excess fresh strong eggs can be washed, if necessary, in a 1% solution of sodium hydroxide or bleaching powder. They can be part-sterilised in bleach water at 14–17°C (57–63°F) for fifteen minutes. Then, eggs can be soaked in limewater, with a little salt, for eighteen hours which reseals the cleansed eggs when re-dried. Eggs are then rubbed with vaseline or petroleum jelly to which a little powdered sulphur has been added (to deter moulds) and can be kept in bran boxes.

Preserving Fresh Eggs

Choose sound fresh eggs:

- Gum shellac is dissolved in methylated spirits or ethyl alcohol to make a thin varnish. Paint each egg and store point down in sand in a box. Keep cool. This is supposedly a very good way to keep eggs for up to twelve months.
- Mix beeswax with a little oil, warm until dissolved, rub eggs with the mix and store in sand, bran, ashes, or sawdust. Warmed suet or bird fat, rendered, can replace beeswax, and a powder (flowers) of sulphur at 2 tsp per pint of oil or wax prevents moulds. Again, this serves for some months.
- On 5 cm (2") of coarse salt, in a box, stand eggs, point down, not touching. Pour 5 cm (2") more salt and another layer of eggs. Salt can be re-used for next year's eggs.
- Boil 1.1 litres (2⅓ pints) water glass (sodium silicate) with 10 litres (21 pints) of water, and beat well together. Using a 36 litre (9½ gal) jar, carefully pack in strictly fresh eggs with perfect shells and pour in cold water glass mixture. Using a plate to keep eggs under, cover and keep cool. Eggs can be added at any time. Okay for nine to twelve months.
- A bucket of unslaked (burnt) lime is mixed dry with 1 kg (2 lb) of salt, and 250 gm (½ lb) sodium bicarbonate. Water is added to this mix until a fresh egg barely floats, then other eggs are added to fill the container, and all are gently weighted to cover. Said to preserve eggs for two years.
- In the Arctic, the Inuit gather the eggs of seabirds, reject the white and fill seal intestines with the yolks. These are then freeze-dried in the Arctic winds for long storage, or cached in pits in the permafrost. (Freuchen, 1961).

Cooling and Storing Eggs

Aboriginal Australians have long preserved eggs in cold sand. Moonbird (*Puffinus tenuirostris*) eggs are gathered on rookery islands, off Tasmania, and the eggs then smeared with waxes or fats from the birds before buried "sharp end down" (air space up) in cool sand pits, below the shade of an evergreen tree behind the beach areas of the islands. Each egg is placed a few centimetres apart from the next, then 4–6 cm (1½"–2⅓") of cold sand covers this layer, and the next egg layer added. Finally, about 20–30 cm (9"–12") of surface sand is returned over the pit.

Eggs preserved this way keep for a full year, although they are most used "baked" whole or scrambled for cakes, breads, or cooking late in the year. The combination of sealing with grease, cool storage, moist atmosphere, aeration, and individual packing is ideal for egg storage.

Dehydration

Blended egg liquid from cleaned eggs can be pan-evaporated in layers 0.6 cm (¼") thick at temperatures of 40–50°C (104–122°F), reducing the weight to 6%. The flakes are ground to powder for storage.

Egg pulp can be de-sugared by adding 0.5% (0.5 gm per 100 gm) dry yeast granules, fermented at 36°C (97°F) for one and a half hours, then pasteurised at 60–61°C (140–142°F) for thirty minutes. The pulp is then acidified (with hydrochloric acid) to pH 5.5 before pan or spray drying, and later vacuum dried to 2% moisture. Sodium carbonate is added to neutralise pH, and the powder sealed in cans under nitrogen. The powder then stores for many months.

Pickled Eggs

Hard-boil and peel eggs, then stack them in glass jars. Eggs boiled for five minutes are hard. Eggs boiled for an hour are mealy and more pleasant to eat.

Prepare the pickling fluid by combining: 5 cm (2") finely sliced ginger; 25 gm (1 oz) black pepper; 25 gm (1 oz) mustard seed; and 4 cloves of garlic, to each litre (2 pints) of malt or wine vinegar. Bring to boil and pour hot vinegar over the eggs. Let cool and stopper well for a month.

Duck Eggs

Duck eggs in a saturated brine eventually are salty to the yolk, and need to be soaked in three changes of water before boiling. They are then used as a condiment for noodle dishes.

Hundred Year Eggs — China

Fresh duck or chicken eggs are coated with a paste made sticky with water mixed with salt, ash from grain stems or straw, tea leaves, lime (burnt, powdered), and sodium carbonate. The coated eggs are then rolled in rice hulls or de-natured (oil free) bran and carefully packed in earthenware jars, the lid sealed with paraffin wax or a mud-salt mix.

The eggs ripen over fifteen to sixty days (summer-winter) at 20–30°C (68–86°F), but can be kept in a cooler place for months or years, when the egg solidifies to green-black and is eaten as a delicacy. These eggs are 45% protein and rich in retinol, carotene, calcium and phosphate.

If eaten at eighteen to twenty days, the eggs are boiled, but after sixty days they are eaten without cooking (Campbell-Platt, 1987).

Balut Eggs

Just as many people sprout seeds, South East Asians "sprout" eggs. Incubated for twenty-one days, then chilled to kill the embryo, these eggs are cooked and eaten this way by preference in Vietnam, Thailand, and Cambodia. Duck eggs are commonly used. I imagine that eggs so treated have better, more complex proteins, less fats, and more enzymes, but I have to confess hesitation. As yet, I cannot bring myself to Balut eggs although my good friend, Sego Jackson, has often tempted me with free gifts. I give them away again. One day perhaps....

Egg Omelette for Sushi

Mix 2 eggs, ½ tsp of sugar, and a pinch of salt. Pour half of this in a pan and fry until cooked, remove this omelette and fry the other half, turning it on top of the first. Cut into rectangles and wrap with a strip of nori, previously lightly roasted.

Hatching Eggs

In the great duck-breeding centres of South East Asia (e.g. Thailand), eggs are not "artificially" incubated, but (once a lot have been selected and tested for fertility) all are first sun-warmed for a half hour or so, then closely packed in a blanket-lined woven basket, with a blanket pad above, and a thick lid. Every four to five days or so, they are spread out, and outer, inner, and bottom and top eggs are changed over; sprinkled with warm water; sunned a little; and returned to the basket. At thirty-one to thirty-three days, they hatch. About 40–80 litres (10–20 gal) of eggs create their own hatching heat. The hatching rate is as high as a modern incubator. Great energy savings, you must admit.

CHAPTER TEN

DAIRY PRODUCTS

10.1 INTRODUCTION

Dairy products account for a large part of the fermented food market and range from simple clotted and fermented milks (yoghurt, kefir) to complex cheeses requiring several bacterial species, finished off with a mould (blue cheese, camembert). See Figure 10.1.

During fermentation, milk lactose is converted to lactic acid by such lactic acid bacteria as *Streptococcus lactis* ssp. and *S. diacetilactus*. Growth of bacteria in milk causes curds to separate from the whey. Draining and drying of curds produces cheese. Many other micro-organisms can become involved, depending on the type of cheese made.

Maturation of cheeses is very important, as enzymes from the lactic acid bacteria are still active and help to produce the many different characteristic flavours of the various cheeses.

Cheese can be soft, with cream (extra fat) added (cream cheese, brie) or hard, made with partially skimmed milk (parmesan). There are over nine hundred different types of cheese in the world today. My great favourites are the "stinking cheeses", Limburger and Stilton. These wonderfully smelly cheeses must be kept in a container in the refrigerator or they will drive everybody else out of the kitchen!

Yoghurts and other semi-solid fermented milks are also major world foods. These, like cheeses, are an important source of calcium, protein, potassium, magnesium, and phosphorus; they also provide A, B, and D vitamins. Claims have been made that fermented milk drinks such as yoghurt and kefir establish lactic acid bacteria in the gut flora and bowel, thus aiding digestion and prolonging a healthy life. This seems highly suspect as *Lactobacilli* rarely survive the trip to the large intestine. Only for *Lactobacillus acidophilus* has it been established that it can survive in the gut. As for the bowels, there seems to be little need to resort to colonic irrigation if enough fibrous food is eaten. I think many people need cranial irrigation to rid their body of putrefied beliefs!

Making clotted milks and cheeses first starts with pasteurising milk to kill tubercular and other bacteria. Tuberculosis is again on the increase due to the demand for unpasteurised milk. This is only safe if batches are tested regularly for tuberculosis bacteria.

Heat the milk directly in a large pan (or in jars in boiling water) to 72°C (162°F) for twenty to thirty minutes. Then cool in cold water quickly to 10°C (50°F) or bottle, cork with waxed corks, and store cool for up to ten days. Milk is best sterilised in jars stood in water or it can easily be burnt. In India, however, it is directly heated to seething and cooled to blood heat for curd-making.

Milk boiled briefly and poured into heated bottles which are sealed with a sunken cork, waxed over, keeps in a cool place for weeks. It can be sealed into jars.

10.2 RENNET

The basic process for cheese production is to coagulate milk, usually with the addition of rennet (obtained from young animals' stomachs). Other coagulants include lemon juice, vinegar, acetic acid, and some types of leaves. A decoction of nettles, strongly salted to keep, coagulates milk with no bitter taste. In Portugal a preparation is made from the thistle (*Cynara carbunculus*) which is not bitter, unlike other known vegetable rennets. The Zulu use the fruits of *Solanum giganticum* for this purpose. In the north of Sweden, the butterworts of marshy areas (specifically *Pinguicola vulgaris*) are used in strainers through which is poured warm reindeer milk. This sets over a day or two and forms a ropy milk. Thereafter, a little of this curd mixed with warm milk produces the same effect. Today, rennin made from the mould *Mucor meihei* is successfully used for cheddar cheeses.

Although we can use junket or rennet tablets for home use, the following recipes give on farm or commercial ways to make rennet.

Rennet—Variation One

The fourth stomach (known as abomasum) of calves, goat kids, or lambs (the stomachs are called vells) are used. Young animals for rennet production must be milk-fed (not grass fed), and should be about two weeks old.

The vells are washed clean and thoroughly dried over mild heat or in air, then chopped into strips. 50 gm (1¾ oz) of minced vell is soaked in 10 litres (2½ pints) of de-albumated whey for twenty hours at 40°C (104°F). The albumin is removed from the whey by first boiling, and acidifying with vinegar or boric acid to pH 5.0. It is then filtered through muslin as a clear liquid and cooled before the strips of chopped vells are added.

The whey extracts the rennet from the vells and also grows *Lactobacilli*. The stomach almost totally dissolves in the whey, and pH drops to 3.8–4.0. Fresh culture is prepared daily or as needed, from dried stomachs, and the whey solution kept cool if albumin has previously been removed. For cheese, 0.2% of fresh rennet whey is added to the cheese milk. All early cheeses were made by soaking a strip of salted or dried stomachs in the milk.

Rennet—Variation Two

In commercial rennet production, the vells are carefully collected; unwashed stomachs are blown up like balloons and shade-dried in clean open sheds. *En masse*, they are block-frozen, or even salted for preservation (dry-salted or in brine). Freezing of stomachs gives almost double the rennet yield of dried stomachs. Dried or frozen stomachs are put in 10% brine with 5% boric acid to bring pH to 5.0, and after several days extraction liquors are strained and clarified. One part rennin coagulates ten to fifteen thousand parts of milk.

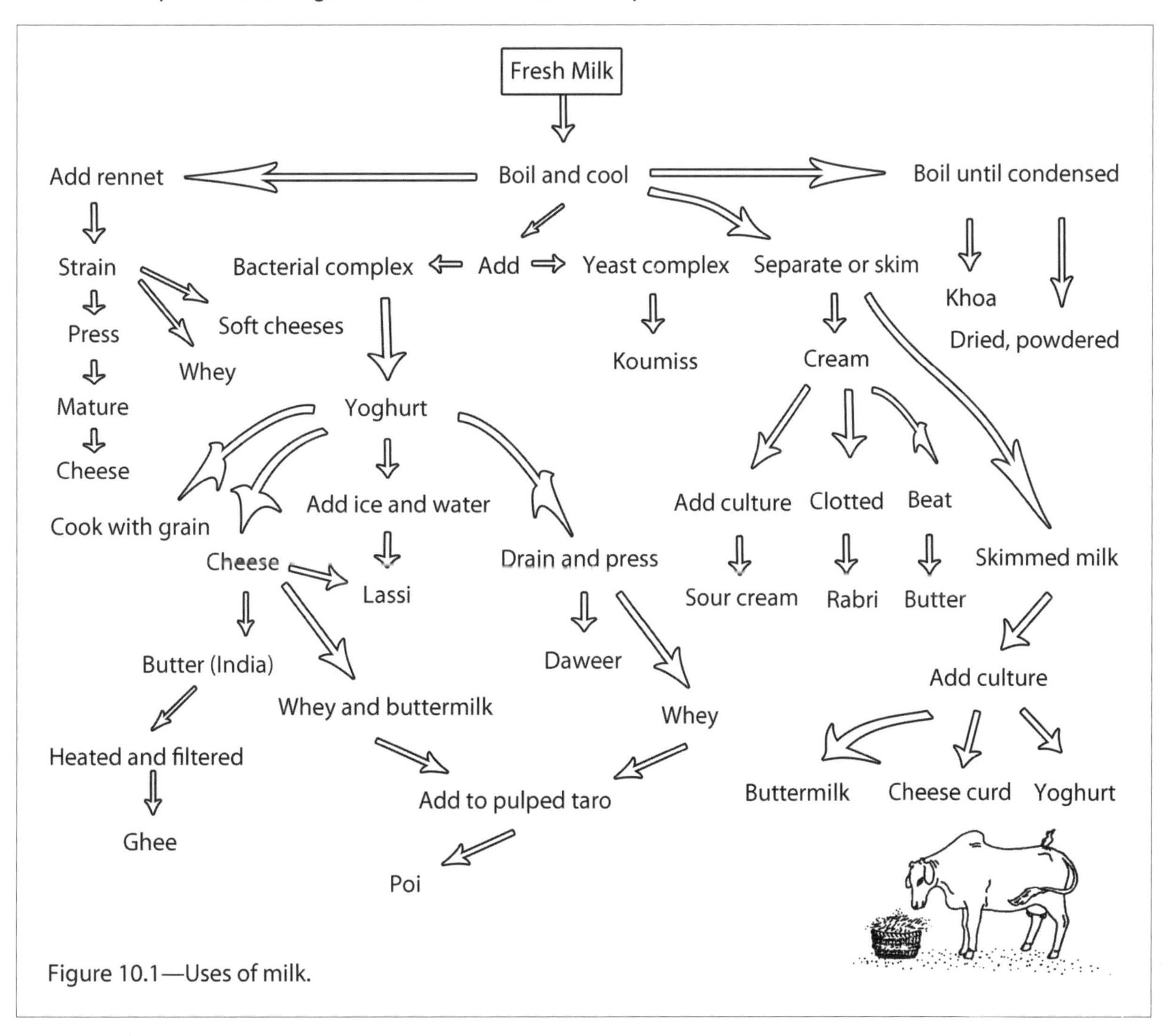

Figure 10.1—Uses of milk.

Rennet—Variation Three

The fourth stomach, is pleated inside, and usually will have curds if taken from a suckling. A good abomasum, stored in brine, is estimated to provide rennet for 65 kg (143 lb) of cheese or nearly 700 kg (1540 lb) of milk (Hale and Williams, 1977).

Rennet—Variation Four

Fresh fig latex (from common edible figs, *Ficus carica*) is thirty to one hundred times more effective in coagulating milk than is rennet prepared from calf stomach mucous. Bead latex is collected early in the morning, when enzyme activity and yield are high. Latex acidified to pH 4 keeps active longer, and best results for storage are by freezing, centrifuging, and vacuum-drying of centrifuged serum of the latex. This dry serum is 95% active at room temperature for months. Ascorbic acid (lemon juice) enhances action and preserves the serum. Used for coagulating milks and milk-like juices of beans for cheese, junket and bean curds.

Fig latex given orally, digests living intestinal worms with a protease (ficin). It acts against *Trichurus* and *Ascaris* worms and must be given with sodium bicarbonate, as it is destroyed by stomach acids.

Ficin may be used to tenderise meat and to help render fats, at optimum pH of 7 (neutral), and up to 60–65°C (140–149°F).

Rennet—Variation Five

A useful vegetable rennet is made from the juice of nettles, pressed out and stabilised by adding salt until the solution is saturated. One tbsp of this solution added to 4–6 litres (8½–12½ pints) of milk produces a curd free of bitter tastes.

Rennet—Variation Six

The dried flower heads of cardoms, (*Cynara*), an edible Mediterranean thistle allied to globe artichoke, is used in Spain as a taste-free rennet, probably as an aqueous extract.

Important Micro-organisms Found in Milk-Based Foods

- *Streptococcus lactis lactis*
- *Streptococcus lactis diacetilactus*
- *Streptococcus lactis cremornis*
- *Streptococcus raffinolactis*
- *Streptococcus faecalis*
- *Streptococcus faecium*
- *Streptococcus thermophilus*—important in starters where high temperatures are used
- *Leuconostoc mesenteroides = L. m. dextranium, L. m. cremornis*
- *Lactobacillus acidophilus*—as a minor factor except when added from a separate culture of bacteria to fresh milk
- *Lactobacillus gasseri*
- *Lactobacillus lactis = Lactobacillus bulgaricus*
- *Lactobacillus casei*
- *Lactobacillus helveticus = Lactobacillus jugurt*
- *Lactobacillus plantarum*
- *Lactobacillus fermentum*
- *Lactobacillus cellobiosus*
- *Lactobacillus reuteri*
- *Lactobacillus brevis*
- *Lactobacillus kefir*
- *Pediococcus* (was *Pediococcus cerevisiae*)
- *Pediococcus pentosaccus*
- *Pediococcus acidilacti*

(*Pentosaccus* and *acidilacti* are not used in any starters but are found growing in mature cheese.)

10.3 MISCELLANEOUS MILK PRODUCTS

Junket (rennet curd)

Prepare ingredients: 1 litre (2 pints) milk, 2 rennet or junket tablets and a little cold water to dissolve rennet. Optional: 4 tbsp sugar and a few drops vanilla to sweeten curd. Salt is added to unsweetened curd.

Heat milk to lukewarm, dissolve rennet in cold water (crush with a spoon). Add ingredients to milk, stir for a few seconds, then set in small moulds in a cool place.

Chhurpi or Dhurko — Nepal and Tibet

Heat buttermilk (from making butter using whole-milk yoghurt) until the solids separate. Strain out the whey and mix the solids with roast and ground ginger, garlic, and cardamom seeds. Add lemon juice, then wrap in cloth and press in a vat with a large stone. After two to three days, cut the pressed chhurpi into strips and press dry for a week, then smoke and fire-dry or sun-dry until very hard. Cut these strips into pieces, thread into a "necklace", and store over the kitchen fire.

Eaten chewed as hard rations or softened in a stew. This is an excellent travellers' food.

Kurani or Khuwa (reduced milk)

Simmer milk in a wide pan until all the water is evaporated and a brownish substance (kurani) is left behind. This is the basis for gravy on roast meats or as kofta balls in sugar syrup. It can be stored for a week or so, and is tested daily to check for freshness. The addition of a little salt or sugar late in the process adds life to the storage.

Rabri (sweet condensed milk)

Simmer whole milk and lift the cream off at intervals. As the liquid is reducing, add sugar, return the cream, and blend the cream clots back in to make a sweet concentrate of doughy consistency. Used for sweets in India, and in syrups.

Smoked Milk

The Lapps hang a reindeer stomach over a fire and add milk daily. This continues for two to three weeks, at which time the stomach is filled with grains of dried milk. The quantity is now enough to feed a Lapp family all winter. The stomach, by then well smoke-cured, is used as a container for the dried milk.

Ghee — India and Nepal

Ghee is clarified or reduced butter; but unlike butter, ghee does not burn easily and is used in lieu of oil. Slow pan-boil (but watch closely so that the butter does not boil) butter made from curd or cream. When the butter is red-brown, smelling strongly, and appearing granulated, strain and store in wooden jars, bottles, or press-top cans. In cans and bottles it lasts a year, in wooden jars up to five years. It is used in cooking, over rice, or on chapattis (bread). The Tibetans add a little ghee, with salt, to their tea.

10.4 FERMENTED AND CLOTTED MILKS

Kefir

Kefir is a mildly alcoholic, slightly acidic milk drink very popular in Russia, the Middle East, North Africa, and the Balkan states. In recent years it has crossed to the USA where it is commonly sold flavoured with fruits.

Kefir grains can be wet (they look like little cauliflowers) or dried (looking like brown seeds). They are composed of milk casein, and a mix of yeasts (*Candida kefyr, Kluyveromyces fragilis*) and bacteria (*Lactobacillus caucasicus, Streptococcus kefir* etc.) bound together by fibrous carbohydrates. Sheet-like structures form to separate yeast colonies from bacterial colonies.

Heat milk to 82°C (180°F), let cool to 43–49°C (109–120°F), add kefir grains (washed, fresh, or dried), and keep insulated for three hours. To use the kefir, pour it into a jug via a household strainer, wash the grains in the strainer with some water, and store under water in a cool place until next used. They can be kept chilled in the refrigerator, in water, for two to three weeks, but then need a fresh culture if not active.

Regular manufacture deserves a wooden ferment pot with a wooden lid, or at least a ferment pot kept warm in a wooden box with a light bulb for cosiness.

In 1991 the South Koreans were making kefir in quantity in Russia, and (in the cities at least) everyone can have kefir delivered for breakfast, which is great for those of us for whom yoghurt is too acid. While in Kursk, I enjoyed a pint daily.

Obviously, the manufacture and delivery of kefir on a smaller or larger basis is possible and profitable. Commercially, a large culture is made from skim milk seeded with grains, after being sterilised at 95°C (203°F) for thirty minutes, and incubated at 25°C (77°F) for one to two days. Then, this starter is strained and the grains washed with water for a new starter. Whole milk is then sterilised (85°C [185°F] for thirty minutes) and the ferment added at 5% of volume, incubated at 18–22°C (64–72°F) for twelve hours, then ripened at 10°C (50°F) for one to three days in bottles, and sold. A good kefir milk foams like beer and has a smooth consistency.

Koumiss—Variation One

Fermented milk is called koumiss, and it may be consumed from open ferment or as a bottled (alcoholic) drink. For open ferment, use whole milk; for bottled ferment, skim fat from milk. Ferment proceeds best at 10–24°C (50–75°F), usually 16°C (60°F).

Use 1–2 parts new whole milk to 1 part sour fresh buttermilk and let ferment for three days. Using yeast, mix well with brown or white sugar, malt, a little salt, and warm water to make up about a cupful. Dilute whole milk with ⅓ boiled water cooled to blood heat and strain active yeast solution in. Ferment two to three days. Add ⅓ of any ferment like those above to ⅔ whole new milk and continue ferment.

When ready to eat, mix to a creamy texture and consume. Start a new ferment every three to five days.

Koumiss was once made from mare's milk; now it is also made from cow's milk enriched with whey protein. Two starter cultures are used, one containing *Lactobacillus bulgaricus*, producing lactic acid, and the other *Candida (Torula) kefir*, producing ethanol and carbon dioxide. The starters are added to fresh milk and fermented at 26–29°C (79–84°F) for two hours, stirring occasionally to let air penetrate.

After the milk settles, it is packed into bottles and sealed. It is left to ferment in the bottles at 18–20°C (64–68°F) for two to three hours or longer, depending on the degree of sourness desired. Sweet koumiss is eaten after one day, while "strong" koumiss is eaten after two to three days. Koumiss can contain up to 3% alcohol and may be further distilled.

Koumiss—Variation Two (an older European recipe worth a try!)

Dissolve 56 gm (2 oz) compressed yeast in 2 tbsp lukewarm water. Heat 1 litre (2 pints) milk until lukewarm (or cool boiled milk), then stir in yeast and 2 tbsp sugar. Fill airtight bottles to within 4 cm (1½") of the top; cork and invert in a stand for six hours if possible at 27°C (80°F). Store chilled.

Matzoon

In a small jug add 0.5 litre (1 pint) of fresh heated milk (at blood temperature) to 2 tbsp of baker's yeast and sit in a warm place until well curdled. This is the starter solution. From this take 6 tbsp and add to 0.5 litre (1 pint) of fresh milk. Do this transfer five or six times, by which time the taste of yeast will have disappeared. The milk curd can be used in cooking, or eaten with malt and fruit.

Thereafter, the culture is used to make matzoon, and each day 6 tbsp of the last curd is used in fresh milk, until it thickens, when it is used in a creamy state, with slight acidity.

The curd is eaten as is yoghurt. Refrigerate to keep when curd is ready; or refrigerate the culture if leaving on holiday.

Yoghurt Madzoon (Armenia)—Laban (Lebanon)—Mast (Iran)—Yogurt (Turkey)

Scald 2 litres (4¼ pints) of milk (preferably whole milk; can even be powdered) and cool to just hot enough to hold a hand in a basin-full to the count of ten. Add 3 heaped tbsp of yoghurt (or yoghurt starter) beaten with some of the milk, and cover the basin with a plate. Wrap the basin in an old blanket or sleeping bag to retain heat, and set it where it will not get disturbed for eight to ten hours. Flavours can be added before and after ferment. A thick wooden calabash is used in India and Africa, and even this is wrapped in winter.

For yoghurt starters, equal ratios of *Streptococcus thermophilus* and *Lactobacillus bulgaricus* are used almost universally; pH is 4.0–4.4. Yoghurts may differ a great deal depending on substrate (goat, buffalo, or cow's milk) and the amount of fats left in or added to the milk.

In India, *Streptococcus lactis* and *Lactobacillus plantarum* or *Lac. bulgaricus* in buffalo milk may all be present. As inoculum is often 20%, only one to three hours is needed for ferment (usually in a wooden pot). Usually, 2% of a previous culture is added, and the yoghurt set overnight.

Yoghurt is stabilised for a meat marinade (replacing cream in many recipes) made by beating 1 tbsp of cornflour in an egg white. Heat the yoghurt to liquid, boiling slowly and stirring in one direction with a wooden spoon at first, adding the cornflour mix. Simmer, uncovered, for ten minutes, until thick and rich. If now mixed with meat, yoghurt will not curdle. No water is used. Spices added to the marinade are ginger, garlic, cardamom, chilli, turmeric, cloves and so on.

In the Middle East, spinach, leek, fine-minced meat and so on are made into a soup, thickened in a butter-flour roux. Yoghurt is added when the soup has gone off the boil and beaten in thoroughly. The whole is carefully re-heated, but not boiled and poured into bowls, dressed with parsley, as a yoghurt soup.

Kishk, Kurut (dried yoghurt foods) Syria, Iran, Lebanon

Yoghurt curds kneaded with spices or herbs, wheat flour, and salt are dried (oven or sun) in small balls for later grating and adding to soups or stews as a protein supplement and thickening. This is an important way for desert or nomadic cultures to save excess milk and reconstitute it later, even after a long time in hot weather.

Lassi

Lassi is a wonderfully refreshing yoghurt drink that has been thinned by the addition of icy water and flavoured with herbs and salt, or sweetened. Lassi is drunk as a cool, fresh summer drink from Greece to eastern Asia, and is available in every Indian or Turkish restaurant.

Lassi—Variation One

Plain yoghurt is thinned with icy water (1 of yoghurt to 3 of ice water). Note—water can be soda water. Blend in either: 1½ tbsp of sugar, treacle, or maple sugar plus fresh mint or mint ice cubes (sweet lassi). For salt lassi, use salt to taste plus mint, and omit sugars.

Lassi (buttermilk and whey drink)—Variation Two

Use whey with some yoghurt, buttermilk, or plain yoghurt mixed with 4 times the volume of water, and add a pinch of salt. At this time you can add some sugar as syrup (to taste) and blend or whisk the drink. When serving in glasses, add grated lemon rind, mint leaves, ice, or frozen banana. Note—that butter is made in India and Nepal by churning yoghurt made from whole milk, thus yielding considerably sour buttermilk.

Buttermilk

Buttermilk used to be produced from the liquid expressed during making butter. As a child, I drank buttermilk from the great wooden churns in my father's factory (Stanley, Tasmania). Nowadays, however, the amount of natural buttermilk never keeps up with demand, and most buttermilk today is cultured from skim milk.

Starter cultures are *Streptococcus lactis* or *S. cremornis*, with flavours added such as *Leuconostoc citrovorim, L. dextranium* and *Streptococcus diacetilactus*, so that the ferment is quite complex. Cream, salt, and citrate or lemon juice can be added to vary the quantity or aroma.

First, skim milk is heated to 85°C (185°F) for thirty minutes (essential) then the culture is added. Inoculate with 0.5% lactic acid starter, and hold for fourteen to sixteen hours at 22°C (72°F) until the pH is 4.5. Violent agitation should be avoided, or the whey separates; gentle stirring to incorporate the culture is ideal. Cool rapidly to prevent further acidity, fill bottles or plastic packs and store at below 10°C (50°F). Salt can be added during chilling. Consume before five days. At home, some of the last batch is saved for inoculating the next, and so on.

Rarely, gelatin is added to increase viscosity, as is butterfat or flaked butter. Most buttermilk today is produced commercially, and often flavoured with 15% thawed strawberry syrup preserve and glucose. (Kosikowski, 1977).

Scandinavian Buttermilks

Filmjolk (Norway and Sweden)—3–5% fat, not "ropy". Replaces fresh milk over cereals, pH 4.4–4.6. Starters—*Streptococcus cremornis, S. lactis, S. lactis diacetilactus,* and *Leuconostoc cremornis.*

Lattfil (Scandinavian)—0.5% fat, not ropy, similar to filmjolk, less rich, starter of same species.

Langfil (Scandinavian)—Ropy, fermented in carton (cannot be pumped, and is not easy to spoon). Incubated at 18–20°C (64–68°F) to produce a protein slime. Traditional on farms. Uses *Streptococcus cremornis,* ropy *S. lactis* strain, and *S. lactis diacetilactus.* Final pH 4.2–4.4.

Ymer (Denmark)—3% fat, pH 4.4–4.6, and pourable. Mixed starter of *Streptococcus Lactis diacetilactus* and *S. cremornis.* Can be thickened by replacing whey with cream.

Skyr (Iceland)—Very similar to ymer, but using ewe's milk, not cow's milk. Fermented from before the 10th century.

Villi (Finland)—Made with mixed starters but has a surface mould growth of *Geotrichum candidum,* giving a slightly musty aroma to the crust of the mould, which is considered essential. The mould is stirred in before consuming. Ropy.

Buttermilk generally is essentially a thirst-quenching drink and should not be too thin or too thick. Thin buttermilks are too high in pH. A good balance of cultures is required as indicated above, for a specific flavour or result.

10.5 CHEESES

There is an almost infinite number of cheeses, but for all these there are only a few basic types or classes of cheese:

- Short-ferment, short-life cottage cheeses, such as panir.
- Acidic cheeses chilled for longer keeping (beyond a week).
- Less acid medium-hard cheese, which can be dried or part-dried.
- Cheddar types, ripened for months, and sometimes years, keep well while waxed.

The "bible" for cheese makers has been written by Frank Kosikowski, 1979 as *Cheese and Fermented Milk Foods.* It gives a great deal of detail on the art or skill of the still somewhat imprecise science of cheese making. So many organisms and enzymes enter into cheese ripening that precise statements are risky. Good procedures, however, consistently produce good cheese.

Figure 10.2 shows the basic equipment needed for cheese making.

Cottage Cheese

Slowly warm 2 litres (4¼ pints) of fresh whole skim milk by placing it in a large container of warm water. Do not use a galvanised metal or aluminium container. Heat the water until the milk reaches 22°C (72°F), then stir in ¼ cup buttermilk or 2 rennet tablets, dissolved in a small amount of cold water. Cover with a cloth and let stand sixteen to twenty-four hours.

To determine when the curd is ready for cutting, insert a knife at the edge of the container and pull the curd gently away. If it breaks away quickly and smoothly, it is ready. Curd must be cut into 6 mm (¼") pieces to let the whey escape. Cut with curd knives; these can be bought commercially or easily hand made with thick wire. If you are only making a few batches of cottage cheese, however, you can cut the curd in the following way:

Insert a knife blade through the curds to the bottom of the pan on the side opposite to you and draw the blade towards you. Repeat every 6 mm (¼") so that now you have a row of parallel cuts. Give the pan a quarter of a turn and repeat. Give the pan another quarter of a turn, but this time insert the knife at a slant (cutting from right to left). Give the pan its last quarter turn and cut from left to right.

Let the curd stand ten minutes to allow the whey to separate and the curds to become firmer. Then start to heat the water in the outer container as slowly and as uniformly as possible. The curd already should have been at 22°C (72°F). Now heat to raise the temperature to 43°C (109°F) in thirty to forty minutes. During this time, stir the curd very gently, about a minute at a time, every four to five minutes.

When the curd and whey reach 43°C (109°F), hold at this temperature for twenty to thirty minutes or until the curd holds its shape and readily falls apart after being squeezed in the hand for a few seconds. If it is not firm enough, slowly increase the heat to 46–49°C (115–120°F). Stir the curd gently, preferably with the hand.

When the curd is firm enough, drain the whey through cheesecloth in a colander for two to three minutes (not too long as the curd particles tend to stick together). Then gather up and immerse in a pan of cool, clean water. Raise and lower the cheesecloth bag for a few minutes to rinse the whey from the curds. Rinse again in ice water to chill the curd, and set back into the colander to drain any leftover whey.

Transfer the curd to a large mixing bowl and add 2 tsp salt for each kilogram (2 lb) of curd. Unsalted cottage cheese will have a definite acid taste. If you want a smooth cottage cheese with good flavour and texture, mix in 6–10 tbsp of cream.

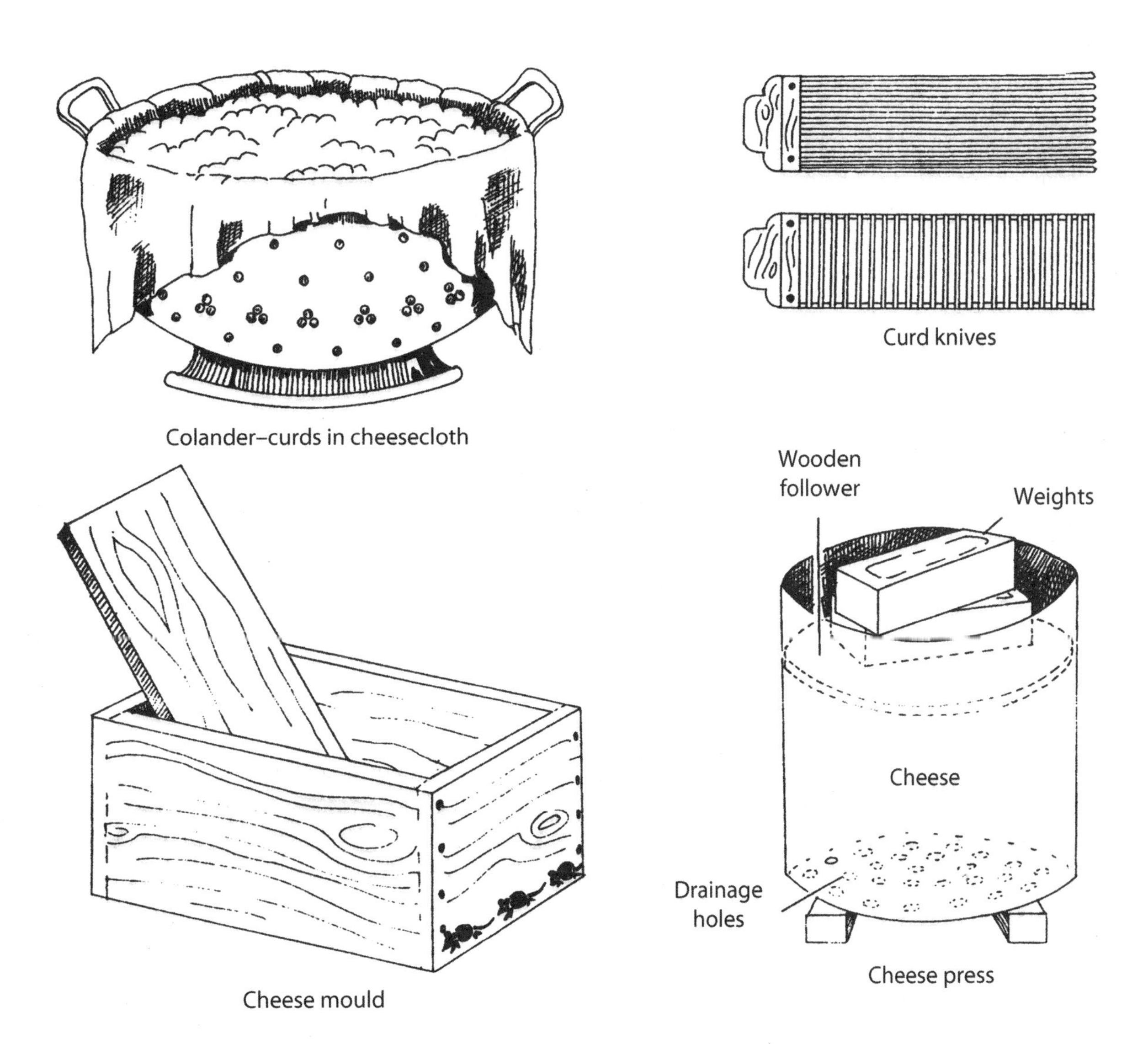

Figure 10.2—Equipment for cheese making.

Panir — India

Panir is a soft cheese, eaten fresh or slightly matured.

Bring to boil 1 litre (2 pints) milk; add 2½ tbsp lemon juice and salt and pepper (to taste); stir and take the pot off heat. Milk will be in curds. Strain curds through a cheesecloth, in a colander, then tie up as in plum pudding, and hang from a tap (to drip overnight). Save whey to mix with yoghurt for lassi, to mix in bread, or mix with poultry food. Flatten curd and press with 2–4 kg (4½–8¾ lb) weight. Drain overnight, then press for five to six hours.

Panir can be eaten as is, as a type of "yoghurt cheese", or cut into rectangles and fried briefly. Store in plastic bags, add to vegetable dishes or stir-fries. Refrigerate to keep a few days.

Ricotta Cheese

Using the whey from making cheddar cheese, combine 41 litres (10 gal) whey with 4.5 litres (9½ pints) skimmed or whole milk, and heat, stirring constantly, to 88°C (190°F). Add 900 mL (30 fl oz) white vinegar and 113 gm (4 oz) salt. Stir briefly then allow curds to settle. If the curds are not settling, re-heat to 90°C (194°F) or add more vinegar in small lots. Dip out curds with a fine perforated strainer or scoop.

Dip off most of the whey, and strain curd remainder through muslin or cheesecloth. Drain in a refrigerator overnight, then hang in a cool dry place for several weeks (watch for mould). This dry curd can be blended with more highly-flavoured grated cheese for flavour before sale.

Haloumi — Greek

Crush 3 rennet tablets in 4 pints milk at 38°C (100°F). Stir for a few seconds and stand thirty minutes to curdle. Break up curds with a wire whisk and leave to settle. Strain out through muslin in a colander. Drain well and shape curds to only 2 cm (¾") thick on a board, place muslin over and press gently over two hours.

Boil all whey to reduce, cut curd carefully into 4 pieces and gently place in the boiling whey. Cook until curd floats, then lift out and dip into salt. Sprinkle with dried mint. Gently press all cheeses together and pack in a sterilised jar. In 2 cups boiled whey, dissolve 2 tsp salt and pour over cheese.

Seal and store cool. Mature six weeks and eat. This cheese can be fried and eaten with lemon juice. The whey, boiling, can be used to curdle another 4 pints of milk, poured in and set with lemon juice. Repeat, lifting off floating curds.

Whey Cheese

Whey, especially goats whey, is boiled down slowly all day (on the side of a winter stove), then cream is added to thicken it, and it is moulded to cool and used as a spread like "whey cheese"! It can be variously flavoured; spiced, or combined with strong cheeses.

Feta Cheese

Heat 45.5 litres (12 gal) whole milk to 32°C (90°F). Turn off burner and add 284 mL (9½ fl oz) lactic acid starter (fresh yoghurt), or 568 mL (19 fl oz) cultured buttermilk. Wait forty-five minutes then add 5 crushed rennet tablets dissolved in 2 cups clean cold water. Stir for two minutes and leave for curd to form (thirty-five to fifty minutes). Cut curds in squares using a simple wire rack frame or cake cooling tray, leave for fifteen to twenty minutes in the whey.

Place curds on muslin, cut into 6 cm (2⅓") cubes and place the bag of curds in a saturated brine (11.3 kg [25 lb] salt per 45.4 kg [100 lb] water); refrigerate overnight at 4–10°C (39–50°F). Next day, place curds in plastic bags, or in clean cans filled with 14% brine (6.4 kg [14 lb] salt per 45.4 kg [100 lb] water). Seal well, and place these containers upside down in cardboard boxes. Keep at 4–10°C (39–50°F) for two to three months. Before consuming, salt can be reduced by soaking in clean water.

Kopanisti (Greek soft cheese mix)

Mix together: 1 kg (2¼ lb) feta cheese, rinsed and crumbed (or whey cheese); 360 gm (12½ oz) blue cheese, crumbled; 120 gm (4¼ oz) softened butter; and 454 gm (1 lb) cottage cheese, mashed smooth. Spices: 1 tbsp each of ground black pepper, oregano and thyme.

Mash all cheeses and butter to a smooth paste with spices, reserve oil to one side. Press down into a jar or small crock and pour 2 tbsp olive oil on top. Store cool for eight weeks to reduce mould. At eight weeks, mash again with the small amount of mould, then screw down in smaller covered jars and refrigerate to keep. Mix again, just before serving. Makes about 1.5 kg (3⅓ lb).

Sheep Milk Cheese

Curd-making in sheep milk, as recorded by Ryder (1983), can be achieved by using fig juice (latex) gathered on wool. The wool is rinsed and the rinsings used. Also, lemon juice and sourdough were used as milk coagulants in the Middle East.

Igt (known as "Milk Cake" in Arabia)

Yoghurt is boiled until thick (about two to three hours) and granulated, then cooled and moulded in flat cakes. Herbs may be added. The dried cakes keep a year and may be mixed with water after pounding.

Laban — Arabic

Yoghurt curds are boiled and drained through fabric, then moulded with salt until hardened. Small balls of this are sun-dried to keep, and later reconstituted with water in which mutton is cooked. Used for rice dishes. (Ryder, 1983).

Whey is boiled at 89°C (192°F) to form a coagulate which can be moulded with gelatinised flour, shaped into balls and dried.

Karish and Mish

In Egypt and Africa, goat skins are used to ferment milk. The skins are de-haired and the openings tied off or fitted with a plug of wood. Fresh milk is added daily.

Butter is made by skimming settled milk in pots. Skimmed milk sours to a sour yoghurt and is used with grains, to make a storable food. In goats' skins, the sour milk is sold for porridges or churned, as in Nepal and Botswana, until the cream forms butter globules and the whey or buttermilk is drained off.

Sour buttermilk has two uses: Conversion to cheese (e.g. Karish); and with grains, as a dried food (e.g. Kishk).

Karish (after El-Gendy 1986, in Hesseltine) — Egypt

A soft cheese for fresh consumption. Cow, buffalo, or goat milk is used, as is sour milk and buttermilk. Milk is heated to 35°C (95°F), rennin added, and the curds transferred to cloth-lined baskets of reed or palm fronds. Amounts are about 200 gm (7 oz) lots, and when the curds are salted and drained, they are stored for fresh consumption under salted whey, or salted skim milk.

Mish (matured Karish cheese)

Mish is simply Karish cheeses stored for long periods under salted liquid. First, the Karish or soft cheeses from curds are sun-dried for several days, washed, and stacked in layers in large earthenware pots, with salt sprinkled on each layer.

This salted mass is then filled with a pickling fluid containing buttermilk, sour milk whey, and the granular residue from boiling butter to make ghee. Spices of ground black pepper, green chilli, chilli, and ginger are added, all in a small cloth bag or *bouquet garni*. The jar is then sealed with mud and stored at room temperature for about a year.

Cheddars

To clean raw whole milk, or pasteurised whole milk cooled to 32°C (90°F) first add 1–3% lactic acid (yoghurt) starter and later (after an hour) rennet. A smooth curd forms in thirty minutes.

The curd is now carefully cut into cubes, often using a rack of wires passed both ways through the curd to allow whey drainage.

Slow heating to 37°C (98°F) over an hour allows bacterial acids to increase (above this temperature ferment ceases). Now the whey is dipped out and the curds stirred to a semi-granular structure (thirty minutes), adding 900 gm (2 lb) to 2 kg (4½ lb) salt per 45 kg (100 lb) of curds (45.5 litres [12 gal] of milk yields 4.5 kg [10 lb] curd). Salt is added in small lots as curd is stirred; 1.5–1.8% of salt suffices.

Salted curd is strained, and enclosed in a muslin "bandage" which is tied to form a tight cover. This is then transferred to a form (a cheese press) and weighted overnight to shape the cheese. Cheeses are usually about 4.5 kg (10 lb) in weight.

The cheeses, in the muslin, are well-dried, in a cool place, over days until the cloth is thoroughly dry. Paraffin is heated to 107°C (224°F), the cheese dipped in, cooled and held at 10–15°C (50–59°F) for two to six months, periodically cleaned of mould, turned, sometimes oiled or washed with salt water to reduce mould. They are then mature.

Cheddars are rich in calcium, riboflavin, and thiamine. They are the common cheeses of small dairies for country areas, or in remote dairy herds. Cheddars enable small producers to preserve milks for distant markets. They were made in alpine summer pastures in Spain and Switzerland, and in thousands of small country centres wherever dairy herds occur.

At home, cheeses may be kept in oil to age, thus excluding air and keeping the cheeses soft, and eventually mild and tasty. Well salted and pressed curds can be aged, or kept for short or long periods, in this way.

If cheeses are made to almost fit a broad tin or earthenware bowl, very little oil need be used.

10.6 BUTTER, MARGARINE AND MAYONNAISE

Butter

The cream from yoghurt, scalded milk or whole milk, or cream from a separator, is left to "set" overnight in a cool place. It is then churned or beaten to butter, salt added to taste, and frequently pressed to extract residual moisture. It keeps, cool, a week to ten days.

Tropical Garlic Butter

Using one part of garlic to 3–4 parts of edible oil (good olive oil), rub or blend until it forms a thick yellow paste. Some stearine can be added, and salt to taste.

This "butter" stores well for a year if refrigerated, or freezes well. It is a popular butter in the tropics, or hot Mediterranean climates.

Margarine—Variation One

Various mixes of butter, edible quality animal fats, and edible oils (olive, sunflower) are mixed for margarines. Turmeric or saffron are added for flavours and colour as needed. Salt is added for taste and preservation.

Some of those mixed in Russia are:

Butter	Fats	Oils
50%	20–25%	25%
33%	33%	33% (equal parts)
16%	59%	34%

Margarine—Variation Two

Mix or blend thoroughly, 5 tbsp of liquid or reconstituted skim milk, with 2 tbsp powdered skim milk and ½ tsp of salt. Gradually add 1½ cups of edible oil (safflower, sunflower, olive), blending until thick, adding a small amount of powdered milk to thicken if too liquid. Fill containers, cover, and chill to harden. (Fairfax, 1986).

Mayonnaise

To this basic mayonnaise, used by many people as a butter today, can be added such spices as mustard, herbs, garlic, chilli or any suitable condiments. Olive, sunflower, and safflower oils are commonly used.

In a bowl, whip or blend 1 egg, a little salt, 2 tbsp of vinegar or lemon juice, and ¼ cup of oil. Blend until the mixture thickens (it will not do this in thunderstorms I am told). Then add ¾ cup of the oil slowly, making one cup of oil in all.

For salads, chopped chives, a small onion, grated ginger, a garlic clove, or a chilli can be added. The basic mayonnaise keeps well chilled, and is in every way equivalent to butter on breads, potatoes, or rolls.

CHAPTER ELEVEN

Beers • Wines • Beverages

11.1 GRAIN BEVERAGES

A stillroom in the corner of the kitchen or basement for ales, stouts, wines, cider, and "fizzy drinks" such as ginger beer, is a must for the serious fermenter.

Brewing and Malting—Variation One

In the first stage, grains have to be converted to fermentable sugars, by sprouting. In the second stage, sprouting is stopped through drying (to 6% moisture at 65°C [149°F], then to 4.5% moisture in an oven at 80–85°C [176–185°F]). This grain, sprouted and dried is called malt and is then ready to produce the beer. It is ground and combined with water to continue the breakdown of unsprouted grains to fermentable sugars. This is called the wort.

Beers of many varieties and of a great variety of tastes are made from all grains. Grain is at first soaked over two to three days, then spread out and allowed to sprout. Sprouting activates a variety of enzymes in the grains, and amylase, peptidase, and glucase are either produced in the grains or released from chemical bonds to act on cell walls, starches, and proteins. All of this can take place at 10–20°C (50–68°F).

Malt will release enough enzymes to ferment a mash composed of 60–70% unmalted cereals and 40–30% malt grains. Both the cereal and the malt can be of any grains or mixture of grains (rice, wheat, barley, maize, sorghum and so on).

The mash or ferment substrate is made by adding water at two to three times the volume of grains and malt, either: As sprays passing down through the grain mass, at 65°C (149°F), until the wort or washings fall below 1.005 SG (specific gravity); **or** by boiling part of the mash and then mixing this with the rest, to raise the temperature of the whole mass or kettle to about 50°C (122°F). The "boil and return" method is done twice over, once at 50–55°C (122–131°F), then at 63–65°C (145–149°F), and last at 73°C (163°F) where starch viscosity is lowered, so that the wort is more easily filtered off.

Before brewing, worts are boiled sixty to ninety minutes, which kills enzyme activity; sterilises the wort; precipitates proteins and tannins; extracts flavours from hops, ginger or other bitters or barks (sassafras); normalises sugars for colour, and drives off volatiles.

The wort is now cooled to 15°C (59°F) or so, aerated, and yeasts are added (*Saccharomyces cerevisiae* or mixes) from a yeast mixture. After from five to ten days, the latter at 6–10°C (42–50°F); the former at 15–22°C (59–72°F), SG is 1.008 to 1.010, pH falls from 5.2 to 3.4. Beers can be consumed unsterilised or fresh, or matured in vats for from four days to four weeks, then bottled and sterilised.

The above process also yields a variety of by-products suited to livestock feeds, but in peasant varieties, beers are not rigorously strained or clarified and are consumed as a thin gruel, rather sour in taste. A great deal of domestic grain may be consumed in this way as a normal sour thirst-quencher, or as a party drink to celebrate some special event. There is a gradation of grain ferments lying between low-alcohol field drinks, beers, sour porridges, and industrial sterilised beer, but really poor people always consume the mash with the beer, and only in industry are large amounts of wastes produced. These are often collected by pig owners, who may mix mash with whey to help feed their pigs, or gardeners who use spent mash and hops with lime to mulch trees and gardens (to feed worms). Note—that some malted grains fed to poultry and pigs greatly assists the digestion of other unsprouted but softened grains.

Malting—Variation Two

Starting from the choice of grains, the production of malt can be very well ordered if attention is paid to the following:

- Grains, especially barley, are ideally at 1.4–1.6% nitrogen (Acland, 1980). Heavy nitrogen fertiliser makes for a very poor malt and "steely" hard grains.
- Grains for malting need to show white, mealy endosperms, and a trial lot should germinate quickly and very evenly. Small or broken grains can be rejected by sieving. Again, 80% mealy grain is an ideal, 50% is just tolerable.

First, soak selected plump grains in a trough, changing water every four to six hours, until the grain can be easily halved by a fingernail. Cease as soon as the grain is softened, then allow the vat or trough to drain thoroughly, and spread the grain out on a cemented, tiled, or hard floor.

The beds (couches) of germinating grain are to be 50–70 cm (20"–28") high. In cool weather or winter, beds need sacking or carpet over them. The temperature of the beds will slowly rise 6–7°C (42–45°F) above that of the atmosphere, and the dryish outer grains will sweat and a small white swelling will be visible. This is the root or germ.

Now the barley or malt grain is turned over every four to five hours and the bed thinned to keep the temperature to 15–16°C (59–61°F). When the root is about half as long as the grain or before the leaf develops, drying should commence, often in a kiln or gently heated drum not exceeding 35°C (95°F) at first. The heat is gradually raised until when nearly dry, 80°C (176°F) is reached.

Ideally, the grain should stay pale; a darker colour means the formation of caramel (and thus loss of some sugar). Hot air rising via perforated base plates is the ideal drying situation. Dry, warm grain can be winnowed to break off the shoots, which are discarded as garden mulch. A good malt now remains. Five parts of ground malt, by weight, will dissolve 100 parts of starch, as cracked or ground flour, in 400 parts of water at 60–80°C (140–176°F), stirred. This is the stage at which the mash is ideal for livestock. Note—that the "starch" can be, in part, of potatoes, cassava, sweet potato, or sago.

For African sorghum beer, malted and unmalted grain is mixed at a ratio of 1:4 and 4 litres (8½ pints) of water are added for 2 kg (4½ lb) of grain, so that a much more porridge-like beer is obtained.

Malting—Variation Three

Soak barley, spread and cover to sprout until the root is 1.3 cm (½") long. This takes ten to fourteen days at an ambient temperature of 10–16°C (50–60°F). Then dry the grain, at first using a low temperature, but rising to 60–63°C (140–145°F), and never exceeding 71°C (160°F) (formation of caramels from sugars). Reduce the moisture to 2–3% of total. The starch in the grains is now converted to maltose, which suffices, at 10% by weight to convert all other starches (from soaked wheat, barley, potatoes etc.) and so on to sugars and via yeast ferment to alcohol. Some starchy ferment values are:

- Potatoes 20% fermentable
- Bananas 18–20% fermentable
- Artichokes 17% fermentable
- Grapes 20% fermentable
- Sugar 50% fermentable
- Rice 78% fermentable
- Barley 68% fermentable
- Sorghum 9–15% fermentable
- Sugar beet 18% fermentable
- Cassava 25% fermentable

Hop Drying

Cut small quantities of hops with stalks attached and dry under shelter, then store in onion bags or paper bags in a dry place. Cut the heads off when small globules of resin are visible, and the leaves are easily rubbed off. Commercial quantities are dried in kilns, on perforated floors in oast houses, where they are piled at 40 cm (16") deep and dried at about 82°C (180°F) for eight to ten hours. A little sulphur is sometimes added to the fire, and the heat fanned through and over the hop floors. The fire is allowed two hours to cool off before the hops are pressed and packed for sale.

The hop vine is perennial. Propagated from stem cuttings with 3–4 buds, they bear in three to four years, generally supported on overhead wires 3–4 m above the roots, with cheap baling twine supports for the vines, cut at harvest.

Hops are used in pillows and mattresses; they are a mild soporific. Historically, children's pillows were often filled using hops as they were thought to prevent suffocation (which other materials were believed to cause). Pillows made of hops can be washed, bleached and re-dried.

Kiesel — Poland

Mix oaten flour and warm water to a batter, then add boiling water to make a thin solution. Mix well, then cool to 20°C (68°F) or so and add yeast (1 tbsp per litre [2 pints] of fluid), or 1 tbsp per 250 gm (8¾ oz) of flour. Stand in a warm place for twenty-four hours. Strain and return to container.

Slightly acid, kiesel is consumed thereafter as a drink like water or milk, or used in vegetable soups. To refresh the mix, add boiled (but cooled) water, flour, and a little yeast. Kiesel can be kept going this way, with frequent addition of water until the new ferment ceases, then again flour and water is added as it becomes exhausted. It is a refreshing field drink.

Honey Mead

Combine 7.6 litres (2 gal) of water with 3 kg (7 lb) of honey (or 4.5 kg [10 lb] of mixed wax and honey from a wild hive) and boil for an hour. Allow to cool, skim, and strain. Remove wax and clean for re-use. Stir in 40 gm (1½ oz) yeast.

Place in a large jar or barrel fitted with an airlock. Siphon off and bottle after a year, or stopper tightly when fermentation ceases. I remember my grandparents arguing about the correct time to drive in the bung. My grandmother always prevailed!

For wines, use enamelled basins, wooden and plastic casks, glass flagons and bottles. Avoid copper and aluminium.

Rice Beverages

These are the "thin porridges" or field drinks of many farmers, who probably "eat" 30% of their rice this way.

Jaand—Variation One — Nepal

A farmer's field drink, taking four days to prepare, and remaining sweet for a further two days. Cook 1.8 kg (4 lb) of rice and spread it in a basket to cool. Crumb marcha (rice yeast, see also koji) mix it with the rice and leave in a warm place for a day, then transfer to a big pot with an equal quantity of water. After a few days the liquid becomes jaand. Dilute with water as a drink.

Jaand—Variation Two — Nepal

Soak boiled rice a few hours in water, drain and transfer to a perforated pot over a pot holding boiling water and boil together for a few hours then cool. Now mix in yeast (koji) with the rice and let the pots stand for four days until the holes trickle out a white fluid ferment. The beer is diluted with water to drink.

Kat Jand — Nepal

Proceed as for poka (next page) by first fermenting and then drying rice, but dry the fermenting steamed rice over two to three sunny days and grind it. Add rice flour (four to five times the ground grain). This mix is returned to ferment (50:50 water to rice). Ferment in a well covered pot for one month. The strong double ferment is often mixed with "sweet" beer. Alone, it is always consumed with the solid residue, as a festive gruel beer. In effect, double ferment uses enzymes and yeasts developed by the first ferment, which is a stage beyond malting, but still like a wort or sponge used in beers and breads. It is probable that the ground first ferment is added to flour at 1 part to 4 of flour or more.

Strong "red" liquors like this are kept for medicines, and herbs extracted in them, or molasses added against women's disorders (providing iron for any blood loss). (Gajurel, 1984).

Jarnd (beer using rice, maize or millet) — Nepal

Rice that has been soaked overnight is steamed over a copper pot in an earthenware container with holes at the base. After steaming for an hour, once stirred and once rinsed, the rice is cooked.

Cooked rice is spread on a clean floor and two biscuits of marcha (koji in dried rice flour) crumbled over it and mixed in. The rice is then pressed down in an earthenware pot and covered with maize husks and cloth in the pot (gyampa). The whole pot is covered with straw, and in winter, cloth also covers the straw to keep the pot warm.

After five days ferment, water is added and the rice strained out; the amount of water (a thin gruel) determines the strength of the beer. The ferment can be watered down two to three times.

The residue of rice is used for jarnd three times. For every 11 kg (24 lb) of the rice strained out (the lees), 11 kg (24 lb) of molasses and 1 kg (2¼ lb) of yeast (marcha) are added. It then again ferments for nine days, and is stirred with a spatula twice a day.

Even after distillation to raksi, the lees are given more molasses and yeast and fermented a further four days. Finally, the lees are mixed with wood ash as fertiliser, or fed to pigs.

Poka (semi-solid rice ferment) Nepal

Whole and broken grains are soaked overnight, and steamed enough to be crushed by the fingernails. The mass can be removed or stirred when half-cooked, for thorough cooking.

The cooked rice, spread and cooled, is mixed with yeast cake (1 yeast cake to 454 gm [1 lb] of rice). The mixture is placed in a clay pot, sometimes with a piece of charcoal and chilli "fire". The mouth of the pot is closed with a clay bowl and sealed with cloths.

It is left for four to five days, when it gives out a brewed smell and tastes sweet. This rice is eaten with chiura by Nepali farmers for the afternoon meal in the field.

The beer, taken to the fields, can be obtained by adding water 50:50 to poka, and after one to two days, pushing a deep woven sieve into the mass and baling out the beer from inside the sieve. This process continues until the poka is exhausted. Residues are fed to pigs when exhausted by dilution (after two to four successive water additions).

Bara Basha Gyampa (twelve year beer) Nepal

This beer has some mythical qualities. I know of no one who has sampled it! Add yeast or koji to cooked, cooled rice in a large earthenware pot and bury in the earth for twelve years. When you dig up the pot, you will find small white grubs (like silkworms) full of alcohol. Split the worms in half and add 2–3 glasses of water. This will intoxicate one person. A big worm split in a large gyampa will intoxicate a dozen drinkers.

(I include this incredible ferment recipe in the hope that someone can confirm or deny that people would wait twelve years just to get drunk with a worm!)

Maize Beverages

Maize Beer Nepal

Maize is ground to flour and boiling water added so that a porridge is made with continuous stirring, then cooled, and yeast added. It is fermented slowly (in a cool area for a month) before using as beer. It resembles granular milky tea, but is stronger than rice beer. Fermented maize with koji is solid, white, and ferments in three to four days (bhyabar is the name). Such beers can be distilled to raksi (grain spirit).

Chicha—Variation One South America

An important maize drink for farmers and field workers is made from cooked maize, chewed and expectorated into bowls. The enzymes and bacteria added from saliva start the ferment to beer. Only in remote villages is it still prepared in this way.

Chicha—Variation Two

(Modern Beer)

Mix 1 litre (2 pints) of corn kernels with 1 litre (2 pints) of barley grain and lightly toast. Mix in another litre (2 pints) of each grain and keep just covered with water for two days, until sprouted. Blend soaked grains.

Blend separately: 1 sweet pineapple, 1 bottle of orange juice, 4 cups sugar, 1 stick of cinnamon, small amount of yeast, and 8–15 litres (17–32 pints) water. Mix with grain mash for two more days, then chill and serve with crushed ice.

Bojwalwa Ja Paleche (beer using maize meal and sour milk) — Botswana

Maize meal is stirred into well-water with sour milk added, cooked, and stirred until it is a thin gruel, then left in an earthenware jar for a week. On the sixth or seventh day pour this pot of fermented maize into a large pot, adding well-water and uncooked sorghum meal. Leave one day and then strain through muslin.

Miscellaneous Grain Beers

Poko (rye beer) — Nepal

Rye is sun-dried, husked, coarse ground, and well-steamed over a boiling pot. Cooking is judged complete when "tears" (condensed water) fall from an inverted bowl over the rye container. This stage is called "bringing the tears". The rye is spread, cooled, and yeast cakes (koji) crumbled over it in a ratio of 1:3 by volume (1 rice koji to 3 rye) and the whole is hand-mixed. The mix half fills vessels which are then filled with water, and an inverted bowl is used to cover them. All pots are well covered with cloths to reduce air flow (and exclude vinegar flies). The mass is stirred at eight to nine days and further fermented for eight days to become beer.

At this stage, poka (see under rice, previous page) is added, again the mix is stirred and kept covered for ten to twelve days. 2.25 kg (5 lb) of sugar is then added, along with some popped rice. This is the "wine" which is distilled to spirits. (Gajurel, 1984).

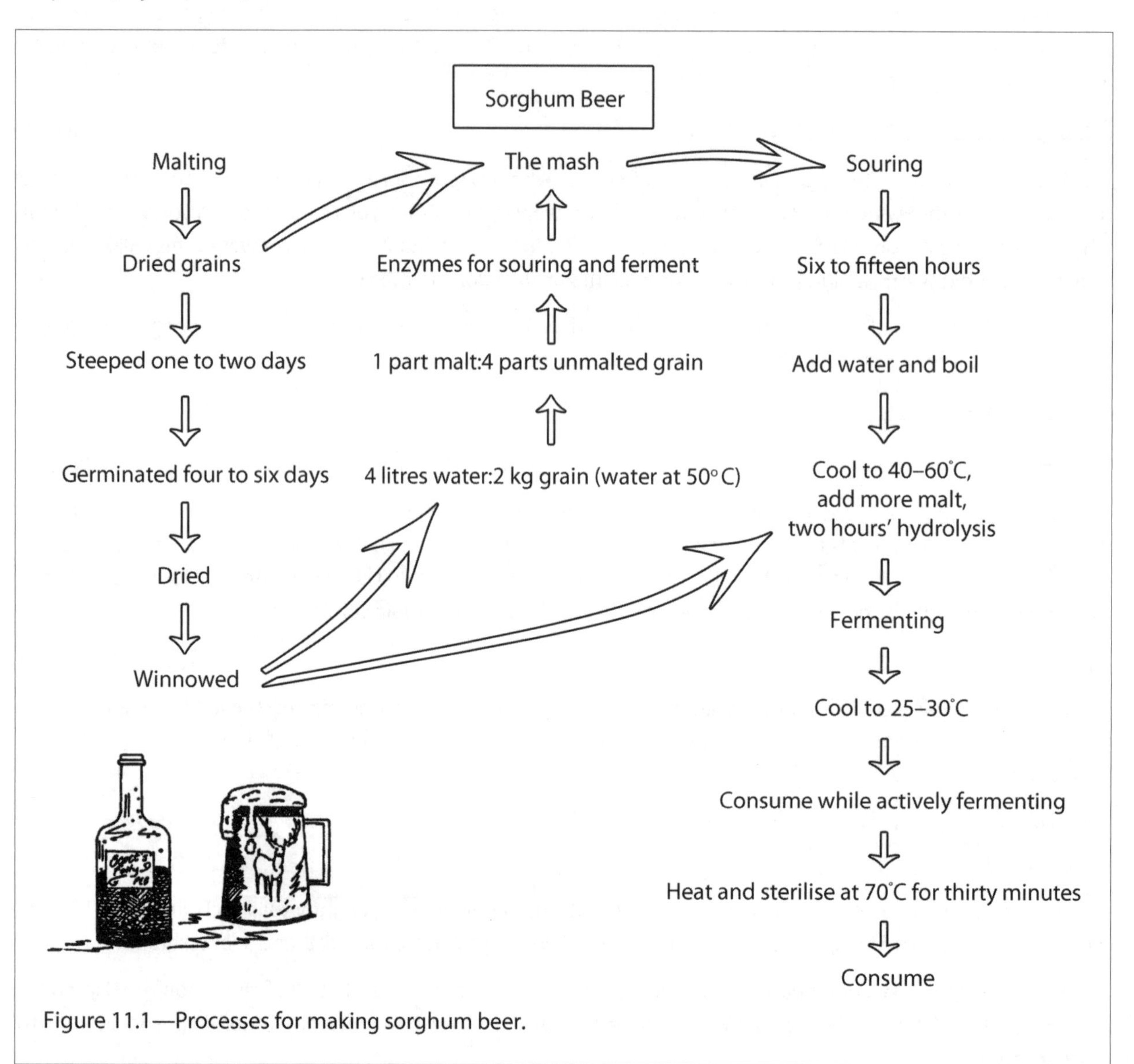

Figure 11.1—Processes for making sorghum beer.

Sorghum Beers

In parts of east Africa, ashes are mixed with the sorghum grains, and the whole mass well wetted, or soaked overnight, then covered for five days or so to germinate, cleaned, and dried as a malt (the flour used in beers). Joined with such malt in brewing is a thick paste made of ungerminated grains left under water four to five days. This paste is lifted out, divided, and fried before being returned to the water with the malt (as above). After two days of this combined ferment, the beer is ready. See Figure 11.1.

The main organisms involved in making sorghum beers are:

- Malting *Saccharomyces cerevisiae*, a yeast.
- Souring *Endomycopsis burtonii*, a yeast.
- *Amylomyces rouxii*, a mould.

Grains used are sorghum, millet, and maize. Occasionally, cooked and mashed sweet potato, cassava, taro, or root starches are added at the mash stage.

Bojwalwa (fermented sorghum beer of the Setswana) — Botswana

In a large pot, water is boiled and maize-meal mixed with sorghum is stirred in until it is cooked. This "thin porridge" is then cooled and poured into an earthenware pot where uncooked sorghum meal is added. Boiled (then cooled) water can be added to the pot to create a thin gruel. An earthenware lid is placed on the pot until it foams (one or two days), and the beer is strained through muslin before being consumed.

Stout (barley beer) — Europe

Boil vigorously, 25 gm (1 oz) of dried hops in 5 litres (10½ pints) of water for forty-five minutes, then add 454 gm (1 lb) blackstrap or "BB" malt, 113 gm (4 oz) of black malt grains (for colour), and 454 gm (1 lb) dark brown sugar, stirring well to dissolve sugars and malt. Leave to cool. At 18–20°C (64–68°F) add a yeast made up in lukewarm malt water.

If water is soft (acid), add a little lye or salt.

Cover with muslin and ferment for five days, skimming if necessary and wiping yeast crust off the sides of the pot. When bubbles cease, leave in a cool place a day or two to still, then siphon into ferment jars fitted with an airlock.

After yeasts settle, bottle, cap or cork, and let mature for three to four weeks. Fresh hops added to the stout on the third day of ferment gives a good aroma. Malt flour (750 gm [26½ oz]) can replace the heavy malt syrup. Instead of bottling, an airlock can be fitted to a small cask (primed with 45 gm [1½ oz] of sugar), and served as a draught beer after two weeks.

Chaang (millet beer)
Nepal

Boil 2 kg (4½ lb) millet in 5–10 litres (10½–21 pints) water for two to four hours. Take care not to burn the grain, by topping up with water as it swells. The mush can be blended or ground, and water added to make a thin gruel. Yeast or lemon juice is added to the cooling gruel, and sometimes a little sugar. It can stand for some weeks and is strained before use.

Chaang or Tongba (millet beer)

Millet is thoroughly boiled, spread on a mat, and yeast culture added (koji or marcha).

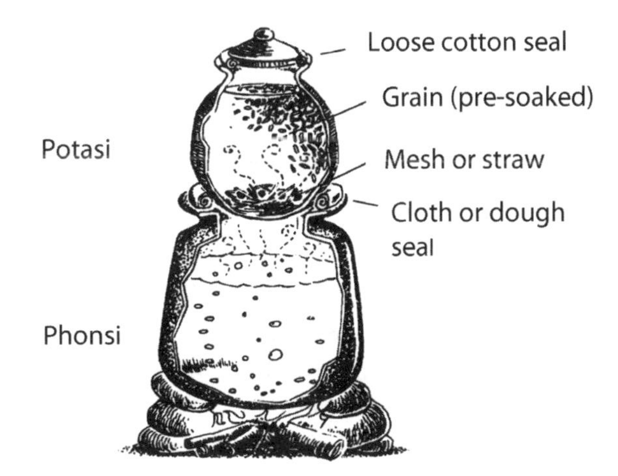

Figure 11.2—Steaming rice, maize or millet for beers in Nepal.

Line a fine-woven basket with banana leaves, spread in the millet and cover with banana leaves. Leave to ferment for three days in summer or six days in winter. The froth is allowed to run to waste. Now the mass is transferred to a jar and thoroughly sealed (air spoils the ferment) for eight to fifteen days. This mass is diluted 3:1 (3 of water) or more and strained to drink.

Chaang can also be made from sweet potato, wheat, maize, rice or a mixture of these.

African Beer (from millet, maize, sorghum or a mixture of these)

Moistened grain, in baskets or covered with green leaves, is sprouted three days and this chimera is dried in the sun and ground into a meal or malt, which is then boiled as a thin porridge (called bota) with water and allowed to ferment for one day. This is the bume or sweet beer (a wort) given to young people or preferred by many adults as a drink.

This sweet beer is again boiled (and is called mangah). This is allowed to cool and ferment. A little uncooked meal is added to the pot on the fifth or sixth day. The pots of mangah are stood with plates covering them as lids for the first five days, when the addition of a handful of meal starts the second strong ferment (strong beer or musumo). After one more day, the ferment is strained and consumed. Strong beer is not preferred by a majority of Africans, and perhaps the greatest benefit comes from a mix of sweet beer with a little strong beer, at least as a food.

Cassava

Juice expressed from the roots and allowed to clear is mixed with molasses and fermented to ouycou in Guiana, and consumed as a beer.

Fast-Brew Ginger Beer

Grate 30 gm (1 oz) ginger, and add this to 454 gm (1 lb) sugar, 30 gm (1 oz) cream of tartar and a grated lemon rind. Over this pour 20 teacups of boiling water. Stir to dissolve sugar and leave to cool to hand-warm.

Mix 30 gm (1 oz) of yeast with the juice of 1 lemon and a little warm water and add to the mixture above. Leave overnight, then bottle the next day. Store in a cool place to mature for at least forty-eight hours. My grandmother added a couple of raisins (natural yeast) to each bottle. Chill before opening.

11.2 FRUIT BEVERAGES

Banana Beers

- In Uganda and Tanzania bunches are overripened by hanging in fire smoke for five days, or are buried in ripening pits. The bananas are peeled and then the juice is expressed by pounding in a long trough, and the pulp expressed through blady grass (*Imperata africana*) wads. To this liquid is added fried and pounded sorghum or finger millet, and the whole mix fermented for two to three days (Acland, 1980).
- Peeled bananas are boiled to a dark colour and allowed to stand in hot water for a three day ferment.

Romany Kini (the wedding champagne of the Romanies)

Peel 24 oranges, and boil the peels in 9 litres (19 pints) of water for a few minutes. Cut up oranges and place in a large pot. Pour the boiling water and peels over the cut up oranges. Add 400 gm (14 oz) of potatoes (cleaned and cut up) and well cover the pot with a folded blanket.

This stands for a week, when the liquor is strained off and the potato and orange removed. Add 2.7 kg (6 lb) of Demerara sugar to the pot, and cover again. The liquor ferments for about three weeks, stirred each time somebody passes. Then a blue mould will form on top, which is removed and the liquor strained and bottled uncorked for another week. A little dissolved isinglass is added to each bottle, and it is tied down. (After Petalengro).

Cider (fruit juice ferment)

Apples are pulped or coarse grated, placed between hessian (burlap) layers, and pressed. The juice, in jars or barrels, ferments vigorously. In casks the bung is kept loose, or a small-necked bottle is used, so that no strain is put on the container.

After ferment ceases (a lighted match is not snuffed out by the carbon dioxide), the cask is filled with some of the same cider fermented in a bottle for this purpose, and the bung driven home. To clear cider, it must be "racked" or siphoned across to leave the lees, three or four times, each time about three months apart. Cider improves with age; age improves with good cider.

Beet

In Russia, a "fortified" beet drink is made from heat-sterilised beet juice. The, culture is of *Lactobacillis plantarum* and *Streptococcus faecium*, and the juice is acidified with lactic acid (to pH 6.0). After ferment, essential and other amino acids are added as a nutritional supplement.

The antioxidant betanine is found as an active form in the drink, and this aids normalisation of the intestinal bacteria. (Kovalenko et alia *Microbiol J.* 1990 52 [6] pp. 53–59, Kiev).

11.3 YEASTS AND ALCOHOLS FROM FLOWERS

Mahua

The creamy-white flowers of *Madhuca indica*, falling in showers in spring, are rich in sugars, vitamins, calcium, and trace elements. Next to cane molasses, these flowers are the most important source for yeast culture, or spirits, in India and Nepal. A ton of flowers will yield about 400 litres (106 gal) of alcohol at 95% ethanol if double-distilled to get rid of the foetid odour and gastric irritants of the first distillation.

The flowers are first boiled down for about six hours as for eating, all water being evaporated. As too much flower pulp causes illness (vomiting, dizziness), modest consumption as a sweet with rice or grains is recommended. Natural yeast ferment occurs in the fresh flowers in boiled, cooled water and this can go to spirits, syrup, or vinegar. Syrups are extracted with hot water from flower masses, clarified by filtration through active charcoal (from bamboo, coconut husk, or willow) and evaporated to a syrup under slight vacuum, when a golden yellow "honey" with the odour of fresh flowers results.

Flower wastes are fed to domestic stock as a concentrate, and are relished by cattle or pigs (*Wealth of India VI* p 214). The berries of the tree are eaten raw or cooked. Obviously, other nectareous flowers can be used to release yeasts, and made into dried pastes or yeast sponges.

Flower Wines

The flowers of elderberry, coltsfoot, dandelion, marigold, and so on (edible flowers) are gathered, and the petals measured (lightly pressed) into a 0.5 litre (1 pint) jar. To each 0.5 litre (1 pint) of flowers pour over 4 litres (1 gal) of boiling water (with or without fine-cut rind of 2 oranges and 2 lemons). Add 1.5 kg (3½ lb) of sugar (or previously boil this in the water and skim it) and 454 gm (1 lb) of raisins. Cool to around 21°C (70°F) and add 1 gm (pinch) of yeast. Leave to ferment, cloth-covered, for five to six days. Strain, and put it into a cask or plastic drum with an airlock. After eight weeks, bottle and cork well, siphoning from the cask to clear.

Rose Hip Wine

Pick ripe hips, de-stalk, and mash well. Leave for a day and then strain, and to each quart of juice add 250 gm (½ lb) of sugar. Cover, then leave one more day, stirring occasionally. Fill a flagon or cask and fit an airlock. Keep dark and when fermentation ceases, cork tightly. Matures in three months.

11.4 DISTILLED LIQUORS

Slivovitz — Yugoslavia

Excess plums are ground-gathered or beaten from the tree onto an old sheet or sail, then left "as is" in an open plastic bin or barrel with a loose-fit lid for five weeks. They are then distilled using a worm condenser. As the slivovitz or plum brandy comes over, toss a little in a fire until it fails to ignite, then stop distilling or fusel oils may come over. You should get a little over 2 litres (4 pints) of slivovitz to 22–30 litres (6–8 gal) of plums. If fusel oil is suspected, add a little black pepper to the liquor in a glass; it will sink and take the fusel oil down. Best consumed in a snowy climate, (Blaz Kokor). This was an annual winter ritual at my house in Hobart (1960).

Raksi

Raksi is a rice whisky distilled from strong beer to spirit. The Nepalese still is a complex, copper vessel (see Figure 11.3).

Sake (rice wine) Japan

Koji culture (see yeasts and moulds) of steamed rice with *Aspergillus oryzae* is where sake culture begins. Sake yeasts (*Saccharomyces cerevisiae*) proceed at low temperatures 15°C (59°F) over a long period to produce up to 20% alcohol. The mash is called moromi.

Rice (50–75% polished, the remainder unpolished), is first washed, then steeped until 25–30% of the dry weight is water (e.g. for 1 kg [2.2 lb] of dry rice you will end up with 250–300 mL (8.4–10 oz) of excess water once steeped), then drained four to eight hours. It is then steamed, thirty to sixty minutes, which also sterilises the rice and raises water content to 35–40%. Steamed rice, cooled to 40°C (104°F) is used for koji inoculation, and the 3–4 steamed rice batches added over the next few days are cooled to 10°C (50°F) for adding to the moromi mash.

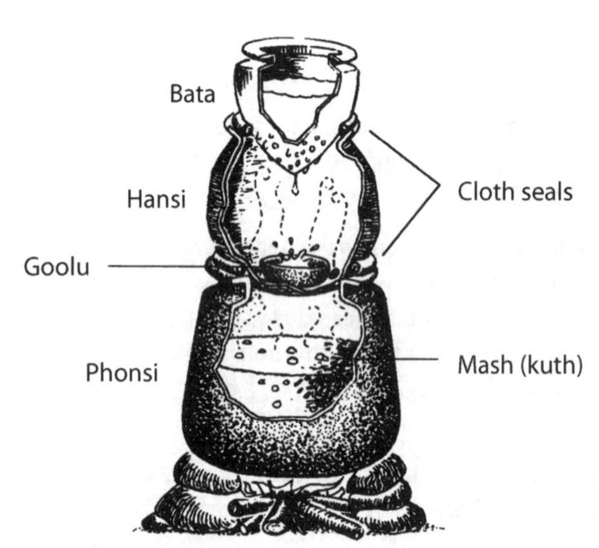

Figure 11.3—Distilling ferments of grain, fruits and cane sugar to raksi in Nepal.

In the koji incubation room, where the first batch is inoculated, temperature is at 26–28°C (78–82°F) and humidity 80–90%. The moist rice is spread on shallow wooden trays (6 cm [2⅓"] deep). Spores of *Aspergillus oryzae* (Tane-koji) are then sprinkled over, and the whole well-mixed, then the trays are covered with cloth, they are now at 31–32°C (88–90°F).

In ten to twelve hours, the rice is broken up, heaped and covered with cloth. At about twenty-four hours, the mass (at 33°C [91°F]) is divided into 1.5 kg (3⅓lb) lots and heaped on small trays, which are stacked one above the other. It now should be at 31°C (88°F), but as it is still fermenting, the temperature will increase to 32–34°C (90–93°F), and the stack of small trays is periodically restacked or changed over. After six to eight hours, the small masses are again turned and spread, and the trays changed over (temperature is 33–35°C [91–95°F]). After six more hours, the temperature is 37–39°C (99–102°F) and a second mixing is made as before. After the piles reach 41–43°C (106–109°F), the trays are again mixed, and a further six to eight hours allowed. After a last stir, the koji is taken out of the cultivation room and mixed with water and steamed rice.

The sake mix is 30 kg (66 lb) of koji plus 70 kg (154 lb) steamed rice added in three to four lots over the first four days into 100 litres (26 gal) of water, at 7–8°C (45–46°F). A heater below the cask raises the temperature to 14–15°C (57–59°F) over a week. 600–700 mL (20–24 fl oz) of lactic acid is now added and the ferment continues for fifteen to twenty-five days. Then a yeast mash is added.

So much for the koji mash. The yeast mash (moto) is prepared with 7% of the rice for koji. A pure culture of the yeast is added at 15–20°C (59–68°F), and for two to three days the ferment is vigorous. After five to seven days' rest, only the beneficial yeast remains and can be kept for a culture. Because of the vigorous yeast ferment, the vats are kept two-thirds full only. All ferment is in open vats. Thus, after twelve days of yeast ferment, the sake is ready for pressing.

Usually, the sake is pressed out of 5 litre (10½ pints) bags under hydraulic pressure. The solids or sake lees are used to preserve fish and vegetables. About 250–300 kg (550–660 lb) of solids are recovered from a ton of rice. The sake is settled at 8–10°C (46–50°F) for one to two weeks, then filtered through activated charcoal and cotton. It is pasteurised at 55–65°C (131–149°F), matured at 13–18°C (55–64°F), and can be diluted to 15–16% alcohol before bottling for sale, after some months. (After Yoshizawa and Ishikawa in Steinkraus, 1989).

Rice wine or mirin is more simply made by crumbling koji, marcha, or ragi over rice fermenting at 28–30°C (82–86°F), and diluting with cooled boiled water. Fresh boiled rice is added, plus malt or sugar cane juice. Yeast (*Saccharomyces*) is added under a covered ferment. Other processes are as for sake, after the yeast mass is added.

Amazake (sweet sake) Japan

Pressure-cook 1 cup of any rice in 2 cups of water. Allow it to cool to 60°C (140°F), then mix in 2 cups of koji and incubate for eight to twelve hours at 60°C (140°F). To serve, combine with an equal volume of water and just as it boils add salt (light), pour into cups and serve with grated ginger on top. Mixed with strong or regular sake, it is strained or decanted as mirin (sweet wine), which keeps well sealed.

A substitute for mirin is made by adding 1 tsp of sugar to 2 tsp of sake or dry sherry.

Sealing Wax

Wax for pouring in or sealing oiled corks, driven in to bottle necks to below the rim, is made from 65 parts of resin (rosin), 1 part of beeswax, in 1 part of linseed oil (all slowly melted together over low-heat). Wax can be mixed with lamp black or vermilion powder for black and red wax respectively.

Such expensive waxes are used only for valuable spirits stored for long periods, or used for sealing legal documents.

11.5 DRINKS FERMENTED BY MOUTH ENZYMES

Couffignal (1979) records several folk drinks fermented by using the enzymes found in saliva, and mouth bacteria or yeasts, as well as natural ferments of many tubers, roots, grains, and milks.

- Masato—is a manioc or cassava ferment where the root is chewed one mouthful at a time then spat into gourds to ferment (South America).
- Kava—is a fermented drink of Fiji, and is traditionally made in this way, but in modern times is pounded in a trough and only a little is chewed.
- Chicha—is a South American beer made from maize, and is a porridge-like beer, first chewed, and fermented for fifteen days. A double ferment can reach 36% alcohol (with sugar added to the first ferment).

11.6 OTHER FERMENTED DRINKS

- Baiga—made in the Gobi Desert, is a millet beer where the millet and a little dried pigeon droppings are first roasted in metal pots, then fermented in skins and bowls.
- Airaq—is the koumiss or fermented mares' milk of Mongolia, and akin to the goat and cattle milks of the Kalahari where long yeasty ferment in skins produces a part alcoholic sour milk.
- Pulque (Spanish) or neutli (Aztec)—where pulque means "bitter", are fermented aguamiel (honeywater) from the stems of agaves (*Agave americana*). Some large agaves produce 125 litres (33 gal) of aguamiel per stem. Stem or leaf bases can be cooked in large earth-ovens to produce sugars, then fermented.
- Jiculi—is a fermented drink from the peyote cactus of the Sonoran Desert Region of Mexico.
- Mescal—is fermented from the flower-stem juices of the grey-blue maguey or agave of the Sonora (*Agave mescal*). Elsewhere, *Agave bacanora* of the Sierra Madre is used to make the local bacanora drinks (distilled spirit). A little dried fruit is added to bacanora for flavour (usually sweet plums).
- Tezquino—combines wild grains (amaranths and grasses) with sprouted maize, both fermented together in the Sonoran Desert Region of Mexico. As it is no doubt the habit to boil grains with lime in Mexico, the basis of this ferment is very close to the "porridge" of ancient Rome.

Tea and Coffee

We often use ferment to free seeds from mucilaginous pulp. Coffee berry ferment, for example, has become sophisticated at the same time. Yeasts used to free or clean or mature the cacao bean, or to ferment tea, give a very specific aroma to the product and subsequent drink.

Processes are essentially:

- Pulp seed and fruit.
- Wash skins and surplus pulp away; skim light seeds at this time. Pulping and washing should be a fast process, as stain substances can develop.
- For coffee and cacao, add sugar to the mucilage to rapidly degrade pectins, preferably using waste water from past pulping to enhance sugars and temperatures. The pH should fall below 4.3 so that lactic and acetic acid form and stop spoilage (or acid can be added).

Coffee Berry Ferment

Ripe berries are pulped by a revolving cylinder with short "music box" teeth or semi-circular projections, crushing them against a fixed beam or steel bar. They are washed through the crusher, fermented two days in a vessel, and the remaining pulp washed off. A mincer with the cutter removed and large holes does the job of pulping on a small scale.

The seed pairs, in their parchment, are well spread and dried. (They are often stored this way, sometimes for more than a year.) Coffee beans are hulled by a heavy revolving wheel which breaks the husk. They are then ready for roasting, grinding, and coffee making. Roasting is in a large iron pan over a fire or gas flame. A little sugar can be sprinkled on for a caramel flavour. This roasting stage takes experience to know when the coffee beans are right for flavour. Try some (well-timed) before committing all the beans (2.3 kg [5 lb] fresh fruit yields 454 gm [1 lb] coffee).

Tea Ferment

The growing tip and first two young leaves of tea gives those teas known as pekoes. The tip and buds are "flowery" pekoes, and the tip and first leaf "orange" pekoes.

Coarser teas (first 3–4 leaves) are "souchong" (smoked teas). Leaves are plucked every eight to twelve days in the growing season, spread out on mats under cover, and withered. Sometimes hot air is fanned over the leaf. It should wither in eighteen hours or so, becoming limp and flaccid. It is then rolled for up to an hour, usually by machine, to form a clinging mass, broken up, and sifted to set aside coarse material.

It is then piled on mats to ferment, and in two hours or so becomes a coppery colour, it also smells peculiar. The fermented tea is dried with hot air draughted through it on perforated floors or mesh trays, and is allowed to become dry and brittle. Green tea is dried as it comes from the break-up directly after being rolled, and is steamed (not withered), then dried. Sieved, grades are allotted, and tea is bulked up or stored dry in sealed packets.

Cacao (cocoa)

Ripe cacao fruits are taken with a sharp, hooked knife on a short pole, split, and the mucilaginous seed removed. The seeds are piled in heaps or vats, and covered with banana leaves, weighted with earth or sand, to ferment. Fermentation takes place over two, or even ten days (the heap turned twice daily) when they rinse clean easily. To dry, they are sunned on mats for two to three hours, and sweated in heaps for the rest of the time (this keeps them plump). In a few days of sun, they will be dry. Husked and ground they make cocoa (chocolate), (Willis, 1909). The husks, as with coffee, make excellent mulch.

Cocoa, and cocoa butters are used in many drinks and sweets, but unsweetened cocoa is also used in stews and sauces for meats. Of course the basic product is a solid chocolate.

Before husking, the "beans" are roasted, and the husked beans are then finely ground (traditionally "conched" in stone troughs). As the heat of grinding builds up, the cocoa becomes a semi-liquid mass. If now pressed, cocoa butter is expressed from the cocoa cake, which is then ground to cocoa powder.

The cocoa butter is combined with milk solids, flavouring, and other vegetable fats to make bitter cooking chocolate, or sugars are added to make sweet chocolates. Some "white chocolates" contain very little cocoa!

Good cooking chocolate may contain rice flour or cornflour, with a little cinnamon and very little sugar, mixed with the mass first obtained by grinding. (See also Grigson, 1991).

Kambotscha ("The Tea Beast") France, Germany, China

The culture for this ferment is (like koji) a yeast-bacteria mix. It is a tough, membranous, fibrous mat used to create excellent cold drinks from waste, cold tea.

My culture came from Germany via France, where it was imported from China! Several people in Australia also use this culture. Three or four wide-mouth jars with muslin and rubberband covers are ideal, as the culture is then visible, and can be watched. The tea culture bacteria is *Acetobacter aceti*, spp. *xylinum* which synthesises cellulose. The drink is sub-acid, a good thirst-quencher, and much liked by children.

To start the Beast, you must already have a piece of the culture, donated by a friend.

Fill a jar with tea, adding 100 gm (3½ oz) per litre (2 pints) of sugar when cool. Boiled and cooled water can be added to strong tea, which will float on the rubbery culture.

Leave eleven to twelve days minimum; more realistically two to three weeks in winter. Bottle, and keep cool, or refrigerate for cold drinks.

My German friends tell me to drink three glasses a day before meals; and that it is called a "health" drink. Travellers carry a culture. The colony-culture itself is never refrigerated, but carried to friends in plastic bags at room temperature. If left for four to eight weeks, a good vinegar is produced. Moulds may grow on neglected cultures. If the drinks are first boiled to kill the culture, and well corked, they may not need refrigeration.

The culture is split up (it separates in sheets) every three to four weeks, and fresh bottles started. Each culture grows to fit the diameter of the container, and floats above the tea in time. Washed, it stores well but is best kept alive and busy in the tea-sugar solutions. I see no reason why ginger or other spices should not be added on bottling. Ferment is at room temperature. This is a useful aromatic ferment culture to pass on to friends and can be quickly duplicated. If at first the culture disc sinks, wait a few days and it will float.

My thanks to George Schweisfurth of Glonn for practical data.

Kargasok (the Tea Beast again!) Siberia

Gently boil together 16 cups of water (or cold tea), 2½ cups of sugar and 5 tea bags for five minutes.

Cool to blood heat before adding a piece of the rubbery culture from the Tea Beast (see previous recipe). Cover the top with muslin and a rubberband. On the jar, write the date and stand in a cool place for fourteen days. Remove and wash the culture, which will split up horizontally. Give a piece of this Beast to a friend.

The liquid is credited with great healthy benefits from insomnia to kidney cures, cataracts to cancer, if consumed twice a day (2–4 fl oz glasses or shot glasses). The liquid is kept cold for consumption, and the new brew is always on the way (don't forget to date each jar!) If you are going on a holiday, float the culture in a weak salt solution in the refrigerator, or freeze it.

CHAPTER

TWELVE

Condiments • Spices • Sauces

12.1 SAUCES

Although this chapter covers most sauces, such important sauces as soy sauce and fish sauce are covered in Chapters 4 and 8 respectively.

Tomato Ketchup—Variation One

(tomato sauce)

Prepare ingredients: 6 kg (13 lb) tomatoes, 454 gm (1 lb) chopped onions, 5–6 chopped garlic cloves, 60 gm (2¼ oz) sea salt, 1 chilli, 1 tsp cinnamon, 15 gm (½ oz) fine-cut ginger, 1 litre (2 pints) vinegar, and 454 gm (1 lb) brown sugar.

Boil all the ingredients, except sugar, together until really soft, strain through a nylon curtain. Re-heat and add sugar, then cork in sterilised bottles. Keeps very well, at least a year in temperate climates.

Tomato Ketchup—Variation Two

(Nelson's Blood)

Prepare ingredients: 30 tomatoes, 10 gm (⅓oz) green apples, 10 onions, 5–10 cloves of garlic, 5 tbsp salt, ¼ tsp chilli or cayenne powder, and ½ cup water.

Cook all together until all are soft. Strain and rub through a colander, removing the skins. Add a bag of cloves, allspice, and a stick of cinnamon tied in muslin. Add ½ cup to 1 cup of sugar (to taste). Simmer for one and a half hours while stirring, reduce to about half the bulk or until thick. Add 1 litre (2 pints) of vinegar and boil for ten minutes. Bottle while hot and seal well.

Tomato Ketchup—Variation Three

Prepare ingredients: 10 kg (22 lb) tomatoes; 1 kg (2¼ lb) brown sugar; 125 gm (4½ oz) white pepper; 3 litres (6⅓pints) vinegar; 60 gm (2 oz) garlic; 1 stick horseradish (grated); 454 gm (1 lb) salt; 60 gm (2 oz) each mace, and mixed spices; 1 kg (2¼ lb) onions; and 30 gm (1 oz) cayenne pepper.

Method as for Tomato Ketchup (Variation Two) above.

Worcestershire Sauce (from Monzert, 1866)

Prepare ingredients: 4 litres (8½ pints) white wine vinegar; 600 mL (20 fl oz) sweet soy sauce; 600 mL (20 fl oz) molasses; 750 mL (25 fl oz) mushroom ketchup; 15 gm (½ oz) each mace, and cloves; 30 gm (1 oz) each paprika, chilli, and cinnamon; 125 gm (4½ oz) salt (omit if ketchup is salty); and 2 cloves of garlic crushed in 1 wine glass of rum or 15 gm (½ oz) asafoetida in rum.

Mix all together and stand for twenty-four hours, then strain and bottle.

Common additions to Worcestershire sauce are:

- Pig's liver at 1 kg (2¼ lb) to 4 litres (8½ pints) water, boiled for twelve hours, blended and strained.
- 2 litres (4¼ pints) sweet heavy wine.
- Walnut ketchup.

Tabasco Sauce

Chillies in brine are fermented three to four years, then mixed 2:1 with white vinegar, blended, and strained. Annatto seed, popped in a pan, can be added for colour.

12.2 JAPANESE SAUCES

Marinades and Dipping Sauces

- Sushi Sauce—Combine dark soy sauce with a dab of wasabi (green horseradish) in a small bowl.
- Tosa Sauce—Mix ½ cup soy sauce, 1 tbsp mirin, and 5 gm (a little less than ¼ oz) dried bonito flakes. Boil together, strain out the bonito flakes, and chill to keep. Garnish with chopped chives when serving.
- Sanabi-zu Sauce—Mix ½ cup of vinegar, ⅕ cup soy sauce, 1 tbsp sugar, and 1 tsp salt. This is the usual marinade sauce for fish fillets. Let marinate thirty to forty minutes before making sushi. After marination, fish are skinned, boned, and sliced for sushi.
- Pon-zu Sauce—Equal quantities of lemon juice, soy sauce, and crushed roasted nori seaweed can be used over tofu or fish. Bitter orange, citron, or green lime juice is also used as a dipping sauce for roast kelp and seafood.
- Sashimi Marinade—Rice vinegar and soy sauce are mixed in a ratio of 6:4 for fish used in sushi. Sprinkle on slices of raw fish.
- Fried Fish Sauce—Heat together ½ cup rice vinegar, 3 tbsp of sugar, and 2 tsp of salt. Pour hot over deep-fried fish (white fish or mackerel) and let marinate one to three days. This sauce can also be used to marinate wilted vegetables and kombu seaweed.

Toasted, dried, and powdered seaweeds can be used in sauces that are not boiled, or in soups poured hot over the powders or flakes. Powders are used with beaten egg and a little mirin to coat grilled fish or scallops until golden (scallops grilled four to five minutes only! Fish grilled to just cooked).

Optional additions to dipping sauces include: grated ginger, grated horseradish, grated daikon (white radish), grated garlic, a little sesame oil, fine slices of spring onion, lemon juice to taste, wasabi, or fine sliced preserved chilli.

It is useful to make up a soy-mirin-rice vinegar mix, which of course keeps quite well, and to add optional mixes as required or as available when dipping sauce is used for fish, eggs, noodles, tempura, etc.

Nerimiso (sweet simmered miso)

Combine in a small pot; 1–2 parts honey, 1 part sweet wine, 5 parts miso; and simmer on low-heat, constantly stirring until it thickens (two to five minutes). This paste is cooled and refrigerated, and will keep well for two to three months.

It is used over porridges, rice, tofu, curries, potatoes, salads, cooked vegetables, and pancakes, or as spreads.

Before serving, you can add any of the following: roasted sesame seed, peanut butter or tahina, fresh or dried herbs, garlic, ginger, grated lemon rind, soy sauce, sake 0.5 parts (e.g. for every 1,000 gm of dry matter, 5 mL of sake is used), hot mustard, chilli pepper, walnut pieces, crushed nuts, or sunflower seed. Lemon juice and grated citrus rind is added over fruits. This sauce, with other dipping sauces, is served in small bowls as a side dish. With shreds of chicken, pork, beef or shrimp, it can be folded in an omelette, or added to scrambled eggs with tofu.

Mirin (sweet wine) — Japan

Mirin is mostly used as a flavouring in cooking. Polished white rice is steamed at 100°C (212°F) for one hour spread on trays, and mixed thoroughly with koji (*Aspergillus oryzae*). After koji mycelia spread (at 10–17°C [50–62°F], for fifteen to twenty days), more fresh steamed rice and 5% alcohol (sake or mirin) is added and mixed well. It is fermented in pots thirty to forty days, filtered, pasteurised at 70°C (158°F) and bottled. Shelf-life is indefinite when well-sealed.

Mirin is widely used in Japanese cooking, in stir-fries, dips, and cold sauces. Alcohol is 14%, but glucose can be added to sweeten after sterilisation.

12.3 SOUP STOCKS

Frozen stocks, each labelled, are the basic starters for gravies, soups, jellies, and stir-fry juices. Gelatin or cornflour are needed as thickening.

To prepare, boil together: 1–2 onions, black pepper, slices of ginger, a few cloves of garlic, a dash of soy, 1 tsp of honey, and a dash of vinegar; with any of the following:

- 6–8 pigs' trotters, or a small pig's head.
- Heads, backbones, and scraps of fish ribs or belly left after filleting.
- Oxtail or kangaroo tail and shanks; shin meat and bone.
- A boiler chicken.

Boil for one to four hours (less for fish), strain, and save any flakes of flesh for potato cakes.

Remove reject bones, vegetables, and ginger. Pour warm stock in about one cup containers or plastic bags, cool and freeze. Label each type (chicken, tail, shin, fish etc). Reserve some smaller quantities of chicken stock (as ice cubes) for stir-fries.

Soup Stock in a Stir-fry

After stir-fry is seared, remove from flame and add 2–3 ice cubes of chicken stock, plus a ¼ cup of warm water in which about a level dessert spoon of cornflour has been beaten smooth with a fork. Return to flame and finish cooking until the "gravy" has boiled and thickened.

Father's Gravy (brown, fat-free)

Melt together: 1 cup meat stock, ½ cup of hot water, and 1 tbsp of miso. When melted, remove from flame and stir in: 1 tsp of Vegemite or Bonox, a dash of soy, 1 tbsp honey, a dash of sherry, and 2 level tbsp of cornflour well beaten in cold water.

Return to flame and just bring to boil until thick. Thickness can be adjusted with hot water (to thin) and cornflour (to thicken). Let simmer for two to three minutes, to cook cornflour. Serve in a jug with roasts.

Gravy Variations

Pitted prunes and dried apricots, cooked soft, go well with lamb (add to frozen stock in first heating), as does fine chopped rosemary. Cream is added for thick smooth gravy for game, do not boil. Add meat juices if carving. Add blood of hares and game birds, livers and gizzards of chickens (blend this gravy) and game. At first boiling, before thickening, blend giblet gravies well. More sherry is needed in game gravy, or a dash of brandy before serving.

12.4 ESSENCES

Basically, macerated or pounded spices and thin peeled citrus skins, powdered or grated ginger, or pounded vanilla pods are bottled and covered with ethyl alcohol (or non-flavoured spirit) for up to three weeks, then the alcohol plus oils and flavours is strained off and bottled for use in cakes, biscuits, etc. A small press aids extraction from orange rinds, lemon rinds (no pith), and for the first twenty-four hours, the maceration pot can be kept warm (tie a paper over and make a thin hole). Seal well once strained.

Dhulo Achar (powdered pickle) — Nepal

This uses seeds of pumpkin, porilla, sesame, and niger. The seeds are roasted and ground with sun-dried chilli, ginger, and salt. The ginger and chilli can be roasted or deep-fried before grinding. A little concentrated lemon juice is added and well mixed. Powder is stored in bottles or plastic press-top containers. It is kept as a condiment.

Sambals (pastes or sauces)

Sambals are Indonesian flavouring sauces eaten usually with rice dishes. Ochse (1931) lists as ingredients: Ginger, and galangal flower buds and juice (ginger roots are bruised with vinegar); salt; sugar (palm sugar or jaggery); chilli, cut fine and pounded; roasted eggplant; turmeric, peeled, grated, and pounded; red onions, often roasted in ashes; shallots (red or green), lightly roasted; tomato, pulped; citrus leaf (pounded), citrus juice; garlic, crushed and pounded; peanut paste (fresh); red and green capsicum (pungent forms), chopped fine and roasted; and shrimp and fish cake (salted, dried, and pounded).

Usually, only 5–7 ingredients make up a specific sambal. Also added are fermented fruit pulp, mango, averrhoa pulp, or tamarind pulp, all pounded to a paste. This paste may be folded in banana leaves and roasted in hot ashes before adding to sambals.

The sambal is mixed, cooked, and sealed in bottles. It is used on vegetables, rice, and fish dishes, or as a side dish.

Sambal Oelek (spicy, thick sauce) Indonesia

Prepare ingredients: 12 tomatoes, juice of 2 lemons, rind from 1 lemon, 20 hot chilli peppers, 1 litre (2 pints) water, 15 shallots, 4 cloves of garlic, 2 tsp sugar, 2 tsp salt, and 1 cup sweet soy sauce.

Scald, peel, and chop the tomatoes into small pieces. Chop the hot chilli peppers. It is wise to use gloves and don't touch your face. I know a chilli enthusiast who one day stirred his huge vat of chilli with his hand and arm. In ten minutes his red and throbbing arm came out in painful blisters. Boil the water and add the peppers, then tomatoes, and simmer for ten minutes. Chop shallots and garlic very fine. Fry these until the onion is clear, in mustard or olive oil, then add to tomatoes, with sugar, salt, soy sauce, lemon juice, and rind.

Boil for fifteen minutes, reduce heat, cover and simmer for four to five hours. The sauce should end up about as thick as jam, and sets in the jar. If in doubt, add more sugar to preserve.

Stored in a cool place, this sauce will keep indefinitely, especially if oil or hot wax is poured on top before sealing.

Choice of other ingredients: 3 tsp of grated ginger, 3 tsp of dry or grated turmeric, peppercorns, or mustard seed (bruised).

Chilli Sambal Indonesia

Grind 4 tsp of hot dried chillies in a blender, then add 2 tbsp of tamarind paste, and ½ tsp of salt. "Dilute" with peanut paste if too hot. Store well stoppered.

Garlic, soy sauce, lime juice, brown sugar, ginger, and tomato paste are optional additions to sambals.

Chinese Hot Red Oil

Heat 100 mL (3⅓ fl oz) of edible oil in a pan, add 1½ tsp of ground hot chilli, stir in then let the oil cool a little before straining it through muslin. Store closely stoppered as a hot addition to soups and Chinese rice dishes.

Basic Fish Sambal

Use for satay and fish grills (freshwater or marine species). The fish to be grilled are lightly slashed and marinated in water, (just covering them) containing tamarind paste, lime juice, and a little salt and sugar. Leave in marinade ten to fifteen minutes, then pat dry and sear the fish in hot oil. A basic paste is then made.

The paste is made by the pounding together of several herbs and spices; fresh leaves of coriander; 3–4 small onions; 4–6 cloves of garlic; a few large red chillies; small pieces of ginger and galingal; about ½ tsp of salt; ½ tsp of shrimp paste; 2–3 tsp of curry powder, aniseed, dill, and cumin seeds; and the tender bases of lemon grass leaves.

The seared fish are either cooked in this paste; coated with the paste and wrapped tightly in banana leaf which is steamed thirty minutes; or slowly grilled over an open fire.

With ground, roasted peanuts added to the above paste, it is used on satays (pieces of squid or prawns grilled on skewers). The satays are covered with the paste, which is bubbling on the table over a flame.

With some water added, the paste is used to baste grilling fish or poultry.

With salt added, the paste is fried and stored as Sambal Ulek, and used as a condiment.

Satay Sauce

For skewered fish, shrimp, or shellfish.

Prepare ingredients: 6 stalks lemon grass; 600 gm (1⅓ lb) galingal; 15 red chilli peppers; 10 cloves garlic; 20 small red onions; 6–8 tbsp curry powder; 3 tbsp shrimp paste; 400 gm (14 oz) dried prawns; 454 gm (1 lb) peanuts; good amount black pepper; and 8 tbsp five-spice powder (star anise, fennel seed, cloves, cinnamon, and pepper).

Pound to a paste, moistened with peanut oil, and fry five to ten minutes in rendered lard. Add water, sugar and salt to taste. Place in a boiler (boiling) on the table and cook skewered fish as needed.

Serve with white bread, rice or cucumber.

Garam Masala (traditional curry powder)

Prepare ingredients: 20 gm (3¾ oz) brown cardamom, 20 gm (3¾ oz) cinnamon, 1 pinch of cloves, 1 pinch of black cumin, 1 pinch of mace, and 1 pinch of nutmeg.

All ingredients are ground in a mortar or coffee grinder, sieved, and stored in an airtight bottle. Good only for two to three weeks, so keep the quantity small.

Other curry powders use ginger, mustard seed, coriander, fenugreek, black pepper, chillies, poppy seed, garlic, cardamom, and cinnamon.

Coriander seed, roasted and pounded, is about 50% of the bulk, and for very hot curries, chillies are brought up to half the bulk! These hot English curries are very coarse and not traditional.

Masala (a Parsi recipe)

Blend or grind together: 6 red chillies, 1 tsp roasted cumin seed, 6 green chillies, 1 tsp roasted coriander seed, 1 clove garlic, 6 cloves, 2.5 cm (1") piece of ginger, 1 piece of cinnamon, and ½ tsp turmeric powder.

Vinegar can be used in blending. Masala is used as a general spice in achars and curries. This mix can be salted to keep.

Alternative additions for curries are grated coconut, or mashed peanuts and grams. This masala spice is added to fried curries before main ingredients (chicken, prawns, fish, eggs) are added to a covered pan. For immediate use, onions can be blended into a masala, and then covered with oil (usually mustard oil) in an airtight jar. Coconut and pulses are added just before frying.

Mustard (a full treatment can be found in Mabey, 1984)

Mustard pungency arises when crushed seed, mixed with water, releases glucosides (sinigrin and sinalbin) which, affected by the enzyme myrosin, forms the pungent vegetable oils which also occur in the fatty oils pressed from heated mustard seed in India.

Commercial mustard is made from the fine-milled seed flour, or from bruised seed, salt, a little cold water (for enzyme activation) some wheat flour, vinegar, wine or beer (yeast and bacteria) and such additives as honey, spices (chillies, peppercorns, fennel seed, turmeric), and horseradish. Green mustards are made with leaves of coriander, chives, nettle, etc. There are many possible extra flavours from citrus and/or spices, etc.

Many modern mustards use the whole seeds, barely bruised, with wine and flavours. These seeds have more volatile oils than the milled seed. A variety of French mustards are made by grinding the soaked seeds into a rich porridge with wine, salt, etc., but fine mustard flours are usually stored dry and made up with water and vinegar as needed, and are far more pungent. This is still the tradition in the UK and Australia.

Chanwa Dahl

Mustard seed and flour in India (especially white mustard) is known as chanwa dahl. Small limes, citrus, or peppers and chillies are covered with mustard oil and stood in the sun for a week or so before screwing down tight in jars. Mustard mills are of thick, heavy, smooth granites, not ridged like flour mills.

Prepared Mustards

- Coarse-grained—Blend 1 tbsp salt, 2 tsp pepper, and 1 cup wine or sherry until smooth and well combined. Coarse-blend or bruise yellow and black mustard seed in 1 cup of oil, and 1 cup of vinegar. Combine these mixtures with a few chopped herbs of choice. Screw down tight in sterilised jars and allow to mature several days in a cool place.
- Smooth-grained—Combine 3 tbsp or ⅓ cup of dry mustard powder, plus 1 tbsp sugar, ½ tsp salt, 2 eggs, and ⅔ cup of vinegar in a small saucepan, (or the eggs can be beaten with the vinegar and slowly poured in) and gently heat and stir until thick. When cool, 2 tbsp of oil is stirred in, and the mix placed in small sterilised jars to cool thoroughly.

Optional additions to cool mustard are roast fennel or dill seed, diced capers, or diced chilli.

12.5 SPICES

Nutmeg and Mace

Mace is the scarlet net (aril) that covers the nutmegs when they fall from the tree. The mace is taken off and dried separately on trays or sheets; sun-dried to yellow over a period of up to six months. The nutmegs are spread out on wooden trays and dried indoors until the shaken shell rattles, when it is removed by a hammer tap and the nuts or kernels graded for size and quality.

Black Pepper

Spikes of fruit are harvested when a few of the green peppers turn red. These spikes are sun-dried, usually on concrete slabs, for four to six days, or are hot water blanched and dried for three to five days. (Blanching is at 82°C [180°F] for two minutes, when the silver skin splits as the green berries turn greyish.)

When well-dried, berry skins are rubbed off and screened and winnowed to remove any stalks, then placed in water where the floating, immature berries and leaves are removed. Re-dried thirty-six to forty-eight hours, they are packed in plastic pillows and stored dry and cool.

There are three species of plants grown to provide seeds for mustard. All are of the family *Cruciferae*:

- *Brassica niger*, black mustard seed (actually deep red-brown) is cultivated throughout Eurasia. This is the basis of very hot and pungent "Hot English" mustards. The plant is large, up to 3–4 m (10'–13') and usually hand-harvested. Traditional mustards are of this plant, often hulled and sieved to remove any husks, hence yellow in colour.
- *Brassica juncea*, brown mustard, is a more easily harvested annual field crop, often 0.5–1.0 m (1½' –3'), and although it is now the basis of many mustards in Europe, and is widely grown for oil in India, it is definitely milder than *Brassica niger*.
- *Synapsis alba*, white mustard, (actually large yellow seeds) has very bulbous pods, and is grown in Europe and the USA. The flours or seeds are used in blends, mostly in the USA mustards, but are forbidden as an adulterant in France (Dijon mustard)! The large seeds of white mustard are sprouted with cress seed as a salad sprout, which are clipped by scissors from kitchen trays containing fine soil.

The mustard oil of India and Asia combines all three mustards also:

- *Brassica campestris*, wild or field mustard.
- *Brassica napus*, rape (grown with grains as forage, or rapini as a vegetable). Usually pressed for canola oil.
- *Brassica rapa*, turnips and radish, etc.

Mustard oil is used to cook vegetables and fish, and also to make preserves of citrus and sour fruits. Mustard itself is a powerful preservative, and aids digestion of fatty foods by the emulsification of fats. Rapeseed oil is called canola oil in modern times.

Five-Spice — China

This is a fine-ground or blended mix of star anise, cloves, Szechuan pepper (*Zanthoxylum simulans*), fennel seeds, and cinnamon. Grind to your taste, and seal well to keep.

Seven-Spice — Japan

A coarse-ground mix of chilli, mustard seed, poppy seed, dried nori, sansho leaves (*Zanthoxylum planispinum*), sesame seed, and dried mandarin peel. According to Grigson (1991) the sesame seed is dry-fried before grinding.

Cinnamon — Sri Lanka, Seychelles

Cinnamon is the stem of the bark of *Cinnamomum zeylandicum*, peeled as quills from the coppice shoots of the trees, which are hand-pruned every two years, and harvested from three years of age (twice a year in monsoon lands). They produce for an indefinite time, reaching full production at twenty years, and are usually re-planted at forty years, or a staged re-planting is made over ten to fifteen years.

Shoots are wilted under bags overnight, then the leaves and twigs removed. The bark epidermis is scraped off to leave a smooth golden under-bark, which is vigorously rubbed with a smooth brass rod about 1.5 cm (½") diameter by 40 cm (15") long, or lightly beaten with slender bamboo to loosen. The shoot (about 1.5 m [4'9"] long) is then slit on both sides, and the two halves of the bark are carefully taken off in one piece (by experts).

These pieces are rolled, fitted inside each other, and telescoped end to end until several "plys" are built up. Again, the bark is covered and let dry for a day, then placed indoors on battens, or in midair on four tight rope supports. When the bark starts to curl into quills, they are brought outdoors for a day or two to thoroughly dry, and are trimmed to the standard 106.7 cm (42") trade size. Broken cinnamon and trimmings are powdered, or distilled for spice oil. (Data from *Geo* 13 [2] pp 34–41.)

Cinnamon is an essential spice, and the oil is still used as a toothache remedy and for perfume oil. As for the peeled stick, twigs, and leaves, farmers in the Seychelles near Mahe assure me that these are preferred mulch for their gardens. The spice cassia (*Cinnamomum cassia*) was also prepared in much the same way, but is less used today.

Tamarind (pulp)

Used widely in curries as a sour spice. The pods are peeled and sun-dried, pounded to free them of seed and fibre, mixed with salt and formed into balls or flat cakes. They may then be steamed to dark brown and kept in pots.

Vanilla

Vanilla beans, picked when the bottom centimetre (⅓") of the pod turns yellow, are variously cured:

- Pods are stored indoors to brown off for ten days. Then the wilted pods are given ten to twenty days' sun exposure until an oily sheen appears. When pods near a moisture content of 30% of original weight, they are given one hour a day in the sun to bring them to correct weight (30%) before grading, bundling, and storing in airtight containers for three to four months. They are then packed in heavy paper and cardboard cartons for sale or storage. The process crystallises dark vanillin in the pods.
- Pods are dipped in hot water at 57–88°C (135–190°F) for two to three minutes, sweated in a heap, and then dried.
- Pods are exposed to sun two to three hours daily, folded in blankets to sweat until next day, and when the beans are dark brown they are dried under cover for two to three weeks.
- The beans are chopped, exposed to artificial heat, sweated, dried and percolated or soaked in 35% alcohol by volume in water to extract vanilla essence. Any cured bean can be chopped and extracted in 35% alcohol.

CHAPTER THIRTEEN

Agricultural Composts • Silages • Liquid Manures

13.1 COMPOST

Composting is a skill common to many gardeners who deliberately assemble all of the totally dry materials for a compost heap, build a pile, add whatever minerals are needed (some to prevent release of nitrogen), water it, and get a good result in ten or so days. The water added should be about the same weight as that of the dried material. A good "container" can be made of straw bales, and about a cubic metre (35 ft^3) is the smallest successful heap.

Some people use the heat of compost in a glasshouse or house, and a few use the carbon dioxide released from the compost pile to increase glasshouse plant production.

The early colonisers of dead plant material are sugar fungi (*Saccharomyces*). Their populations flare up in compost piles, competing for oxygen, nitrogen, and minerals. Their metabolic heat cannot easily escape. Most other organisms are killed by heat (60–75°C [140–167°F]) and the exhaustion of oxygen and nutrients by the few thermophilic (heat-loving) fungi who can stand these conditions. Both the mycelia and spores of other fungi are destroyed over ten to twelve days. Only if the heap is allowed to thoroughly cool can other micro-organisms like earthworms and large insects invade.

Fungi require nitrogen (about 0.5% [e.g. 5 gm per 1,000 gm] of dry matter), phosphorus, and potassium to do their work. Nitrogen is essential for the fungi to increase their own growth, so that the rate of decomposition of plant wastes depends on nitrogen supplied to the micro-organisms. Green crops contain more than enough nitrogen for their decomposition to soil, whereas dry materials or old leaf take their nitrogen from the soil itself. So we "turn in" green crop, but leave dry plant wastes and cellulose (carpets, cardboard, old clothes) as a mulch on top of the soil.

13.2 SILAGE

In cold countries such as the UK and northern Europe, grass is cut and stored in the autumn as silage. Silage is the anaerobic ferment of forage greens for fodder, chopped and compacted, and sealed in earth pits, plastic bags, or concrete silos. Microbial action preserves the fodder by conversion of sugar to lactic acids. In the process, proteins and vitamins are also produced.

The green fodder can be wilted before making silage, to reduce liquids. There are plentiful natural yeasts and bacteria plus plant enzymes to attack the sugars. If molasses is locally available, farmers may sprinkle the layers with a strong solution of molasses, and of course there is no harm in sprinkling salts and trace elements for nutrition. The right bacteria are encouraged by anaerobic (no air) conditions and the absence of free water (hence the advantage of wilting). Careful sealing is essential to good silage, which is an ideal winter fodder for livestock from rabbits to milk cows. (*New Scientist*. 22/29 December 1983. pp. 918–923.)

It is essential in the preparation of silage that any drainage liquids, which are evil-smelling and which can badly pollute streams, are diluted and sent via a long drainage channel through reed beds and trees for disposal.

In recent years, farmers have mixed microbial inhibitors in silage grasses or legumes in order to preserve more of the sugars, which may be 15% of the dry matter, and which form an easily digestible energy source for ruminants and bulk feeders. This has resulted in more milk, fats and body weight in animals fed silage. Milk yields have increased by 11 kg (24 lb) per day.

Thus, silage can be variously manipulated to produce just vitamins, or more sugars to increase milk yields, or sugars can be added in the form of molasses. As silage is now made in sealed plastic bags in the field, a variety of mineral, vitamin, and food additions can be tried. Urea-molasses mixes have long been used for fodders high in fibrous materials or in lignin and cellulose (paper, sawdust, bark, crushed cane, sorghum and grain stover, wood chips from poplars etc).

Feeds

Cattle fed on protein-free, grain-free diets can build their bodies and produce milk in no way inferior to cattle fed special concentrates of high-protein foods. They just grow a little slower and produce 25% less milk than pampered cattle. All ruminants need nitrogen as urea, uric acid or ammonium salts. A friend (M. C. Pereira from Nepal) provides cellulose and nitrogen to his milk buffaloes by soaking rice straw in the sludge tank of his biogas unit, where urea, bacteria, and yeasts increase the digestibility of straw. Rack-dried, it is fed to buffalo with coppiced legumes as their main ration.

Cows can be fed on molasses, urea, and cellulose and will remain in good health even if the cellulose is from sawdust, cardboard or paper. Sucrose and starches provide energy for the animals, but many agricultural wastes contain cellulose: citrus skins (15% of feed), cocoa residues, coffee husks, overripe bananas, straws of all legumes and grains, wood bark, and grain husks, bagasse, dry sorghum leaves and palm wastes. Sorghum, maize, and sugar cane stalks in India are fed to cows. Sugars at three to five carob pods per day are provided to cattle in Afghanistan and goats in Western Australia, and pod legumes of honey locusts and *Inga edulis* (ice-cream bean tree) produce masses of sugary food excellent for cattle fodder, as do sweet oaks, chestnuts, even mango trees and avocado on range, as starches and oils.

It is simply a waste of energy and good food to feed cattle or pigs on edible seeds and fish cake. It is better to enrich the range and to grow backup forages for hard times.

Fish Silage

Fish silage is prepared, often at sea, from viscera, trimmings, heads, and blood wastes. An anaerobic process, it uses low pH either from inoculation with *Lactobacillus plantarum* or from the addition of formic, hydrochloric, or sulphuric acid. The wastes are minced to 1 cm (⅓") and frequent stirring is advised, and the pH is reduced to below 4.5 to prevent spoilage. At from 10–21°C (50–70°F), viscera or bones decompose in three to four days or two to three weeks respectively. Often, the floating oils are separated and sold if the silage is not stored.

The silage is dried (often using heat from engine exhausts or cooling water) and neutralised with lime, for sale as fertiliser or animal foodstuff. The dried materials can be 80% protein. No salt is used if rapid acidification is achieved after mincing.

With *Lactobacillus*, molasses is mixed to rapidly lower the pH to 3.0 or 4.0 in three to four days (molasses at 5% weight, adjusted for temperature). Solar drying of fish silage is possible, but may need gas-fired back up.

The high-protein fish silage is a valuable mineral and protein source for domestic animals, game birds, and (when neutralised) even humans. However, it is not ethical or necessary to degrade good pelagic or fresh fish stocks to protein powders for raising pigs or chickens. In fact it is a degradation of food via an energy-wasteful process in large factories, where the oils are a major pollutant (Mexico).

Such processes should be outlawed to save fish stocks and energy. But silage based on true wastes and dried via solar collectors or waste heat is a good idea even for small boats at sea. For feed mixes with grains, 15% of salt can be added as a safety factor to prevent *Botulinus* occurring, or proliferating (plus low pH).

Using Fish Wastes from Processing

Fish wash-water, scales, bones, or fins are all excellent garden manures or good for watering crops such as melons or cucurbits. For human food, a small proportion of fish backbone can be thoroughly oven dried at 66°C (150°F), ground and blended with dill or celery seed, dried chilli, and salt as a dry condiment for fish or meats, giving us calcium and phosphates as well as trace minerals. Larger quantities, boiled, pressed, dried and ground, can be added to chicken food, or stale fish hung to feed chickens on fly larvae.

Unused fish offal (guts, skins, scales, heads, and fins) can be buried under garden mounds for rhubarb, or fruit tree fertilisers. Crab, and shrimp shell can be crushed, or simply buried, as can the shells of oysters, abalone, or scallop.

In short, every edible scrap can be used (with some bones) for human food or gardens. Spoiled fish or viscera can go to vegetable production; and boiled, pressed, dried, and ground meal can amount to 5–10% of feed for small livestock. End products feed people.

13.3 LIQUID MANURES

Liquid manures, whether from strong manures of pigeons and chickens (which need dilution for gardens) or from a ferment of wet manures and leaves from such plants as comfrey, are made in water-filled drums. The method consists of placing 9–18 litres (19–38 pints) of leaf, manure, or seaweed in an open mesh bag, and hanging it in a drum or barrel containing about 180 litres (48 gal) of fresh water. Lime, trace elements, sulphates, and phosphates can be added. The pH needs to be about 5.0–7.0. Every three to four days this "tea" is diluted three to five times and watered on young plants. The drum is topped up with fresh water until the solution is weakened, when a fresh lot of bagged material is added (about twice a month). The old material is mulched or added to the compost pile.

13.4 USE OF SEAWEEDS

Liquid seaweed extract is made from the brown intertidal seaweeds (*Macrocystis, Fuscus, Ascophyllum*) by first chopping and then subjecting the mass to hydrolysis with dilute hydrochloric acid and then a sodium carbonate wash under a few kilograms (pounds) of steam pressure. Like any agricultural extract, additives suitable to the product (urea, chelated iron, or manganese) are added before the concentrate is bottled. Such liquid extracts are used as leaf sprays on high value crops.

Seaweed meal, of the same species, is dried first in air, then in gas kilns, and hammer-milled. Its nutrients are not so quickly available, and so the meal is added as a longer-term soil amendment under mulch, in compost, and as a surface dressing for high value crops. Use about 100 gm (3½ oz) per square metre (10 sq ft) or only 25 gm (1 oz) per square metre (10 sq ft) in good loams.

Liquid seaweed is used at a ratio of 1:100 to 1:1,000 with water. Both liquid and powders supply almost every trace element, and therefore act to make many forms of enzymatic process efficient.

For long storage of extract, 1–2 parts per 1,000 by volume of formalin is added to prevent ferment (1 tsp per 4 litres [1 gal]). Alginates in seaweed extracts form gels which assist soil crumb structure and water-holding capacity. Fresh and powdered seaweeds provide vitamins, and assist ferment processes in nuka boxes, fermented mushrooms, and so on. Formalin can also be added to containers of dry seaweed powder for agricultural use.

Seagrasses

The seagrasses, all estuarine or in shallow, sheltered bays, are not algae but true flowering plants. *Zostera, Heterozostera, Amphibolus* and *Posidonia* are typical genera. They wash ashore in autumn gales, and I have collected them for mulch when I lived by the sea. Washed and composted over ten to fifteen days, and the outer layers turned in four times, they form an excellent fire and pest-proof insulation, loose, packed, or bagged, and have been so used in Scotland and Tasmania in walls and ceilings. It is of course, expensive to purchase, but very cheap to make.

The desert Amerindians of Baja, California gathered the seed of *Zostera* for grain, and cattle eat the fresh or part-composted fronds. Like seaweeds, the larva of kelp flies bred in seagrass masses on beaches provide excellent fish food, as do the amphipods who swarm in the beach masses in autumn. The undersea seagrass beds support many fish species, and are refuges for fish fry.

13.5 SEED RECOVERY

For tomatoes, tomatillos, aubergines, in fact any *Solanaceae* fruits and many cucumber family fruits the seeds are best collected after ferment. Spoonfuls of fruit pulp from several mature plants are just covered with water in an open bowl over three to seven days (cold to hot water) until bubbles and a sour smell result.

Fermentation enzymes thin the pulp, and the acids destroy disease organisms. Gelatine on the seed is dissolved, leaving the seeds dry and non-slippery. Floating seeds and pulp are carefully washed away; seeds are stirred occasionally to prevent sticking and the heavy seeds sink, are sieved and dried on newspaper.

Be sure to overripen all cucurbits for six to eight weeks before removing seeds, as these mature in the fruit even if that fruit was originally a little immature. Much of the cucurbit seed collected from fruit is eaten, and often field-dried in the shell of the fruits.

13.6 FIBRE RETTING

With flax, retting (rotting) can take place in the field or in tanks. As with all fibres, the idea is to decompose unwanted plant tissue between fibres, so that pectin and lignin are removed. Bacteria are anaerobic, butyric-acid producing, and must be watched so that the material to be rotted is removed from pits or tanks before the fibres are attacked by cellulose-degrading bacteria.

Jute

A few jute plants (*Corchorus capsularis, C. olitorius*) can provide garden "string". Commercially, jute is grown in the rainy season of the tropics (24–32°C [75–90°F], humidity 90% or so). It is sown and harvested annually, and is sown in the pre-monsoon showers by broadcasting on prepared fields. Early sowing is essential for good fibre as only adult plants stand flooding. Lime, cow-dung, and ash can be used, but no artificial fertilisers, especially not nitrogen. At full height and with half the pods formed, the plants are cut and harvested.

They are bundled, stood at first upright in water for two days, and later tied as bundles in two to three layers, but a little apart. Then they are sunk by poles and weights, and fermented. At twelve to twenty days, when the fibres are not rotted but the stem is, the bolts are beaten, and the stem shaken in water by the base fibre until leaf fibres are free.

Fibres are washed, cleaned, hung to dry for two to three days and later bundled for sale, or made into rope, hessian, sack-cloth, etc. The woody centres are used for stick fires. Flax, sisal hemp, and even banana leaves are all rotted in much the same way for fibre.

13.7 DYES—WOAD

- Indigo—(*Indigofera* spp.) and woad (*Isatis tinctora*) have traditionally been used to dye cloth. Both plants are bundled and sunk in vats to oxidise, when the dyes are released in quantity. Agitation or stirring after twelve hours is usual, and next day the dyes are drawn off and the fibres sieved or filtered out. Dyes can be allowed to settle, and pigments drained and dried for sale. Indigo cake is fixed with urine as a mordant and cloth dipped and dried. It is this process that produces the oft-noted smells of the dye fields.
- Purple—as favoured by the Romans, was extracted from up to seven species of lichens (e.g. *Roccella tinctora*) and rare whelks. (*New Scientist,* 2 February, 1991. p.14).
- Red—from the madder (*Rubia tinctorum*) and the shield louse (*Kermes vermilio*) (the latter an insect source) were both used by Vikings.
- Yellows and browns—came from various plants (tannins from oak, and acacia) and onion skins. Many of the dye extraction processes are lost, but most involved (as they do today) simple boiling for dye extraction, staining, and soaking of cloth or fibre first in the dye, and secondly in a fixative or mordant, which can be aluminium salts (alum), urine, or iron salts for tannin dyes. The mordant fixes the colour, hopefully permanently, in cloth or rope.

Certainly, woad was produced by ferment, but for many plant dyes, boiling suffices.

13.8 FERMENT FOR GOURDS AND LOOFAHS

The loofah (*Luffa cylindrica*) is grown on a trellis until mature, when the pod feels light and yellows at the flower end. They are then rotted in tanks of water or in changes of water until the skins and flesh disintegrate. Then they are washed and kneaded until the seeds and pulp are gone. They are bleached with hydrogen peroxide or domestic bleaches, and left in the sun to whiten. Widely used as bath sponges, they can also be used for insulation and shock absorbers, or as fuel and oil filters, even tough sandals.

13.9 BUG JUICE (WHOLE BUG FERMENT)

A strange but apparently effective insecticide (for any insect) is to gather the target species, especially those appearing sick or diseased, but in any case all those caught, and to blend them in water. Allow them to ferment there two to three days, then use a very dilute solution (ratio 1:1,000 bug juice to water) to spray on plants. It is said to be effective for any insect, and of course would spread pathogens widely, plus any decay organisms. Bacteria, viruses, and parasitic fungi or nematodes could all be involved. I am sure we could not survive such treatment!

13.10 INOCULATING PLANTS

The root and soil associates of higher plants are algae, fungi, and bacteria. In rice paddy culture, where 7–8 cm (2¾–3") of water covers the soil, nitrogen fixation and soil nutrition may depend on two associated plants, blue-green algae, and a specific algae associated with a water fern (*Azolla*). *Azolla*, with its associate algae *Anabaena azollae*, is a floating fern used in rice and taro, with the coppicing tree *Sesbania* as a pond side mulch. However, see notes on nitrogen fixation versus nitrogen scavenging in section 14.8.

There are many genera of nitrogen-fixing blue-green algae in paddy fields, especially when there is a good phosphorous status and the pH of the soil is adjusted to 7.0–7.5 using lime. Algae, raised in flat trays with phosphate and lime, are dried and the flakes stored in bags for field inoculation. There they are credited with 20–30 kg (44–66 lb) per hectare of nitrogen if phosphate is sufficient.

As the plant is a phosphorous accumulator, the cultured pond can be very high in phosphorous. Unlike algae, *Azolla* may die out if the soil is dried and culture ponds must always contain water. Excess heat kills *Azoll*a, and it may be supplanted by blue-green algae in hot periods.

The field can be kept "friendly" for algal production by adjusting pH, avoiding herbicides, and keeping good levels of phosphates in the soil and water. In the field, heavy biocide use kills algae and root associates that fix the nitrogen for plants, a big factor in incurring fertiliser costs in rice crops, so that only natural methods allow soil and water microbes to fix crop nitrogen.

In many areas, *Azolla* is not used due to a lack of knowledge about it, but in actively researched or well-informed areas, *Azolla* is well established. Where ducks and fish are also grown it is valued as a food for these livestock, who of course manure the fields. *Azolla* acts as a shelter for amphipods, snails, and shrimp eaten by fish and waterfowl.

In the family *Dioscoreaceae* (the yams), the long pointed leaf tips are folded to shelter glandular chambers lined with hairs, in which live bacteria secrete a gel or slime. Infolding of the tissues enables the plant vascular system to carry away nutrients produced in these factories, and plants deprived of bacteria produce a very few yellowed leaves (cf. the rampant green foliage of most yams). Just what nutrients are involved is not yet known. (See *New Scientist*, 29 October 1987, p 38.)

Similarly, leaf chambers or "domatia" in *Coprosma* leaves, at the base of veins on the underside of the leaves, contain colonies of bacteria and algae, again not proven to fix nitrogen, but inoculated plants produce glossy foliage, and other plant groups may prove to have similar leaf domatia or leaf glands hosting colonies of symbiotic organisms aiding the nutritional health of the host plant, either as bacteria-fixing nitrogen or organisms scavenging nutrients from atmospheric sources.

It is therefore beneficial to "leaf inoculate" new or cultivated plants, using the leaves of healthy field plants to do so, by washing them in a water bath applied to the plants to be inoculated. This may be important for yams.

Legumes can produce nodules at their roots which contain rhizobia (root inoculants). Nitrogen is "fixed" by the rhizobia, and is absorbed by the plant. If the rhizobia are missing, legumes do not do as well, so it is important to make sure the seed is inoculated with its particular rhizobia. Rhizobia can be bacterial or fungal or both. Many, if not all, plants foreign to an area benefit from inoculation by rhizobia (soil from below healthy plants).

Insofar as general yield is concerned it is rare nowadays to find legume seed not supplied with inoculum in new areas, and few areas now seem to call for inoculants (insofar as nitrogen fixation is concerned). The introduction of really new crops (blueberries into Australia, macadamias into India) may call for an initial inoculation, but for most micro-organisms it is better to supply lime to alleviate acidity and phosphate (not in excess) for nutrition, and to rely on the numerous wild or field strains to find and colonise new plant species.

Almost all authors agree that there is no miraculous effect on plant yields from inoculants, and scientists often find new inoculants do not compete well with active field species anyhow. However, genera of plants completely new to an area may not thrive at all unless inoculated, and this does not apply only to legumes.

Many higher plants (*Casuarina, Banksia*, etc.) can accumulate both nitrogen and phosphates at root nodules (in this case the nodules are woody, not soft as are the nitrogenase nodules). The *Frankia* root associates are invaluable to *Casuarina*. Thus, most plants have a specific leaf, stem, root, or intracellular associate (fungal, algal, bacterial) that is active in bringing nutrients to that plant. The importance of inoculation of "new" plants with their microbial associates is now realised, and many firms supply cultures. Enquire from your local department of agriculture for supplies of inoculum for a specific legume or crop. (*New Scientist*. 18th June 1987, p. 59, 29th October 1987, p. 38, and 7th May 1987, p. 29.)

13.11 INOCULATING ANIMALS

Esko Nurmi (Finland) has developed the idea of inoculating artificially reared chickens with cultures from the gut caecae of healthy free-range adults, thus resisting harmful *Salmonella* infestation by "competitive exclusion" (the Nurmi effect). Cocktails of up to forty-eight strains of caecal bacteria, especially the *lactobacilli* and gram-positive cocci, and harmless strains of *Salmonella* can be used. Aviary owners, rearing birds, often "inoculate" chicks of new species by mixing a little of the droppings of adults in the food of their chicks.

Effectiveness is greater if such inoculations occur before infection challenges. The simple methodology is to spray watered cultures of healthy gut contents on day-old chicks, and add the harmless mixes to the drinking water of older chicks. Treated animals have a very low rate of *Salmonella* infection, and there is a general 40% reduction in infected poultry in the UK (*New Scientist*, 17 September 1987, p 55).

Similarly, calves inoculated with the rumen contents of healthy free-range calves show better resistance to infection and a better ability to digest plant foods. At least four rumen micro-organisms are now spread through sub-tropical Australian herds to enable them to detoxify mimosin, an alkaloid in *Leucaena leucocephala* shrubs that otherwise causes severe photosensitivity in cattle. The inoculants were isolated from Hawaiian cattle, who showed no such sensitivity. These results suggest that a wider medical investigation of our gut flora may explain why some of us react strongly to milk, yeast, or grain foods. Children raised in "hygienic" high-rise buildings may lack many inocula for digestion of foods.

Like cockroaches and termites, people and their domestic animals need an appropriate gut flora. Even in the case of sudden changes in food and water quality, we can suffer illness such as diarrhoea or malabsorption of foods. Koalas and many other animals feed their young or allow them to feed on adult faecal pellets to achieve the same result. Caged animals may need such inoculants to resist disease, to utilise a variety of foods; and to detoxify some foods.

In Germany at least, some preparations are offered for travellers to Asia which are said to contain beneficial strains of human gut flora for that region. Because tribal people smear their food with the stomach contents of game animals, and even marinate or cook in stomach juices, they inevitably take in a range of digestive organisms. We, of course, cannot live long in a sterile environment, and normal (not excessive) house hygiene usually ensures that we become eco-compatible with the local organisms.

Sheep Tick Dip

Soft soap and potash or a little lye, nicotine water, and neem oil are all able to kill and dislodge ectoparasites, and are not dangerous to operators if carefully mixed (unlike the chemicals and halogenated hydrocarbons now used). Agricultural sheep dip stations have set up hundreds of pollution points in Australia, and have badly affected the people dipping cattle and sheep. Most cattle dips are near streams used for drinking water downstream. Arsenic, BHC, and DDT are common pollutants.

Rubbing hoops (made of 7.5 cm [3"] pipe) for cattle and rubbing triangles for pigs and sheep can supplant dips if bags soaked in natural oils are bound on to these rubbing posts.

13.12 ORES

The bacteria *Thiobacillus* spp. and *Ferrobacillus* spp. are used to leach copper, nickel, molybdenum, zinc, arsenic, antimony, and germanium from their sulphides, producing sulphates in solution. *Thiobacillus* needs good aeration, a large inoculum in good culture, a range of 20–28°C (68–82°F), and a pH of 2.5–4.5 to rapidly oxidise ores. *Thiobacilli* occur in hot springs. In nature, bacteria concentrate metals such as manganese as metallic nodules on the ocean floor, and may cement minerals around or in roots in sands and soils. Bacterial leaching of low-grade ores is now widely used in mineral recovery programs.

CHAPTER FOURTEEN

NUTRITION AND ENVIRONMENTAL HEALTH

14.1 NUTRITION AND LIFE-ENHANCING ACTIVITIES

The very basis of health is clean water, clean air, and healthy soil, hence a variety of plants and animals also in good health. Enough exercise comes from caring for soil, plants, and animals.

That said, the more we know about nutrition, the better. Many allied sciences are involved in nutritional sciences. See Figure 14.1.

Fish, game, birds, cheese, natural meats, garden and wild foods, fermented food, unfiltered beer, low salt diet, an active life, gardening, and walking certainly plays a part in a healthy lifestyle, as does a quiet life and not too much travel! Losing the traditional way of life, buying most of one's food, and too much comfort seems to bring on early death and diseases. If we want to be healthy, there are things we can do about it. Good practices in gardening, an active interest in life, a balanced diet, and some exercise all contribute to health. The steps are relatively simple.

Gardening

For soils, add mulch and trace elements that are lacking. Bring the pH to 5.5–7.5 for optimum growing conditions. Slight acidity is not harmful. Mulch to retain soil moisture, and add compost to ameliorate the soil.

Hygiene

Hygiene is particularly important in areas where faecal material may contaminate water supplies or food. Boil water if necessary, and use fly-proof pit toilets.

Test soils and water rigorously to make sure neither are contaminated with toxic elements. This is especially important for drinking water. If necessary, collect water into rainwater tanks from the roof. Check bore waters for excess salts and radioactive minerals.

Nutrition

A balanced diet should include fresh fruit and vegetables. This is only possible with a garden, as most fruits and vegetables lose vitamin content soon after picking. Overcooking also leaches out valuable nutrients.

Exercise and Life Interest

This is a natural adjunct of gardening. Building compost piles, trellising, picking vegetables, and mulching are all active processes, and looking after the garden and the health of the family or community provides a varied interest in life.

Every gardener has the best chance of being healthy. To garden is to be patient, optimistic; to live expectantly as if tomorrow matters; and to teach and share with others. Gardeners sleep well, get mild but regular exercise, eat very well of many foods, and are generally good cooks. They browse or snack often from their gardens and eat simply of good food. They live in fresh air and with natural things, and daydream rather than meditate. They are practical, sceptical, and hardy people, and feel self-reliant and secure in themselves. All this leads to good health and a long life. Make your garden and live in it, and help others to do so. This will return power over their lives. Above all, gardeners are content to die, as resurrection from dead material is a daily event in a garden, not a rare miracle guaranteed by some guru! I imagine that the ultimate torture would be eternal life!

14.2 SOURCES OF ESSENTIAL FOOD GROUPS

See Figure 14.2 and Figure 14.3 for essential food groups and good combinations of food.

- Enzymes—ferments, livers, fresh fruits and vegetables, and pawpaw.
- Vitamins—fresh fruit and vegetables.
- B vitamins—milk, meats, and ferments (thiamine is essential for using carbohydrates).
- Vitamin A—fish oils, and livers, yellow vegetables, seaweed (nori). Add a little oil or fat to yellow vegetables.
- Vitamin C—fresh fruit and vegetables.
- Minerals—co-enzymes for many digestive processes. Add to soils as seawater or foliar sprays.

- Iron—green vegetables, iron pots, and liver.
- Zinc—cucurbit seeds, testicles, and oysters.
- Trace elements—ash, dried and roasted bird manures, seawater, and bitterns.
- Fibre—whole grains, brans, fresh fruit and vegetables (**not** juices).
- Fats and oils—milks and butters, animal fats, fish oils, oil seeds, oily fruits (olive, nuts, avocado, and coconut meats).
- Starches and sugars—root crops, grains and nuts, honey, palm and cane sap, and aguamiels.
- Water—coconuts, fresh fruits, saps, clean rainwater, and clean springs.
- Protein—eggs, fish, meats, complementary foods, e.g. pulses and grains; milk and root crops; meat and vegetables.

14.3 WATER AND TRANSMISSIBLE DISEASE

It is not necessary to drink water to be infected by it, as many pathogens (*Giardia*, inner ear infections, viruses and worm eggs) survive in streams, wet sands and mud, or seawater, and can infect us there.

Nor are these the only dangers. Disease organisms breed in air-conditioning towers (Legionnaires disease, cold virus) and are also infective; and to take a shower in Louisville (Kentucky) is like a chlorine gas attack in the First World War. Medical geographers correlate chlorinated water with cancers in their distribution maps.

The more centralised water systems become, and the more well waters or borehole waters are used, the more likely it is that a great variety of contaminants from wastes off farmlands (biocides, nitrates, and fuels) or industry (radioactives, heavy metals, solvents, reagents, aluminium, and various chemical cocktails) will enter the water supply, together with pathogens from animal wastes, processing, or sewage.

Most cities of Europe, the United Kingdom, and the USA have public water supplies that contain a variety of pollutants, many thirty to fifty times recommended health levels (London, Frankfurt, San Francisco). There are up to 1,500 such substances in badly polluted rivers!

As we need such a small amount of water to drink and cook with, then that water should be caught as rainwater. Public supplies can be used mainly for toilets, showers, and (alas) gardens. As I have caught rainwater from my roof all my life, I see no problem with this; nor does any Australian, but we antipodeans are almost alone in the world in doing this. Change will only come about when people realise that city authorities cannot guarantee good supplies, and clouds can!

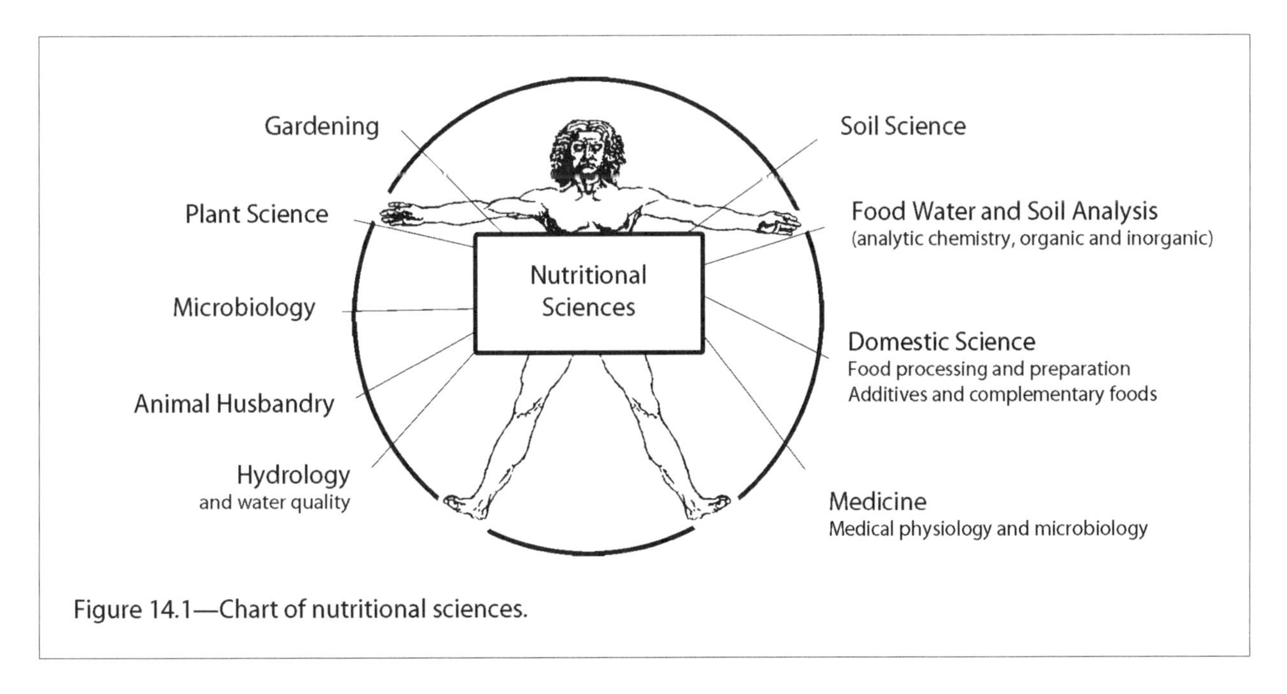

Figure 14.1—Chart of nutritional sciences.

In deserts in Australia, 18% of bore waters (often warm and salty) dissolve radioactive ores, and deliver uranium, thorium and radon to water supplies. In other deserts (California, Botswana), excess fluorine and selenium are dangers, and in old leached soils in Africa and India, excess aluminium is common from leached soils. Thus an analysis of any well or bore waters in deserts is **essential**. Millions of people are dying in India, poisoned by arsenic from deep bores often provided by aid groups.

Clarifying and Sterilising Drinking Water

Whether from silty rain ponds, polluted wells, or muddy and suspect rivers, water for drinking has long been clarified (the process of flocculation and coagulation) by adding various substances to water containers.

These fall into the following categories:

- Earth and clays with high aluminium; kaolin, alum and most morillinite.
- Leaf, stem, and seed juices of land plants (e.g. *Moringa*), and legume seed starches, crushed.
- Standing water for three to four days in settling tanks for silt and organisms to settle. This means using several storage vessels in sequence, and decanting the clear water.
- Filter beds of sand and charcoal, cleaned and roasted when clogged.
- Boiling water for twenty to thirty minutes, which kills most infective organisms.
- Passing water slowly through a bed of such reeds as *Scirpus, Typha, Acorus,* and *Eleocharis.*

(It is probable that this last process gives greater cleanliness, as the roots of water plants are actively imbibing micro-organisms and worm eggs).

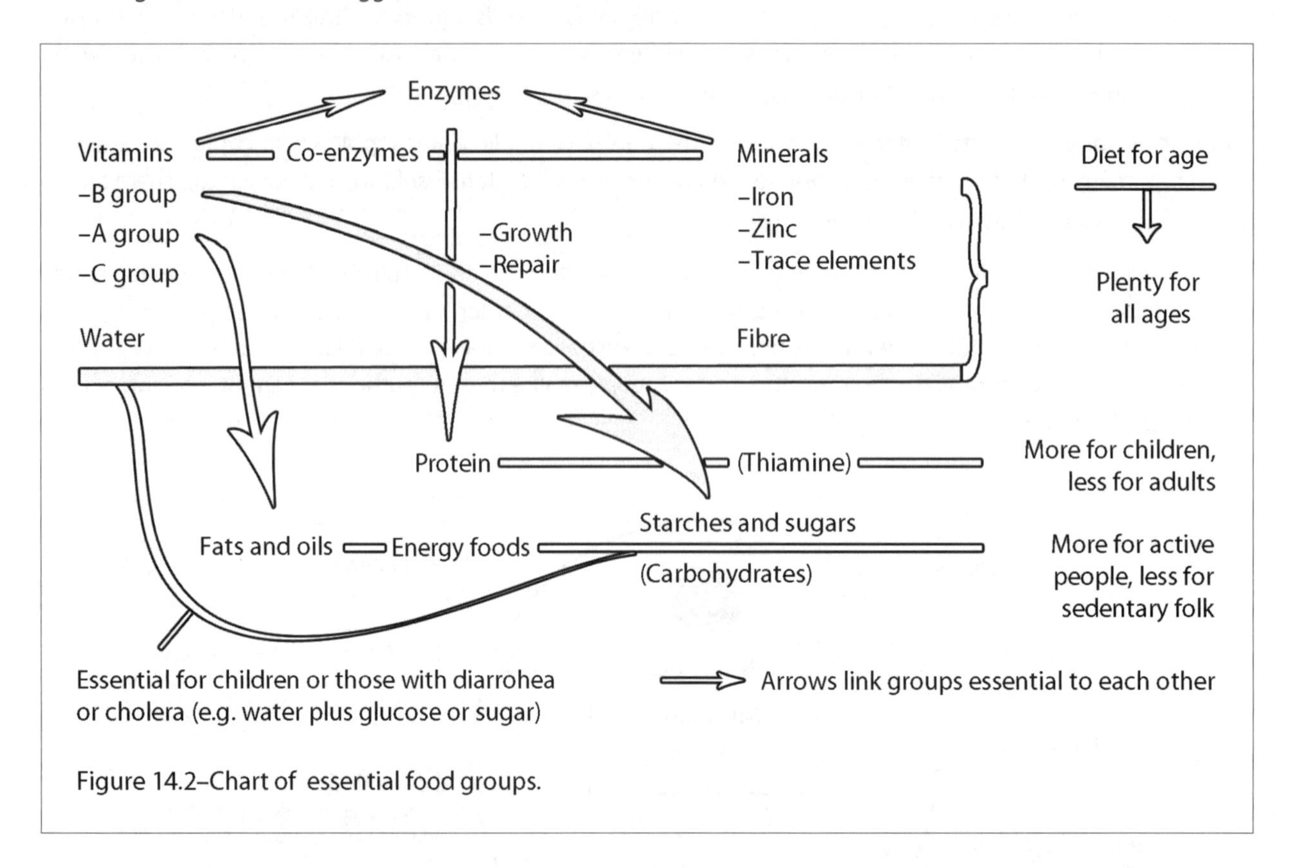

Figure 14.2–Chart of essential food groups.

14.4 HYGIENE IN FOOD PREPARATION

Boil to sterilise milk from untested cows; many of my teenage friends died of tuberculosis. Thoroughly cook pork, beef, and freshwater fish to avoid tapeworm diseases. Fish tapeworm is world wide in ponds and can come to your area, given poor hygiene in sewage. Keep your dog and cat wormed and outside. If you have no work for them, don't keep a dog or a cat, as they eat too much of the world's protein today.

Avoid unwashed or even non-chlorinated salad vegetables from gardens manured with human wastes (as a leaf spray). Avoid watercress where domestic animals have access to the water. Human manure in deep soakage pits is useful (and harmless to us) if applied to fruit trees, bananas, and even tall foliage plants below ground level, as a soak through gravel trenches or pumice.

Several diseases, some very disabling and long-term, pass from animals to other animals or to humans. Tuberculosis in poultry can spread to cattle, thence to milk and humans. Dogs and cats carry some serious diseases and should be regularly wormed by owners, and kept outside. Worming pens may be infected for years. No uncooked offal should be fed to pets, nor indeed to any domestic animals.

Good hygiene should apply to all our domestic species. A careful household buys and maintains disease-free stocks of plants and animals, and takes care with new stock (isolating them until sure of their health).

Cleanliness is essential to all food preparation, and especially important in ferment if unwanted organisms are not to be included. Note—that correct ferment procedures can actually inhibit harmful organisms and reduce poisons in plants. At the optimum temperature on a sterile base, beneficial ferments overgrow any chance contaminants. For this reason ferments are not known as especially risky food. Compared to airline food, and commercial chickens and eggs, ferments are benign!

14.5 COMMON FOOD TOXINS

Natural

- Solanins—found in green potatoes, exposed to light in storage. Keep them in the dark. Rarely causes death or chronic illness.
- Aflatoxin—throw out any old peanuts or nuts before they become rancid or mouldy. *Aspergillus flavus* is the mould responsible. This is a common danger in kitchens (liver cancer can result from eating rancid peanuts or nuts).
- Tetraodontin—a deadly poison in the organs of puffer fish. Avoid these, or go to a very good fugu restaurant in Japan, where your chances of survival are 3,999 in 4,000. (Better odds than travelling in cars!)
- Lathyrism—do not grow, or allow others to grow, the gram *Lathyris sativus* (a small legume seed) for human consumption (as is done in India, where landlords may grow them for workers). Eating the gram consistently over time causes permanent paralysis and eventually death.
- Botulism—most common in poorly-preserved sausages and preserved meats; not common today, but has always to be guarded against. It comes from *Clostridium botulinum*, which can also occur in drying salt lakes. A deadly poison.

Food poisoning is actually on the increase in the western world, due to mass production systems and careless feeding of stock, poorly preserved fast foods (e.g. airline foods) and "agricologenic" (caused by modern agriculture) diseases, in both humans and livestock.

Minor toxins exist in many foods. Some people are now developing allergies due to immune-system damage or genetic intolerances. Factory farms may use biocides, antibiotics, and hormones, and create strains of disease organisms immune to drugs. To avoid this, eat organic and free-range foods.

Added Pollutants

- Sewage and biocides in water—Test water frequently or buy a rainwater tank for your drinking and cooking water.
- Radioactives—Get rid of sources and avoid foods from fallout areas (such as Europe, USA, Russia). Always test for radioactives in milk, water from deep wells, and meats.

  Hand-held geiger counters are available online, starting from approx $150. One suffices for a large area, if on hire. I wouldn't be without one. Many rocks, and granite houses, have high radioactivity levels, as do some bricks made from aluminium smelter wastes, and beryllium sands used in sand-blasting (thorium contamination).

- Agricultural—Spray residues in meat and vegetables, or in seed. Demand organic produce or (better), tested food, as organic practices may not be enough on polluted soils or farms, and labels are just labels, whereas tests are definitive.

14.6 COOKING AND STORAGE LOSSES OF VITAMINS

- Riboflavin—Do not store milk in sunlight (two hours destroys 50% of riboflavin).
- Vitamin B6—Heat sensitive.
- Folate—Heat and alkali destroy folate rapidly. More stable in meat and eggs.
- Vitamin B12—Stable in liver or meat, especially if cooked rare. Can dissolve into cooking water, so make gravies from this.
- Ascorbic acid (vitamin C)—Readily oxidised and is quickly lost in storage of fresh vegetables. Alkali and copper utensils can destroy; minimal loss in stir-fry or fast boil cooking.
- Vitamin A (carotenes)—Stable in mild heat, oxidises at higher temperature. Not easily lost in water.
- Vitamin E—Heat and light sensitive.
- Minerals (trace elements)—Most are fairly stable if least water is used. Iodine can be lost, even in long storage.

Finally, consider making a garden and browsing. This is often the best way to eat asparagus, or carrots, or peas. Also, fast-cook with stir-fry systems.

Cooking has the benefit of releasing niacin (with alkali) in maize, and destroying the trypsin inhibitor in legumes (like soybeans), as well as destroying many poisons in legume seed and cassava, taro, (oxalic acid).

14.7 NUTRITION

"I saw few die of hunger; of eating, a hundred thousand." Benjamin Franklin.

"The only way for a rich man to be healthy is ... to live as if he were poor." W. Temple.

"If you want to know if your brain is healthy, feel your legs." Bruce Barton.

Rather than lacking food in any absolute sense, many societies reduce their access to perfectly good food as a result of religious or cultural bans. Overcivilised societies reject organ meats and some insect protein or products (although rarely honey), due to niceties taught in schools, or in "polite society". Mankind has nevertheless survived and spread because of an eclectic choice of many foods, and the intelligence to apply careful microbiological processes to low-quality but hardy foods, to render inedible foods harmless, or to try any food in emergencies.

Some modern western habits (e.g. the rejection of fish livers and roes) may not only reduce good food by 30% of total fish weight, but also deprive the consumer of vitamins A and B, some essential oils, and proteolytic enzymes.

Malnutrition—The Disease of Affluence

Diabetes and obesity, dental caries and often avitaminosis are the common result of a sedentary life, a diet of refined foods, and low-fibre diets rich in carbohydrates and fats. The more serious results are, for diabetics, a risk of peripheral nerve decay, blindness, and contraction of gangrene. Some 80% of non-accident (traumatic) amputations are caused by diabetes. Impotence in males is a frequent result, as is early death. Alcohol consumption exacerbates vitamin C loss and diabetes. Much of diabetes is diet-generated, a lesser part is genetic.

Obesity not only causes a poor self-image, and often bad complexion and vein enlargement, but also greatly increases the risk of heart and circulatory system disease, strokes, and exhaustion at physical tasks. A fat-mobilisation hormone (FMH) is collected from the urine of starving people (e.g. in Bangladesh) and used as a slimming injection in the USA and other western countries. It is much cheaper to send fat people in exchange for starving people until both are "normalised". The obese would be paying a fraction of the costs they already payout on retreats, treatments, massage, physical workout centres, special diets, medications, and FMH, to fly exchange trips to Bangladesh.

In most situations where food is overmilled, overrefined or scarce, several disorders may occur, as they do in diets of refined plain white rice, white flour, or with too much starchy root foods as staples.

Malnutrition—The Lack of Food

We need a basic sufficiency of calories and proteins daily, of about 2,800 calories and 58 gm (2 oz) of protein, for a man; and 2,000 calories and 36 gm (1¼ oz) of protein for a woman. Excess proteins consumed are converted to energy or eliminated, and excess calories (including those from protein) are turned into fat or burned up by strenuous exercise. Recently, athletes have been allowed more latitude for calories as they need this energy more than sedentary groups. Many poor people manage to work on as few as 1,500 calories. A calorie to protein ratio of 60:1 is recommended by Lappé. Many popular foods (such as potato chips), have a 179:1 ratio, and can create excessive calorie (fat) problems.

Primarily, we should ensure that enough food is being eaten, and this means that enough starchy foods and some protein (potatoes, taro, cassava, bananas, grains and grain legumes, or meat) are available to supply energy at from 1,600 to 2,500 calories per day per working adult. Fresh food and especially fruit is also essential or there will be a lack of vitamins C and A. (Symptoms are called scurvy and pellagra respectively.) Selected green vegetables are also essential.

Lack of food as calories is called marasmus and this may be accompanied by any number of other diseases such as measles, malaria, severe parasitic infection, and anaemic cirrhosis. Marasmus is typified by a cadaverous (skeletal) appearance of the face and body, oedema of the feet and legs, and cessation of menstruation.

Lack of protein is called kwashiorkor. This includes retardation of growth, diarrhoea, dyspigmentation of skin and hair, skin rash over buttocks, thighs and forearms, hair straightening and a steely grey appearance to hair. See Figure 14.4.

In the fourth world we can find a terrible mix of the diseases of affluence (obesity, alcoholism, diabetes) and the diseases of poverty (mineral, vitamin, protein deficiencies, cold or insufficient housing, homelessness, hopelessness). Sometimes the very worst health conditions lie within the first world, as with the Aborigines in Australia. Their misery becomes the source of income to internal bureaucrats, who are never likely to want to solve the problems, and so reduce themselves to poverty!

14.8 THE AMINO ACIDS

For a billion years, bacteria had the earth to themselves, then slowly cells with nuclei and chromosomes developed. These cells themselves developed for about two billion more years, and from them all plants and animals arose. All living things are made of cells, and in every cell, protein is built from the same twenty amino acids. In our bodies are sixty trillion cells, each containing one hundred or more similar cells (the organelles and mitochondria). These are bacteria and provide the energy for our bodies. So we are made up of a cooperative of cells fuelled by independently-breeding bacteria; and so is a plant. All life is essentially similar, holding the body patterns in DNA and RNA, and all of the cells built from just twenty amino acids. Although there are hundreds of amino acids, only these twenty build proteins.

John Portgate, Professor of Microbiology at the University of Sussex, UK asserts that until recent times (*New Scientist*, 4 February 1988), **only bacteria of a few dozen genera** have been proven to have the ability to fix nitrogen. He instances the bacterial genus *Rhizobium* attached to many legume roots in symbiotic association. No other invertebrates can fix nitrogen, nor can yeasts, insects, algae, termites, fungi, or higher plants.

Thus, it is necessary for many plants to enter into a symbiotic relationship with these bacteria. Few animals can enter into any such relationship, and no algae have been proven to "fix" nitrogen; that is, to convert atmospheric nitrogen to organic compounds and thence to protein. Thus, all amino acids are built on a few genera of bacteria, plus a few very efficient scavengers.

The enzyme nitrogenase is peculiar to the bacteria. Although other organisms can efficiently scavenge very low levels of ammonia or nitrates from the air, and a low (incomplete) level of nitrogen can be provided by gut bacteria, most nitrogen must be found in foodstuffs; but about 14% of the dried weight of any bacteria is nitrogen. However, as almost all dust is rich in nitrogen from such sources as: body fluids, skin flakes, microbes, microbial spores, and so on, and dust commonly contaminates water, agar, air, and chemical reagents, such sources as ammonia and nitric acid are "banned" from bacteria testing labs for this reason. "Scavengers" can be mistaken for "fixers" if any such contamination occurs. Note—that lightning and photochemical reactions such as the effect of Iron-Titanium (TiFe) catalysts in coastal or desert sands produce ammonia for scavengers.

To build our body proteins (tissues) we must eat food containing the twenty amino acids that we need. We digest proteins to peptides in our stomachs and to amino acids in our small intestine. They are then absorbed into the bloodstream, which transports them around our bodies so that our cells can take these building blocks and make the proteins of our body. We eat proteins to make amino acids to build proteins. Leaf juices, fruit juices, meats, and fermentation all supply protein to our bodies. If we eat vitamins, we digest them in fifteen to forty-five minutes. If we eat proteins, we digest them in thirty to one hundred and twenty minutes. Thus, if proteins are made by enzymes (a protein working with a mineral or vitamin co-enzyme), it is wise to include minerals, vitamins, and proteins in our meals so that we can digest our food. One without the other will not work. We need about thirty thousand different proteins in our bodies to function properly, and these must be built from our food. This is the essential basis of nutrition. (Data from Kapuler, A.M. and Gurusiddiah, S. 1988, *Peace Seeds Research Journal*, Vol. 4.)

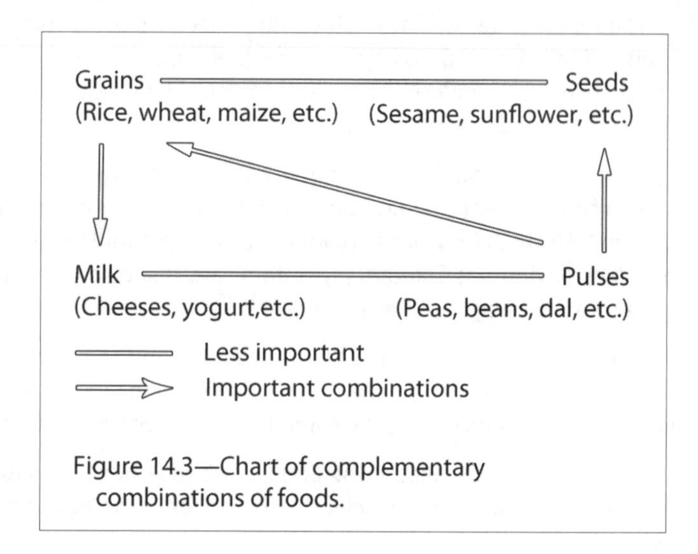

Figure 14.3—Chart of complementary combinations of foods.

Where more than ten amino acids are linked by peptide bands, these are called polypeptides, and large polypeptides are called proteins. We eat plants or meats in combination, so that in total all the essential amino acids are provided. Generally a grain like rice or wheat needs to be consumed mixed with a legume to give our bodies a complete protein. Such foods are said to be complementary. See Figure 14.3.

The amino acids have been likened to an alphabet from which an infinite number of sentences (the proteins) can be synthesised. It is the sequences of amino acids that give proteins their unique characteristics.

14.9 VITAMIN DEFICIENCIES

Water-soluble vitamins—(vitamin C and the B group) are easily lost in washing, soaking, rinsing, and long water-cooking of foods. Thiamine (a B vitamin) is particularly important to enable utilisation of carbohydrates (starches and sugars). As yeasts form thiamine, ferment is important with starchy foods. Local field beers are good sources, as is poi, root-crop ferment, coconut and palm toddy, liver, pulses, eggs, meat, yeast extracts, or whole grains. Fresh fruit and vegetables are the main source of vitamin C.

Fat-soluble vitamins—(vitamins A and D). Natural fats and oils in cream, butter, eggs, fish fats, and oils (particularly fish livers), and some seaweeds (nori) contain large amounts of vitamin A. Children, in particular, need a little of such fats with their vegetables (dark green and yellow vegetables) in order to utilise carotene. Vitamin A forms strong skin cells and assists good sight. It occurs as retinol, found in animal oils and easily absorbed, and as carotene from vegetables, which needs conversion to vitamin A in our livers, hence it is called a pro-vitamin, and is easily absorbed. Cod liver oil supplies vitamins A and D.

Vitamin A deficiency causes ulceration of the cornea of the eye (keratomalacia). Lack of vitamin A is very common in poorer areas, especially in children, who often get little meat foods or oils and fats. It is the small amount of fat that carries these vitamins into the bloodstream, through the walls of the small intestine.

Good combinations are:

- Dark green salads with a little oil dressing.
- Sweet potato, bananas, pawpaw, and mango with a little oil, fat, or butter.
- Pumpkin cooked with a little butter or coconut oil.
- Fish with vegetables, or a little seaweed (mackerel, sardine, or mullet).
- Eggs in salads or with meals once or twice a week.

An excess of vitamin A is poisonous. High concentrates are found in the livers of fish, dogs, bears, and sharks; seabird oil, and carrot juice. Such foods should be eaten in small doses.

An ideal tropical garden for good health would contain:

- Yellow fruits; dark green vegetables; a few chickens (for eggs and meat); a family cow, milk buffalo, or access to whole milk or butter; and coconut trees and oil-seeds like mustard or olive oil, sunflowers, and avocado (to help with vitamin A absorption).
- Ground covers of taro, sweet potato, ginger, and a good many berry-like fruits, for children to browse. This ensures enough vitamin C.

Basic Deficiency Diseases

Rice and Thiamine

Milled white rice lacks thiamine and as a staple food this can cause beriberi in people, chickens and domestic animals. Parboiling or composting techniques before milling tend to offset this loss. Unmilled brown rice in the husk is soaked in warm water for three to five days, then steam-heated and dried. In this way minerals and vitamins from the husk and bran of the rice (including thiamine) are dissolved in the warm water and taken up in part by the endosperm or starchy seed as it gelatinises or begins to germinate.

If this (dried) grain is now milled, it contains twice as much thiamine as the normally milled and polished grain. About 50% of the rice crop eaten in India and Nepal is treated in this way, and other ferments or moulds are used to increase the vitamin or protein content of the remaining rice.

Maize and Pellagra

Maize as "mealies" is widely consumed in southern Africa, and where it is a staple, people suffer from pellagra (a lack of niacin—a B vitamin). However, in the lands of its origin (the Americas), maize has no such bad effect as it is always cooked in lime or ash water, or limewater is used to mix tortillas (pancakes). This alkaline water frees the niacin and proteins in maize.

Similarly, light ashes are cooked with grains as a source of trace minerals and as a way to create an alkaline condition in the cooking water. It follows that corn (maize) grits, cornflour and corn breads are best mixed with limewater. In the Roman empire, soldiers cooked their porridge grains with alkaline earths as creta. A great many peoples use ash in cooking, as ash may be the main source of rare minerals in deserts. Modern mass production of flours or maize products omits the alkaline soak, and produces a vitamin-deficient product.

Trytophan and nicotinic acid (nicotinamide, niacin) are very low in milled flours of maize. Wheat diets seldom show up as pellagra, but in India some sorghum eaters develop it. Sources of niacin are yeast and liver extract (niacin is heat-stable), and milk rich in trytophan. Fermented milk is a great help in prevention (amasi or yoghurt). Pellagra can be associated with vitamin B12 deficiency, always with a poor diet such as mealie flours. Diets should include milk, eggs, meat, or fish. Rest and gradual improvement in diet is necessary for recovery.

Deficient Protein Plus B Vitamins

The "Burning Feet" Syndrome is common in famine, poverty, or in prisoners on poor diets. The feet ache, burn, and throb, and this may be followed by stabbing pains like an electric shock of agonising intensity. Pains are worse at night and come in waves. Patients lose sleep and become exhausted.

Yeasts, grain germ, high-protein diet, or yeasty fermented foods may bring relief over a few days.

Beriberi

Beriberi is the chief cause of infantile death in rice-eating areas. It is a thiamine deficiency. Muscles in adults become wasted, walking very difficult, high blood pressure and oedema develop. Especially suspect in the diet is white rice, refined cereal, fast food intake, or alcoholism. Treatment needs to be fast or the heart may fail; 10 mg of thiamine per day is needed, or 25 mg injected twice daily for three days (Davidson, 1979). Improvement may be dramatic, in a few days, or a few hours after injection. Diets should include a daily intake of beans or lentils (125 gm [4½ oz]), parboiled or beaten rice, or other home-milled grains (unpolished).

Rickets

Rickets is a disease of poverty and darkness; it is a lack of vitamin D (sunlight). Bowlegs and gross bone deformities may occur. Rickets may be found in Muslim female children kept indoors by the purdah system, and those with limited access to sunlight, milk products, and eggs. It also occurs in smoggy and smoky cities and shaded tenements. Rickets is treated with cod liver oil, and normal sun exposure, but most of all by health education.

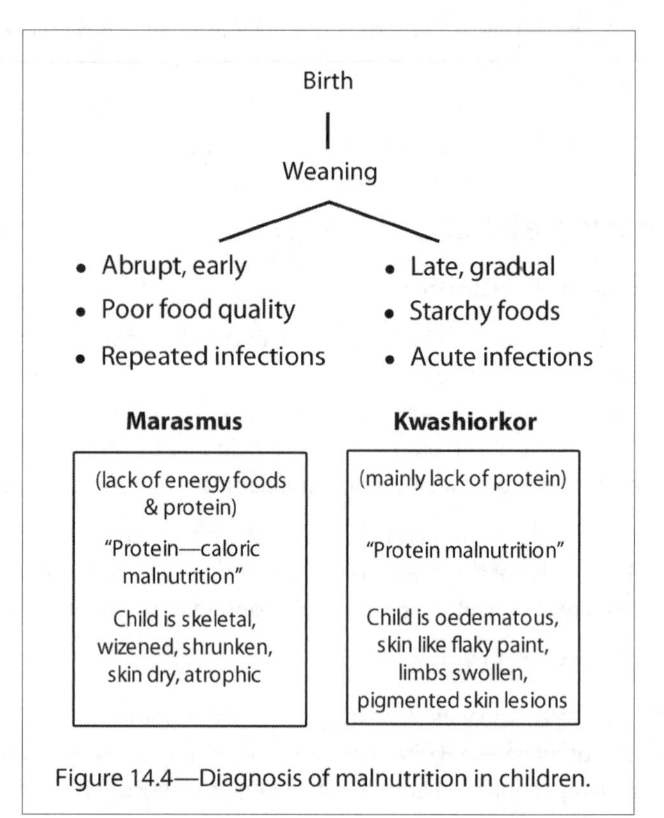

Figure 14.4—Diagnosis of malnutrition in children.

Scurvy

Scurvy is a lack of fresh fruit and vegetables (a vitamin C deficiency). Seamen and explorers were early sufferers. Today, they are the lonely old people, remote desert dwellers, and those on unsprouted or unfermented grain diets. Symptoms are putrid and bleeding gums, loose teeth, weakness, dizziness, and eruptive boils on the thighs and arms. Stress other than dietetic stress can affect vitamin C levels. Aborigines are often sufferers in Australia, and migrant labourers in Africa.

Often scurvy is associated with vitamin A, iron, and folic acid deficiencies. Wounds fail to heal, and patients are very susceptible to infection (often the inner ear in Aboriginal children). The cure is vitamin C tablets, 500 mg or more per day for a week or more, fresh oranges, lemons, greens, even unpeeled cooked potatoes, sprouted legumes and grains (thirty hour soak), pine needle extract in cold areas, ferrous sulphate for anaemia. This disease is early diagnosed by thick yellow nose discharge in children, earache, and listlessness.

An alarming trend in the West is the rise of illiterate and poorly-educated young people, many of them credulous and anti-scientific in matters of health. The basis of good nutrition is not known and diets may be vegetarian but very much insufficient for health or indeed any stress condition, and fermented foods are not often consumed.

Vitamin A

The deficiency of this vitamin in children first affects the eye as conjunctivitis, then as ulceration of the cornea of the eye, finally as softening of the cornea (keratomalacia) leading to permanent blindness, often in malnourished children. Since 1,500 BC, liver has been known as a cure in Egypt and China. If milk and butter costs rise, even European children may suffer. Dried or condensed milk is not a sufficient replacement source for fresh milk.

Today, problems are widespread in poor and poorly-nourished communities. Pumpkin flesh, carrots, liver, milk, fresh vegetables and fruit, and breast milk are all easily available cures. Children from two to five years are most susceptible if artificial foods are fed, or if the child is weaned early. Cod or halibut liver oil can be given if suspected, without harm, at 30 mg per day.

It is estimated (*New Scientist,* 13 October 1990) that twenty to forty million children in the world suffer a deficiency of vitamin A. 46% are more likely to survive if this vitamin can be supplied over a year. Mortality rates fell by 89% in those already malnourished who were given vitamin A. The youngest benefit most, so it is obvious that this vitamin is a key health factor and a major factor in child mortality.

Vitamin B12 is essential in the body for making amino acids, and for maintaining DNA. Deficiency produces pernicious anaemia, slowing down the formation of the blood cells, and (if severe) produces paralysis. Hydrogenobyrinic acid is the B12 precursor, and this acid needs to bind to cobalt to become B12, so that cobalt is, in this case, an essential micronutrient or trace element for creation of the vitamin (*New Scientist*, 27 October 1990). It is supplied to livestock as heavy ceramic pellets, which are very slowly digested.

14.10 ENZYMES

The Role of Enzymes in Nutrition

Enzymes are proteins, and they act like chemical scissors to cut up the long-chain peptides of food proteins (animal or vegetable) into simpler, digestible (assimilable) amino acids or food proteins in our bodies. Enzymes also build needed body proteins from simple peptide chains or amino acids.

All proteins are made up of amino acids, used by the body to build structural tissue such as muscle, skin and hair, also to make enzymes which are excreted into the digestive tract by e.g. salivary and pancreatic glands and by the digestive tract itself, in the liver. Enzymes also make hormones, neurotransmitters and other non-structural proteins found in the blood or in the endocrine glands.

Heat cooking and coagulation by acids (vinegar) inactivate enzymes in food. Vitamin B6 and vitamins (niacin, riboflavin) plus the small molecules (e.g. calcium, magnesium copper, molybdenum, zinc and hydrogen ions) activate enzymes and the "co-enzymes" of vitamins or metals may join with the protein portion of the enzyme to increase or initiate activity.

Enzymes are themselves proteins, which plants make using chlorophyll and sunlight, but which animals like ourselves obtain by eating plants or other animals. The linkage of some twenty-two amino acids by peptides into shorter or longer sequences creates millions of proteins and when we eat them our enzymes break them down (snip them off) into the amino acids that they then use to create those special proteins needed by our bodies. These "new" proteins may encode memories, carry vitamin A in the blood, or they may be enzymes specifically shaped to cut apart or assemble other amino acids, or they may form insulin or other hormones, etc.

Enzymes then are mostly derived from our food and are of four main types:

- Made by micro-organisms in ferments.
- Activated in fruits and grains by ripening, heat, cell wall injury, etc.
- From animal body fluids, organs, or glands.
- Made in our bodies (synthesised) from sequences of ingested amino acids.

Although destroyed by boiling, enzymes are merely inactivated by cold or freezing, and are re-activated by thawing. Enzymes are not only initiators of food assimilation, but they feed the single cells of the ferment organisms (fungi, bacteria, and yeasts).

Aubert remarks on modern fastfoods and their "de-natured" character, and compares this with the traditional breads, cheeses, pickled cucumbers and kvass of the Russian peasant. These are always prepared and always tasty and nutritious "fastfoods", and are still very much used in Russia. He showed weight gain in rats (and gain per one hundred calories of food ingested) to be much greater in fermented rice (idli) than in plain rice, in both cases by a factor of more than double the gain.

Non-Microbial Enzymes

Malted cereals (barley), sprouted, boiled, and dried, produce amylases, proteases and glucanases for the liquefaction of porridges and mashes, in food for people and animals. In brewing, malt extracts sugars and nutrients from grains, especially ground grains. Malts or sprouted mash can liquefy cooked grains, as can saliva.

Crab livers contain proteolytic enzymes used in making salt-stabilised fish sauces, which in leached tropical soils can provide an iodine source for consumers.

Saliva, added to such bases as chewed, sprouted grains is used to liquefy porridges or to make more efficient brewing or food fermentation. Often the substrate itself is chewed (or part is chewed) to release salivary enzymes, usually by mothers feeding their babies.

Papain from the fruits of the pawpaw (*Carica papaya*). It is used for meat tenderisation, brewing, and food processing.

Chymosin from the fourth stomach of unweaned calves is used for curd production in milk (rennet). The stomach is washed, cut into strips, and stored in brine (see section 10.2).

The Role of Enzymes at Weaning Time

Babies are normally weaned to some form of porridge or gruel, but such foods are stiff or sticky for small children, and may be difficult to digest. Here, sprouted grains, ground to a paste or ground malted grains can be added to the porridge pot and thinned to a gruel with sprout enzymes. Such sprouts are often chewed to a paste by the child's mother in Africa and are "...rich in semi-digested carbohydrates and ideal to a baby." (*New Scientist,* 2 January 1986), see Figure 14.4.

Saliva "is a complex mixture of vitamins, lipids, ions, immunoglobulins, polysaccharides, and a variety of peptides, proteins and glyco-proteins dissolved or dispersed in water." Just as saliva provides a good medium for ferment in the mouth, and stomach, so chewed materials spat out into dishes or vessels are provided with all they need for a ferment, and enzymes for the digestion and synthesis of proteins. At pH 6.7, saliva also buffers acidity in foods in the mouth (Marsh, 1980). Sprouted grains chewed and spat out into porridges, hydrolyse the porridge for infants.

As enzymes and safe milk ferment do not allow pathogens to breed, babies are safe on home ferments, and beneficial micro-organisms help to build a beneficial internal ecology.

14.11 PROTEIN VALUE AND COMPLEMENTARY FOODS

Complete proteins are fish, eggs, meats, and a range of fermented foods such as kawal, miso, tempeh and so on. All essential amino acids are present in the protein of such foods. Many foods lacking or having low levels of amino acids present, can be combined with those that do, to make up all essential amino acids. A little complete food and fresh vegetables, for minerals and vitamins, should keep one healthy. Fresh fruit and vegetables, however, are just those items missing from the diet of the poor, as are complete foods.

Complementary proteins for use with foods where amino acids are low, are most easily represented by the simple chart , see Figure 14.3 (after Lappé). In this chart, the arrows mean that many items in the group provide good protein if consumed together, whereas the double lines mean that only a few do so.

Thus, milk porridges, bread and rice puddings, rice-dal, "succotash" (a corn, bean, pumpkin dish), soy-rice, soy-wheat (miso), sorghum-pumpkin seed with milk, potato mashed with milk, and dozens of other traditional foods or dishes are, when combined, far more able to supply whole protein than if each is consumed alone.

Thus many traditional diets maximise the available amino acids in one meal by usually combining two or more foodstuffs on one plate.

14.12 ZINC IN DESERTS OR ARID LANDS

The high alkalinity of desert waters and soils may be expected to create severe mineral deficiencies in plants and in foods, as is the case with zinc or iron. This is a serious mineral deficiency problem, widespread in deserts. Natural sources of available zinc can be found in oysters, (although on desert coasts even these are not necessarily high in zinc), and in the seed of cucurbits (more common in deserts).

More commonly, desert peoples eat the testicles and the ground ear-bones of animals for zinc. They save the seeds of squash, melons, pumpkins, and gourds, which are ground with grain porridges as a staple food or side dish (Kalahari), or are eaten after being cooked in a pan with a little salted water. The water is evaporated and the cucurbit seeds roasted for keeping or grinding. In the Turkish or Afghan deserts, or in south west Asia, special seed pumpkins are grown with papery seed coats, and these are widely eaten as food. Low zinc levels create high copper levels in the blood, high lead levels, poor wound healing, late maturation, and eating disorders.

14.13 DUST-BORNE DISEASES AND AEROSOLS

Many viral diseases and allergens travel in winds, and on dust particles. Asthma, sinus problems, encephalitis, respiratory infections, and viral diseases are typically chronic during and after dust storms in arid and semi-arid areas. Part of good planning is to keep traffic dust, plough agriculture and its sprays, and unsealed roads well downwind or outside settlements, and to erect deep defences of forest belts between settlements and dust sources.

Near ore refineries, lead, cadmium, fluorine, and toxic gases may all cause long-term problems in settlements; and near atomic power or research establishments, leukaemia or hyperthyroidosis is an ever present risk. Even feed pellet mixing centres are a grave danger to employees and residents where arsenicals, antibiotics, hormones, and growth stimulants are used in powder form.

Rooms or water supplies where mineral fibres (asbestos, rock wool, fibreglass) are used can be extremely dangerous to lungs, and cancers can result. Incinerators (especially those dealing with agrochemical disposal) release dioxins to the atmosphere, and no levels of these are safe. As for herbicide sprays; the latter may break down quickly, but their breakdown products are persistent and very toxic, which in part accounts for high rates of leukaemia in chemically-minded apple growers.

14.14 GEOPHAGY

Geophagy is the addition of earth materials to food and water for detoxification, digestion, and enhanced nutrition. It is widespread in all climates and practised by many peoples. Sweet (calcareous) and sour (acidic) soils are classified by taste by many farmers, and a few people eat "meals" of raw or cooked clays, which possibly have a beneficial effect on acidic foods and poisons in such foods as cassava, taro, and zamia "palm" nuts (*Macrozamia*). Earth eating is not uncommon among children, and many of us taste soils or clays in our childhood as part of foraging.

Curiously (or perhaps aptly) many of the clays consumed by people (clays with illites, montmorillites, aluminium and talc contents, and the clays of termite mounds) are just those clays used to clarify polluted water. These particles flocculate (aggregate) silt and entrap bacterial and virus strains. In addition, the illite clays swell up by water absorption to dry out fluids. All these clays are usually taken, even today, by tribal peoples in the case of diarrhoea, in order to prevent fluid loss, and it is probable that when they are eliminated, their entrapped microbiological loads go with them.

About eight species of primates, including man, eat clays or earths in association with bitter leaves, roots, or seeds (Harris & Hillman, 1989). Quite specific clays are normally eaten with bitter potatoes, leaves, acorns and so on, and are so used to make poisons or anti-nutritive substances physically or chemically inert in the food (detoxification). They "bind" alkaloids and prevent illness or poisoning. Many bitter potatoes resist disease and are cold-tolerant.

Grain and Lime Ash (Alkali)

The following mineral and earth substances and salts are added to foods or involved in food processing:

Burnt Lime (agricultural lime or mortar lime)

Limewater is poured off a lime base. In the East Indies, India, Nepal, and Sri Lanka a common post-meal digestive is called pan, which is the crushed nut of the *areca* palm, combined with burnt lime and wrapped in a betel vine leaf. Sugars (jaggery or palm sugar), silver metal foil, tobacco, chilli, spices, and other additions may be combined to produce a great variety of pans.

Alica porridge was commonly eaten by Romans. It was made from coarse (unrefined) grains cooked with lime-earths (creta). Chalks and limes soften grain husks or the husks of grain legumes, making the protein available. They also free niacin in maize.

Ashes

Ashes are sourced from wood fires which are sieved to remove charcoal. Lye is also derived from ashes. Texquite, an alkaline pumice from ash cones in New Mexico (USA) is ground, soaked, pressed to strain, and the liquid used as leavening (like baking powder) in doughs and cakes. (Romero, 1970).

Corn cob, cornsilk, juniper, and other plant ashes were mixed with corn mush by Amerindians, as were tepary and lima beans in succotash or porridges. Succotash is boiled in kernel milk or fresh corn.

Ashes are used in a variety of ways in cooking, nutrition, and food storage. Firstly, the light ash of open fires is often used as a condiment (hence, a mineral and trace-element source) added to the water of boiled grains or vegetables. Eggs, cracked or pierced, are baked in ash (and dipped in ash). Peanuts, popcorn, and many grains, meats, and nuts are roasted in hot ash. Meats that are to be dried or stored are often rolled in ash before hanging to dry (as a surface seal and fly repellent). Peanuts are village roasted in India by mixing them with straw and burning in an open fire.

Lye (water dripped through ash) is used to de-hull maize, corns, beans, and to soften the hulls on sunflower seeds. Ashes are used (dry) to store dried products, as a source of potash in gardens, and as a source of lye in soap making.

Potatoes and root crops dusted with ash, store well in straw, and cut surfaces for planting are dipped in ash or ash-cement mixes to seal the cut and to deter such pests as pill bugs (wood lice).

Maize cobs, sheaths, and silks are burnt, the ash placed in a container, and water added. After the ash sinks, the water is poured off and used in cooking maize kernels and dried beans, spinaches, okra, and taro leaves.

Ashes are also made from the whole dried plants of *Amaranthus* and from *Espalter gariepina*, the latter first soaked in water and then burnt. The salts from evaporated lye (dripped slowly through ashes) are used to cook or preserve meat. *Cyperus esculentus* tops are also used for ash and the bulbs are eaten. These were the usual ways to produce salt inland. Cassava root rind peelings are used for a strong potash (not the outer bark) for gardens.

Earths and Clays

Termite mound clay—Eaten directly in spice as a snack before meals (Zimbabwe). In Africa, the clay from termite mounds is given free with dried caterpillars (Zimbabwe) to improve the appetite, and after a meal as a digestive.

Iron clays—Mixed with high tannin foods such as acorns to make insoluble ferric tannates.

Calcareous clays—May be directly added to porridges, and chalks eaten to relieve acid stomach conditions.

Swamp clays and peats—Butters, live acorns, eggs, and barrel foods are stored by burial in anaerobic swamp pits.

Clay alone—Traditional or special clays with calcium, silica, iron, or alumina are often eaten locally. Sometimes, clays are eaten as roasted cakes (East Indies, northern Australian Aborigines), and clays are also added to fruit jams to reduce acidity or to make iron or zinc available.

Talcs—Soft, white, leached, alumina-rich clays are eaten to absorb fats and poisons, reduce diarrhoea, retain body fluids, and as carriers for other medicinals. They are also used in face powders, and as a polish.

Salt—Rock salt from deep mines, surface salt from farms, sea salt from splash pools, and salt evaporated from seawater ponds are all collected and some are purified as table salts.

Soils licked by animals (salt licks)—Are dug up and sieved by washing repeatedly through grass sieves in perforated pots, as are the ashes of water plants or the bark of grasses and trees known to contain salt. The pure salt solution is boiled until a whitish residue remains (Zimbabwe, New Guinea).

Perhaps the most common mineral addition to food that is familiar to all of us is table salt. Too much, and too little raises blood pressure and although the modern sedentary life demands less salt (there is less loss of salt by sweating) it is also true that a deficit of salt leads to renal tubular necrosis and haemorrhage (*New Scientist*, 30 August, 1984, p. 5). Unfortunately, refined salt also contains aluminium powders so that it "runs" freely.

14.15 TRACE ELEMENTS

- Aluminium (Al)—Suspect in acid soils, or acid foods cooked in aluminium. Accumulated aluminium may cause brain problems, so alum (aluminium sulphate) is not recommended in food or for cleaning water! Be cautious and avoid ingestion. Common in clays.
- Arsenic (As)—Minute quantities may be essential but toxicity is well known!

- Boron (Bo)—Essential for plants, especially the cabbage family. Toxicity is rare, but do not use as a food preservative as it is poisonous in excess. Beware of excess from boron-containing washing powders as these end up in streams, gardens, and soils.
- Cadmium (Cd)—Does not seem essential. It accumulates in the kidneys and can become toxic. Cadmium is found in tyres, some plastics, superphosphates, and shellfish (abalone). Poisoning is very painful. Acid water can dissolve and carry this mineral. It can be high in vegetables near mine leachate (e.g. Canberra, Australia). Beware (as for fluorine, selenium, uranium) of deep bore waters, hot salty water from springs or wells, and vegetables grown in annual commercial dressings of superphosphate. Note—that the use of superphosphate as an annual dressing will prevent the uptake of zinc.
- Calcium (C)—We use calcium as lime, dolomite, chalk, or seashells in our gardens, and to make mortars for stones and bricks. We often burn calcium rock or shells to powder them, and in this way also use it for cooking. My friend Ranil Senanayake (*pers. comm.*), wrote the following about lime (burnt lime or agricultural lime) in about 1988:

  > "When two populations of corn (maize)-eating people, one from Mexico and one from Africa were compared, it was found that while both had the same diet, the African population was suffering from protein deficiency while the Mexican population was not. Further enquiry revealed that the use of lime and ashes as an additive to the water used to boil the Mexicans' corn was a traditional practice. Everyone in Mexico "knew" that you put some lime or ashes into the water you boiled your corn in".

  The effect of lime was to soften the seed coat and to increase access of enzymes to maize protein. As maize was developed in the Americas, the use of lime had co-evolved and had passed into tradition. Only the seeds of maize, not the associated traditions, were taken to Africa, and the result tells us that to take the seeds, and not take the cooking or ferment recipes, can be a very serious mistake indeed.

  I believe, however, that a much more serious case is that of rice, as outlined herein. To have promoted polished rice as an unfermented food was to promote lowered nutritional value, especially a thiamine deficiency.
- Chromium (Cr)—A rich source of chromium is brewer's yeast, also black pepper, liver, even natural brewed beer. This helps regulate glucose levels in the body. Protein malnutrition may cause chromium deficit.
- Cobalt (Co)—Cobalt is essential in vitamin B12 synthesis. Copper and phosphorus are also contained in the B12 molecule. In cattle rumens, B12 is synthesised by bacteria. Excess cobalt causes goitre, or heart failure. Enough B12 and cobalt comes with meat in the diet, but animal blood should be periodically checked for cobalt deficiency, especially in leached lowland soils.
- Copper (Cu)—Lack of copper produces anaemia. Copper, which is bound to proteins and blood in the body, is often in excess where water is boiled in copper utensils or iron is unavailable. Lack of liver copper means many enzymes are inefficient. Copper acts as a co-enzyme and regulates iron use, so that many anaemias arise from lack of either copper or iron. Protein deficiency may lead to copper deficiency. Copper excess chelates with D-penicillamine and is excreted in urine. Often desert people show blood copper excess due to the lack of iron or zinc.
- Fluorine (Fl)—Most tea drinkers get plenty of fluorine from tea. Fluoride is a bone-seeker and is deposited in bone and dental enamel. One milligram per day is considered a sufficient dose. Excessive fluorine, often from deep wells or under fallout from mineral processing industries, causes mottled teeth, bone thickening, and outgrowths of bone.

  Poisoning occurs over 8–9 parts per million in water. The backbone may fuse due to the calcification of ligaments. Several localities have fluoride excess due to minerals like cryolite, and it is prudent to test all well-water sources for excess, and to drink tea if deficient.
- Iodine (I)—Lack of iodine has been known since 1820 to cause goitre or hyperthyroidosus. In 1991, lack of iodine was responsible for about 30% of mental impairment in children, mainly in China, the Himalayas, and South America. This lack is especially severe in high tropical mountain areas, like the middle Himalayas.

Lack of iodine causes the thyroid to overdevelop. All vertebrates require iodine rapidly taken up in the body, combined with tyrosine, and released in the blood as hydroglobulin to regulate cell metabolism, and a lack of iodine lowers all metabolic function. The only rich source is seafoods, hence the importance of fish preserves and sauces, seaweeds, and the use of kelp. Today, sodium iodine is routinely added to table salt where people are living in leached soil areas, tropical uplands, or areas of deep volcanic soils.

- Lead (Pb)—Known only as a toxic substance with severe nervous system effects from overdose. Can affect up to 80% of city children near highways, also inhaled from traffic fumes. Continual monitoring is essential.
- Lithium (Li)—Used in cases of nervous excitation. Usually found in vegetables, often in warm volcanic spring waters.
- Manganese (Mn)—We usually get enough from tea, whole cereals, and leafy vegetables; it is rarely in excess. Can be added to soils as dolomite, as it is usually deficient in leached tropical soils.
- Mercury (Hg)—A severe metabolic poison, destroying brain and nervous system function. Large fish, weighing in excess of 6–8 kg (13–18 lb), should be avoided on islands and coasts near industrial processing.
- Molybdenum (Mb)—No deficiency disease is known, but bacterial nitrogen fixation depends on molybdenum as a soil trace element. It is low in acid sandy soils, relatively high in neutral or alkaline soils. Forms part of several enzymes, xanthine peroxidase for example.
- Nickel (Ni)—Function not known. Some adverse nickel skin reactions occur. With excess copper, it forms a synergistic poison.
- Selenium (Se)—Excess selenium may occur in deserts, especially in over-irrigated areas, and can cause severe birth malformation, and improper development of the eye and braincase in animals. Deficiency causes white muscle disease in lamb, and liver necrosis in rats. Brewer's yeast is a source; as is vitamin E (pumpkin family, seed). Selenium also binds with enzymes (glutathione peroxidase) which guard against oxidative damage to body tissues (lipid membranes). Most normal levels come from cereals and meats. Fish is a good source: vegetables are a poor source. Severe poisoning may occur in desert USA, and Botswana, in irrigated soils.
- Silicon (S)—Cartilage and connective tissue formation. Plentiful in bamboo, hard grains, horsetail, and Portland cement. May be lacking in alkaline areas with warm rains, and can be increased by bamboo mulch, or 1 bag of cement per acre in grains and gardens.
- Tin (Sn)—Toxic only in high doses, from poorly canned foods or too much canned foods.
- Vanadium (V)—Radish is a good source, as are some mushrooms. Known to be catalytic in enzymes but other functions are not well known. The fungus, fly agaric, (*Amanita muscaria*) accumulates vanadium selectively, and grows in pine forests, so that pine "duff" with *Amanita* hyphae should be good mulch for nitrogen-fixing bacteria. The mushroom itself is poisonous (to flies and people).
- Zinc (Zn)—Low in modern diets, due to overuse of superphosphate. Essential to health and an essential garden trace element.

  Dr Derek Bryce-Smith (1986) further proposes that zinc is an essential co-enzyme in thirty or more digestion processes, and a deficit therefore leads to a great many nutritional disorders. Zinc may be more generally missing from modern diets than in the past. We need 25 mg per day of this element, and people eating the foods of modern agriculture may obtain only 10 mg per day.

  Bryce-Smith identifies two reasons for this deficit:

  - Soils dressed with too much superphosphate block zinc uptake in plants and animals.
  - Short-ferment breads may not have sufficient of their inherent phytates destroyed, (as they are in long yeast ferments) and we then lose body zinc because it binds to phytates in breads, brans, wholemeal and legume flours. Thus, since 1959 the shift away from traditional food, and traditional farming and food preservation, or ferment, has meant that our diets may lack essential minerals.

APPENDIX A: GLOSSARY

acetic. (acetic acid) Specific to vinegar.

achars. A mixture of pickled fruits and vegetables.

acidic. A general term for low pH.

aerobic. Living or occurring only in the presence of oxygen.

aflatoxins. Any of a group of toxic compounds produced by certain molds, especially *Aspergillus flavus*, that contaminate stored food supplies such as animal feed and peanuts.

aguamiel. The sugary saps or nectars derived from plants. In its fermented state it can be enjoyed as a beverage.

albuminous. Relating to, containing, or consisting of albumin.

albumin. A class of simple, water-soluble proteins that can be coagulated by heat.

ameliorate. To make better or more tolerable.

anaerobic. An organism or tissue, living in the absence of air or free oxygen.

anhydrous. A substance is said to be anhydrous if it contains no water.

antipodeans. The antipodes of any place on Earth is the point on the Earth's surface which is diametrically opposite to it. Two points that are antipodal to one another are connected by a straight line running through the centre of the Earth. In the British Isles, "the Antipodes" is often used to refer to Australia and New Zealand, and "Antipodeans" to their inhabitants.

antiscorbutic. Efficacious against scurvy.

aqueous. A solution in which the solvent is water.

barm. A solution or cake of yeast.

Bass Strait. A sea strait separating Tasmania from the south of the Australian mainland.

beriberi. A nervous system ailment caused by a thiamine deficiency.

biocides. Any chemical that destroys life by poisoning, especially a pesticide, herbicide, or fungicide.

biotamines. An amino acid.

bush fallow. A type of subsistence agriculture in which land is cultivated for a period of time and then left.

caecae. A saclike cavity with only one opening or the large blind pouch forming the beginning of the large intestine.

calabash. A vine grown for its fruit, which can either be harvested young and used as a vegetable or harvested mature, dried, and used as a bottle, utensil, or pipe.

centrifuged. To rotate (something) in a centrifuge or to separate, dehydrate, or test by means of this apparatus.

chiangs. Chinese ferments.

cilantro. Coriander.

clamps. The means of storing (mainly potatoes) in an earth mound.

coagulation. A complex process by which blood forms clots.

coppiced. A traditional method of woodland management which takes advantage of the fact that many trees make new growth from the stump or roots if cut down. In a coppiced wood young tree stems are repeatedly cut down to near ground level. In subsequent growth years, many new shoots will emerge, and, after a number of years the coppiced tree, or stool, is ready to be harvested, and the cycle begins again.

credulous. Having or showing too great a readiness to believe things.

cyanides. Any of various salts or esters of hydrogen cyanide containing a CN group, especially the extremely poisonous compounds potassium cyanide and sodium cyanide.

dahl. Small, highly nutritious seed of the tropical pigeon-pea plant.

decoction. A method of extraction, by boiling, of dissolved chemicals, or herbal or plant material.

deity. A divine character or nature, especially that of the Supreme Being; divinity.

de-naturing. A process in which proteins or nucleic acids lose their tertiary structure and secondary structure by application of some external stress or compound.

diastatic. A component of malt, important in the brewing process.

diatomaceous. Relating to, consisting of, or containing diatoms or their fossil remains.

dosas. A fermented crepe or pancake made from rice batter and black lentils.

encephalitis. Irritation and swelling (inflammation) of the brain, most often due to infections.

epithelia. One of the four basic types of animal tissue, along with connective tissue, muscle tissue and nervous tissue.

estuarine. Relating to, or found in an estuary.

estuary. An estuary is a partly enclosed coastal body of water with one or more rivers or streams flowing into it, and with a free connection to the open sea.

expectorate. Cough or spit out phlegm from the throat or lungs.

flocculation. Flocculation, in the field of chemistry, is a process where colloids come out of suspension in the form of floc or flakes.

flushes. Short periods of fruiting.

foliar. Foliar feeding is a technique of feeding plants by applying liquid fertiliser directly to their leaves.

formalin. Is an aqueous solution of the chemical compound formaldehyde.

fusel. A mixture of amyl alcohols, propanol, and butanol: a by-product in the distillation of fermented liquors used as a source of amyl alcohols. Poisonous.

genera. In biology, a genus (plural: genera) is a low-level taxonomic rank used in the classification of living and fossil organisms, which is an example of definition by genus and differentia.

geophagy. The eating of earthy substances, such as clay or chalk, that is practised as a custom or for dietary, medical or subsistence reasons.

glut. To fill beyond capacity, especially with food; satiate or to flood (a market) with an excess of goods so that supply exceeds demand.

gram. A small legume seed **or** a unit of weight.

hydrolysate. Substances produced by hydrolysis. These vary from yeast extracts, marmite (yeast protein), soy sauce, miso, and fish sauces. Hydrolysis (and therefore hydrolysates) are often more effective in the presence of salts, and enzymes, and at specific pH and temperature ranges.

hydrolysis. Water reacting with other substances to "dissolve" them into a variety of compounds. Used mainly for the dissolution of protein into ferment sauces herein.

hyphae. (In a fungus) one of the threadlike elements of the mycelium.

idlis. A south Indian savoury cake popular throughout India.

inoculate. To introduce a serum, vaccine, or antigenic substance into (the body of a person or animal), especially to produce or boost immunity to a specific disease.

integuments. An outer protective covering on organisms, such as fur, clothing, shells, skin, and seed coatings.

isinglass. A pure transparent or translucent form of gelatin, obtained from the air bladders of certain fish, especially the sturgeon: used in glues and jellies and as a clarifying agent. Dried fish swim-bladder or sounds.

kotero. A fermented sweet potato yeast. Maori origins.

koumiss. A fermented dairy product traditionally made from mare's milk in Russia.

kumara. Sweet potato.

ligno-cellulose. A compound of lignin and cellulose that occurs in the walls of xylem cells in woody tissue.

lye. A corrosive alkaline substance, commonly sodium hydroxide (NaOH, also known as "caustic soda") or historically, potassium hydroxide (KOH, from hydrated potash) **or** ashes.

mace. A cooking spice obtained from the dried covering of the nutmeg fruit seed.

macerated. Softening or breaking food into pieces using a liquid.

malt. A sugar derived from sprouted barley or grains by enzyme action on starch; and malted barley sprouted and dried is added to beers to provide sugars and enzymes.

mealy. Resembling meal in texture or consistency; granular.

mealies. An ear of corn, Africa.

metabolic. Relating to metabolism, the whole range of biochemical processes that occur within us (or any living organism).

milled. Ground or hulled in a mill.

mucilaginous. Sticky, viscid.

must. The skins, seeds, and fruit pulp of grapes left after the juices are expressed.

mycelia. White threads, a binding substance produced by fungus.

necrosis. The premature death of cells and living tissue.

nigari. Bittern; a concentrated solution of various salts remaining after the crystallisation of salt from seawater.

old timer. An elderly person.

okara. A white or yellowish pulp consisting of insoluble parts of the soybean.

paddy rice. Rice that is still in the husk, either as cut heads or on the stem, is referred to as paddy rice.

parboiled. The partial boiling of food as the first step in the cooking process.

pelagic. Relating to, or living in open oceans or seas rather than waters adjacent to land or inland waters.

phytates. Interfere with the absorbance of several nutrients.

plasmids. Small chromosomes (genetic material) from 3-20% of the size of cell (fertility) chromosomes found in bacteria. Plasmids can go with other "fertility" genes from one bacterium to another, and it is the plasmids that carry a variety of abilities to resist diseases, biocides, or to make antibiotics and toxins. They also act as the metabolic moderators of sucrose. Thus plasmids for specific jobs can be artifically transferred to other cells or other organisms to make special strains of bacteria, or to inoculate yeasts to make antibiotics (*New Scientist*, 16 February 1991. p 31).

polysaccharide. Polymeric carbohydrate structures, formed of repeating units (either mono- **or** di-saccharides) joined together by glycosidic bonds. Examples include storage polysaccharides such as starch and glycogen, and structural polysaccharides such as cellulose and chitin.

pomace. Specifically, the crushed apple pulp left after juice is extracted, but more widely used for fruit remains e.g. "pear pomace".

potoroo. A kangaroo-like animal about the size of a rabbit, native to Australia.

purdah. Purdah is a curtain which makes sharp separation between the world of man and that of a woman, India.

ragi. A cereal grass, cultivated in Africa and Asia for its edible grain.

ropy. Resembling a rope in being long, strong, and fibrous. Refers to "ropes" in milk.

roux. A cooked mixture of wheat flour and fat, traditionally clarified butter.

ruminants. A ruminant is a mammal of the order *Artiodactyla* that digests plant-based food by initially softening it within the animal's first stomach, then regurgitating the semi-digested mass, now known as cud, and chewing it again. The process of re-chewing the cud to further break down plant matter and stimulate digestion is called "ruminating". Found to occur in kangaroos (first discovered by Bill Mollison).

saltpetre. Potassium nitrate, or the mineral niter, the critical oxidising component of gunpowder, and a food preservative.

sambal. A condiment or side dish of Indonesia, Malaysia, and southern India, made with any of various ingredients, such as vegetables, fish, or coconut, usually seasoned with chili peppers and spices and served with rice and curries.

saprophages. Eaters of decayed material.

seething. Bubble up as a result of being boiled.

seraglio. Also known as "serail", it is the sequestered living quarters used by wives and concubines in a Turkish household.

smut. A group of plant parasitic fungi.

sponge. The bubbly gruel made by bakers from compressed yeast which is used briefly to disperse the yeast through bread. That "spontaneous yeast" in a large bottle provided by your grandmother when you are making bread—"Here, take the sponge."

spontaneous yeast. Yeast "caught" in a basic flour-salt-sugar-hops liquid exposed to air (usually in the kitchen) until is shows bubbles and smells yeasty, or goes sour and is rejected.

stearine. A component of fats and oils.

steep. Saturation in a liquid solvent to extract a soluble ingredient, where the solvent is the desired product.

stook. Also referred to as a "shock", it is a circular or rounded arrangement of swathes of cut grain stalks placed on the ground in a field.

styptics. A specific type of anti-hemorrhagic agent that works by contracting tissue to seal injured blood vessels.

substrate. The underlying and main ingredient.

suet. Raw beef or mutton fat, especially the hard fat found around the loins and kidneys.

symbiotic. A close, prolonged association between two or more different organisms of different species that may, but does not necessarily, benefit each member.

synergism. The joint action of agents, as drugs, that when taken together increase each other's effectiveness.

synergistic. Pertaining to, characteristic of, or resembling synergism: a synergistic effect.

tenement. A substandard multi-family dwelling, usually old, occupied by the poor.

thorium. A naturally occurring radioactive chemical element, found in abundance throughout the world.

threshed. To beat the stems and husks of (grain or cereal plants) with a machine or flail to separate the grains or seeds from the straw.

tinder. Material that catches fire on contact with a spark.

tocopherols. A class of chemical compounds of which many have vitamin E activity.

turn in. Cause to rotate or revolve. To spade or plow (soil) to bring the undersoil to the surface.

unslaked. Hasn't been disintegrated with the treatment of water.

vasoconstrictor. To prevent excessive bleeding.

vasoline. Petroleum jelly is a mixture of mineral oils, paraffin and microcrystalline waxes that, when blended together, creates something remarkable—a smooth jelly that has a melting point just above body temperature.

wallaby. A wallaby is any of about thirty species of macropod (family *Macropodidae*). It is an informal designation generally used for any macropod that is smaller than a kangaroo or wallaroo that has not been given some other name.

winnowed. Blow a current of air through (grain) in order to remove the chaff.

wort. Provincial wort. The sponge (see sponge).

APPENDIX B: ABBREVIATIONS

Unit of Measure	Abbreviation
acre(s)	ac
bushel	bu
Celsius	C
centimetre(s)	cm
confer	cf.
cubic feet	cu ft
degree Celsius	°C
degree Fahrenheit	°F
dozen	doz
Fahrenheit	F
feet	'
fluid ounce	fl oz
gallon(s)	gal
grain	gr
gram(s)	gm
hectare(s)	ha
inch(es)	"
international unit	IU
kilogram(s)	kg
kilolitre(s)	kL
kilometre(s)	km
kilometres per hour	kph
metre(s)	m
mile(s)	mi
miles per hour	mph
milligram(s)	mg
millilitre(s)	mL
millimetre(s)	mm
ounce(s)	oz
parts per billion	ppb
parts per million	ppm
pound(s)	lb
quart(s)	qt
revolutions per minute	rpm
square	sq
tablespoon(s)	tbsp
teaspoon(s)	tsp
ton(s)	T
yard(s)	yd

APPENDIX C: CHRONOLOGY

Some Chronological Events in Food and Ferment

BC Before Jesus Christ (approx 2,000 years ago).

AD Anno Domini. After Jesus Christ.

BP Before Present.

About 40,000 BP. West Australian aborigines develop extensive yam fields over many hundreds of acres of alluvium. They dwell seasonally at the gardens, have firm ownership, well-developed paths, permanent substantial hut sites, wells, drains, and conspicuous yam mounds (Harris & Hellman, 1989).

30,000 BP. Cultivated varieties of yam and taro are present in New Guinea uplands; traces of food are found on stone pounders (*Pacific Magazine*, December, 1991, pp. 41-42). There is no doubt that this is an independent agricultural evolution of great antiquity.

11,000 BC. Domesticated sheep present in Iraq.

Prior to 10,000 BC. Production of wine in the Mediterranean. Moulds are used in folk medicines as anti-infection agents. Natural ferments of aguamiels (nectareous waters or saps) are used.

8,000 BC. Elaborate irrigation exists in New Guinea highlands, associated with yam and taro fields.

7,000 BC. Neolithic sheep herds in Europe.

6,220-5,520 BC. In Egypt there are no domesticated crops, animals or any pottery. (Quaranian period).

6,000-5,000 BC. Beers produced in Egypt (barley doughs soaked in water). Milk is changed to curds and cheeses by storage in calves' stomachs.

About 5,000 BC. Grains, some from productive grasses such as Shoshone Indian rice grass, *Oryzopsis hymenoides,* appear in the diet of mankind and over the next centuries, an increasing proportion of grains shows modern selected characteristics. Maize is grown and in process of development in southern Mexico, Peru.

4,391-3,070 BC. Cereals and domestic animals have developed in Egypt (cattle are not an important item) fayum neolithic period. From herein, permanent settlements developed, and by the Dynastic period there is a great dependence on agriculture.

4,000 BC. The Chinese cultivate garlic and are developing improved plants. Brewer's yeast is in use for bread leavening in Egypt. Horses are domesticated and ridden. *Spirulina* algae is harvested by the Aztecs.

3,400 BC. Excavated urns reveal yeast cells (Egypt).

3,000 BC. Rice, in the East, and wheat, garlic, and leeks in Egypt and the Middle East, are well established. In the Andes, quinoa and amaranth are the grains cultivated, as is maize in the Sonora and the sunflower is developed in Canada and midwest North America. Most modern grains and vegetables have now been locally evolved, although many people have grain only as a minor crop, and very few use it as a staple. Exceptions are larger settlements in the eastern Mediterranean, Africa, and the Americas, in the Andes, and the cities of China and South East Asia. Soybeans, rice, Japanese millet, and barley are used in the Orient. Peas are found in Swiss lake dwellings and probably came from the Middle East. Unleavened bread is usual in Europe, yeast breads less common.

2,800 BC. There are Babylonian cuneiform inscriptions for barley wine or ale.

2,400 BC. Lentils and chick peas are common in Egyptian tombs. Broad beans evolve in the Middle East, and lima beans in Peru.

2,000 BC. Guinea fowl are kept in aviaries in Egypt and have long been kept in Africa along the Niger, Volta, and Gambia rivers.

1,475 BC. The chicken is domesticated in South East Asia, and eggs are artificially incubated in Egypt (mud ovens, heated with camel dung).

1,300 BC. Rice is cultivated in India.

1,100 BC. The guinea fowl is kept on Greek farms.

1,100 BC. Yin Dynasty. Chinese begin fish culture of common carp.

1,000 BC. Fermented foods are certainly established in settled areas. Cucumbers are cultivated in the Middle East. Plutarch later mentions pickled cucumber in "Acetaria" (acidic ferment). Pliny states that almost 600 years after Rome was founded, the Roman armies brought Grecian balms into Italy and yeast breads are thereafter common. 1,000 years after this Britain was conquered by Julius Caesar, yeast bread was "seldom used except by the higher classes of inhabitants" in Europe. Distilled spirits recorded in China, as are soybean ferments.

800 BC. Sourdoughs are in use in Europe and Eurasia for rye bread.

400 BC. Chard is grown in the eastern Mediterranean (Aristotle). Ginger cultivated in China.

322 BC. Alexander plants bananas in the Indus valley. Later these are spread by Arabs when slaving in Africa.

200 BC. Asparagus cultivation spreads to Rome from the Middle East.

100 BC. About 250 yeast bread bakeries operate in Rome.

200 AD. Fermented milk products are produced in Rome; also fish sauces (garum).

600 AD. Japan takes soybean ferment techniques from China.

618-907 AD. Fish polyculture practised in China.

800 AD. Frequent Asiatic and Polynesian boat voyages spread breadfruit, yams, sweet potato, plantain, and taro to remote islands of Polynesia and Pacific coasts, and many plants travel to new continents.

900 AD. Arab-Africa contacts bring maize to west Africa from South America.

1435 AD. Maize-like seeds are found in excavations of this date in Assam.

1492 AD. Columbus is in America. He is pre-dated by Native Americans, Polynesians, Chinese, Vikings, Celts and Basques. Portuguese find bananas on the Guinea coast of Africa and transplant them to the Canary Islands, thence to South America.

1493 AD. Peppers (capsicums) are described, after Columbus' voyages. Padres and priests rapidly spread American seeds via monasteries and postings to India and China.

Columbus in the Canary Islands takes on board seeds of grains, vegetables, oranges, and lemons. Gardens are established on Haiti at San Domingo, and from there to South America.

1494 AD. Maize recorded in Spain, from America.

1516 AD. Bananas arrive in South America with the help of the Spaniards.

1518 AD. Oranges in Vera Cruz, Mexico, from Haiti or Spain.

1530-1550 AD. Guinea fowl and turkey introduced to England, replacing peafowl and swan as favourite table birds.

1542 AD. Germany gets its capsicums from India, and believes India to be the origin of these fruits. They are certainly from seed taken to India in the previous fifty years!

1547 AD. Ginger introduced to the West Indies and is re-exported to Spain.

1550 AD. onwards

Empire-building spreads European grains and methods to many other countries. As the consolidation of empires proceeds over the next one hundred and fifty years, many merchant classes arise, busy in the storage, transport, and sale of grains. Grain, as a crop, is used to enrich landowners and entrepreneurs, and is seen as a major bulk trade item and in fact as an accessory weapon in war and conquest. By 1800 this trend is well established.

About 1556, artichokes are cultivated in Italy.

1565 AD. Spanish bring citrus, figs, grapes, and pomegranates to Florida. Jesuits bring oranges, figs, grapes, and olives to California.

1575 AD. Maize in Japan, China, and Iraq.

1780 AD. Europeans mistakenly believe that only yeast (wheat) bread is fermentable, and that rice, maize, sorghum and so on are used unfermented. However, only the European grains are still eaten (for the most part) as unfermented breads, while rice already has a complex series of ferments in Asia. Theodore Schwann discovers that living organisms are responsible for ferment. He calls the yeast he discovers Zuckerpilz or "sugar fungus", which later is Latinised as *Saccharomyces*, the generic name for brewer's yeast.

1781 AD. Pressed baker's yeast is produced (Dutch method).

1811 AD. Iodine discovered (in kelp) by Coutois, in France. Trace elements are being isolated.

1812 AD. On the Island of Skye, Jacob's or Norwegian sheep, with coarse multi-coloured wool and frequently carrying four or even six horns were being replaced (by the lairds) by larger Linton (or Blackface) sheep from the lowlands, and the "elegant" Cheviot sheep "of superior wool but less resistance to bad weather". A very large wool industry is being developed based on large sheep properties.

By the 1820s, Blackface and Cheviot were dominant breeds; black cattle and "improved" horses and pigs were being introduced; clansmen displaced by lowlanders and Anglophiles. By 1820, such improvements had led to starvation as crofters divided their small three-acre plots among their families. They became "wage slaves and scavengers".

1819 AD. Andrew Fyfe finds iodine in animal bodies.

1820 AD. Coindet, in Geneva, treats goitrous patients with iodine (later, overdoses show toxic effects and research is discontinued).

1830 AD. New South Wales, Australia, the focus for Scottish emigrants displaced by large landowners and their sheep.

1836-1837 AD. Researchers opine that yeast is a living thing (Cagniard, Latour, Schwann, Kutzing) but this is ridiculed by others.

1837-1839 AD. A series of ships sail for Australia. Large land grants repeat the European model. There is no notice taken of extensive aboriginal plant and aquatic agricultures.

1840 AD. Skye has many families in Australia. The highlands are "cleared" for sheep by the military and by forced emigration.

1845-1846 AD. The Irish potato famine occurs, due to potato blight. The first European example of famine due to reliance on a staple crop. Staple crops are also the most likely to produce malnutrition due to deficiency diseases. Such famines have long occurred in the Middle East, Polynesia, Africa, and the Americas as staple foods are developed.

1846 AD. The Vienna process yields the yeasts used in processing starchy substrates to spirits.

1847 AD. Ignaz Semmelweiss in Germany reduces mortalities from "childbed fever" (the infection of mothers at childbirth) by having surgeons wash their hands between operations. Primitive hygiene is being practised in the West, although such hygiene was already common in China and India.

1850 AD. Missionaries bring the soybean to Europe from Asia. Bananas are brought to the USA.

1856-1857 AD. Louis Pasteur studies specific ferment organisms, products, processes, and distinguishes microbial species.

1865 AD. Joseph Lister uses phenol (carbolic acid) to sterilise surgery operations. The "microbe theory" of disease is becoming accepted.

1870-1876 AD. Pasteur settles the question of yeasts and bacteria as living organisms and the origins of vinegars, and "ferment diseases" from impure cultures. In fact the "germ theory" of disease is born and eventually gives rise to vaccines, and compulsively clean hypochondriacs. Pasteur and others suggest that there could be therapeutic applications of microbial products.

1879 AD. Marquardt introduces aeration of grain mash to increase spirit yields.

1881 AD. Robert Kock pioneers pure culture techniques in use today. He sets the use of modern sterilised media and aseptic conditions. Commercial production of lactic acid commences (*Lactobacillus delbruckii*).

1894 AD. Goccichi Takamine patents a method for producing enzymes from moulds (diastase). He is a naturalised Japanese living in the USA.

1895 AD. Metchinkoff proposes *Lactobacilli* as a longevity factor. Subsequent investigations throw doubt on his theory of "colon putrefaction".

1900 AD. The soybean reaches the USA.

1906 AD. G. Thom produces cheese fungi (*Aspergillus* spp.) in culture.

1907 AD. The Nobel Prize is given to Harden and Eule-Chelpin for elucidating the role of enzymes in fermentation.

1908 AD. Nikolai Ivanovich Vavilov in Russia begins to collect the world's domesticated seeds and vegetative crops, fruits and nuts.

1913 AD. Vitamin A (fat soluble) is extracted from butter. The elucidation of vitamins begins.

1915 AD. Vitamin B (water soluble) is extracted from whey, yeast, rice polishings, and rice water.

1914-1918 AD. Germany produces yeasts for food (protein) in molasses due to war shortages of food.

1920 AD. Masine and Kimball publish a classic experiment showing that iodine treatment reduces goitre. The role of trace elements is recognised.

1926 AD. J. B. Sumner crystallises urease, the first enzyme isolated from jack-bean meal. Today, there are probably in excess of 1,000 enzymes identified.

1934 AD. Gautherer pioneers plant cell culture, now widely applied to reproducing plant species in laboratories.

1939 AD. Vavilov has completed sixty-five plant collecting expeditions, and has collected, in Leningrad, 165,000 different varieties of economic plants.

1939-1945 AD. Germany uses *Candida utilis* as a protein source from paper industry wastes and sugar from the acid hydrolysis of wood during the war period. The siege of Leningrad; lasts nine hundred days. Stalin arrests and imprisons Vavilov, and he dies of starvation in prison. At the Vavilov Seed Institute, twenty scientists (fourteen of them leading Russian scientists) die of starvation amongst thousands of bags of rice and crop seeds. They believe in preserving this seed for the future, rather than preserving their lives. Vavilov and his fellow scientists had established the core centres or origins of our domestic plant varieties on a world basis (the Vavilov centres). (See modern maps in Fowler and Mooney, 1990.)

1940 AD. USA produces yoghurt commercially.

1950 AD. The sweet corns begin to replace fodder maize on Australian farms and fodder maizes become uncommon over the next decade. First animal (chicken) cells are grown in a medium devised by Morgan and his colleagues.

1953 AD. The structure of DNA is discovered by Watson and Crick. Corporate "think-tanks" in the USA devise strategies of patenting main crop seeds and set up international seed-breeding institutes to collect local varieties and to breed patented varieties. The seed trade corporations begin to replace governments, elected by the people, as the real power. The process is largely completed by 1980, and the international seed trade has passed from governments to corporations. Thousands of local cultivars have been supplanted and have become extinct. No finance or credit is given to the poor farmers (mainly women) who developed the crops for thousands of years, and who now prove to be the only reliable repositories of such seeds, the living plants grown by small farmers. Genetic diversity is being reduced by corporate greed, and as a method of food control.

1955-1960 AD. Eagle and co-workers devise nutritional bases for the culture of animal cells. Viral vaccine production becomes possible. Enzymes are used in washing powders!

1961 AD. Agricultural production peaks on a world basis. Soils are being rapidly degraded and salted due to monoculture cropping, export cropping, over-production of subsidised food staples, and refined and "fast" foods.

1968 AD. Several single cell protein (SCP) plants are built to produce protein from wastes. The multinationals plan to feed the world on wastes of soybean products (they own 100% of the soybean patents). Most soybeans are grown to feed livestock or to paint cars (duco). AIDS is starting in New York.

1970 AD. Diet in Britain derives 42% of energy from fat, 20% from sugar. Coronary heart disease, rickets, gallstones, diverticulosis, and some cancers are partly determined by diet. Malnutrition and obesity begin to shorten life expectancy.

1972 AD. Life expectancy peaks in Western countries and begins to shorten. Truly sedentary lifestyles have become widespread.

1973 AD. The Boyer-Cohen team at Stanford succeed in transferring genetic material between individual bacteria, and eventually from cell to cell, across species.

1981 AD. ICI (Imperial Chemical Industries), in England, produce Pruteen for animal feed at three thousand tons per month. Rank Hovis McDougall, in England, use the fungus *Fusarium gramineum* to produce a textured meat analogue approved for human use.

1982 AD. Ira Herskowitz and Janet Kurjan discover that in the sexual reproduction of yeasts, the alpha factor, a pheromone-secreting protein, is sent beyond the yeast cell wall. Using this factor as a carrier, yeast vats can produce extra-cellular medicinals in vats, leaving the yeast mother cells to continue manufacture. This has led to mass production of specific products from yeasts.

1985 AD. Mitsui Petrochemical Company produces a metabolite (shikanin) by using plant cell culture. Monoclonal antibodies and other medicinal products are being produced. AIDS has a "doubling time" of eight to ten months, and is established world wide.

1988 AD. "Gene shears" are beginning to be used to manipulate RNA/DNA. They are, in effect, a manufactured enzyme tailored to specific genes.

1990 AD. Fast foods of poor quality have largely supplanted regional and domestic foods. The stage is set for total corporate control of the majority of the world's populations. Many modern families exist only if the fast foods system exists. Land and resources in the Third World are destroyed for money alone. But the official seed banks are losing seeds faster than they can collect them; and long-term seed storages fail. A growing army of home gardeners, local and regional seed banks, and small farms are using open-pollinated (unpatented) seed, and there is a much greater variety of perennial plants, organic and conservationist soil treatments, and a world wide series of data-banks, meetings, and training across frontiers. People begin to assert regional independence. The old empires are gone or crumbling and the serious fight against corporations begins. Many corporate and political figures are gaoled or fined for dishonesty and theft or misuse of public monies. Many regions seemingly become self-reliant for food. The Russian empire crumbles suddenly. Trade wars, based on subsidised wheat and beef, are being fought by government-corporate groups. The tool is government-subsidised surplus food traded by corporations. Food for people is no longer the main concern of these groups. Famine, soil erosion, salting of lands, patented seed, poisoned water, and atmospheric ozone loss accelerates. Pet foods become the main grocery item of food chains in the cities, and carbonated drinks the main beverage. Blue-green algae poisoning of major rivers in Australia spreads to dams, and water supplies. Causes are over-clearing of river banks, over-irrigation, excessive phosphates from sewage (40%), and fertilisers (60%). Sewage includes detergents.

1991 AD. Pat Roy Mooney visits the Leningrad Institute where Vavilov had worked and where his co-workers have died. There are mice, rats, and neglect. "It was the finest collection [of seed] the world has ever known, in fact will ever know. The collection was simply killed by Russian bureaucracy over the last few decades." And so it goes on ... but for how long can war and tyrants be allowed to destroy this earth? The very future of mankind on earth has been gravely imperilled by greedy, dishonest, power-hungry politicians and corporate or finance managers. The battle is everywhere joined for survival against great odds. Wars themselves cause enormous environmental damage, resources are laid waste, forests now destroyed, and biological and chemical weapons are spreading across the earth based on corporate agricultural "research". The stage is now set for the biological time bomb. AIDS is found in up to 40% of African and 20% of Western city babies. If scientists succeed in making nitrogen fixation possible for pasture grasses, or one of many such projects, the world will collapse from soil acidity. Ethics of life preservation have long been lost by corporate groups. The future is, to say the least, uncertain. Money is poured into consumer projects, recreation, tourism, inefficient agriculture and industry, and cities. The ratio of "consumers" to "producers" is at breaking point. Homelessness, illiteracy, and disease, are on the increase. Genetic resources are lost and laid waste. No peoples any longer believe in political or financial systems, and the empires are going or gone. A truly dark age of environmental degradation is upon us.

1994 AD. Australian scientists warn of collapse of coastal ecologies due to excessive development, sewage, and sugar cane fields releasing sulphuric acids from old mangrove stands. Rain causes acid release (pH 1.9). Politicians try to minimise information and action.

APPENDIX D: RESOURCES

Sources of Cultures (also enquire from your local Department of Agriculture)

* Indicates that text is unchanged from the original 1992 print and individual research may be required.

Pure Culture Collections

UNITED STATES of AMERICA

American Type Culture Collection www.atcc.org/
Catalogue on request, prices given.

JAPAN

Japan Institute for Fermentation
4–54 Juso-Nishinoch, Higashigodu Garvaku, Osaka
Catalogue on request, prices cited.

AUSTRALIA

Australian Society for Microbiology
Suite 23, 20 Commercial Road,
Melbourne, Victoria 3004, Australia
Ph: +61 3 9867 8699
www.theasm.org.au/
Can advise on sources of cultures.

Dairy and Kefir

UNITED STATES of AMERICA

Chris Hanson's Laboratory
Ph: +1 559 485 2692
www.chr-hansen.com/

DENMARK

Chris Hanson's Laboratory
Bøge Allé 10-12, DK-2970, Hørsholm, Denmark
Ph: +45 45 74 74 74
www.chr-hansen.com/

POLAND

*Institute of Dairying—Kefir
Olsityn Rostowo, Poland

FRANCE

*G. Roger & Co.—Cheese moulds
La Ferte-Sous-Jouare (S&M),

AUSTRALIA

Ken Buckle, Graham Fleet
School of Food Technology,
University of NSW.
www.chse.unsw.edu.au/about/school_history.html

*Australian Rennet Manufacturing Co. Ltd.—Rennet and cultures for cheeses
23 Fuller St. Walkerville, South Australia

Koji Starters

JAPAN

*Mr. Amono Yaichi—Misos
Soto Kanda 2-18-15, Chiyoda-ku, Tokyo.
Ph: +81 03 251 7911

*Nihon Jojo Kogyo, The Japanese Brewing Company—Soy & misos
Koishigawa 3-18-9, Bunyo-ku, Tokyo 112.
Ph: +81 03 816 2951

UNITED STATES of AMERICA

Northern Regional Research Centre (USDA/NRRC)
1815 N. University Street, Peoria, IL 61604
http://nrrl.ncaur.usda.gov/

G.E.M. Cultures
Miso types and shoyu, imported from Japan.
P.O. Box 39426, Lakewood, WA, 98496 USA
Ph: +1 253 588 2922
www.gemcultures.com/

*Westbourne Natural Foods
4240 Hollis St., Emeryville, California 94608

Mushroom Cultures

FRANCE

Agri-Truffle
http://www.agritruffe.eu/

GERMANY

*E. Mangelsdorf
Post fach 1680, 435, Recklinghansen, Germany

*Burbaker Pilzfarm
Am Denkmal 14, 5909 Burbach-Holz Lansen

*Hornberger Pilzlabar
Werderstrasse, 7746 Horn/Schwatzwald

HOLLAND

*Les Mix Holland
B.V. Knapenstraat 15, Amsterdam Holland

UNITED STATES of AMERICA

MushroomPeople
560 Farm Road, P.O. Box 220, Summertown,
TN 38483, USA
Ph: +1 931 964 4400
www.mushroompeople.com/

Rick Kerrigan, Far West Fungi
541 Poplar Avenue, San Bruno, CA, 94066
Ph: +1 650 871 0786
www.farwestfungi.com/

APPENDIX E: SPECIES LIST OF LATIN NAMES

A

Acacia
A. senegal
A. seyal
Acanthopeltis
Acanthosicyos
A. horrida
Acer
A. saccharum
Acetobacter
A. aceti
A. xylinium
A. xylinum
Achatina
Acorus
A. calamus
Actinomucor
Agaricus
A. bisporus
A. brunnescens
Agave
A. americana
A. angustifolia
A. bacanora
A. mescal
A. tequiliana
Ahnfeltia
A. concinna
Alaria
A. esculenta
Albizzia
A. procera
Aleurites
Aloe
A. castanea
A. ferox
Alpina
Amanita
A. muscaria
Amaranthus
Amomum
Amorphophallus
A. konjal
Amphibolus
Amylomyces
A. rouxii
Anabaena
A. azollae
Analipus
A. japonicus
Anchoviella
A. comersonii
Anthrobacter
A. terrigens
Arachis
A. hypogaea
Arenga
Armoracia
A. rusticana
Artemisa
A. apiacae
Artocarpus
Ascaris
Ascophyllum
Asparagopsis
A. taiformis
Aspergillus
A. flavus
A. oryzae
A. rouxii
A. sojae
A. soyae
Astralagus
A. gummifer
Atta
Attacus
A. ophimacus
Auricularia
A. polytricha
Azedarachta
A. indica
Azolla

B

Bacillus
B. megaterium
B. natto
B. subtilis
Banksia
Bauhinia
B. malabarica
B. tomentosa
Boletus
Botulinus
Brassica
B. campestris
B. juncea
B. napus
B. niger
B. rapa

C

Cajanus
C. cajan
Calmaria
C. palmata
Canarium
Canavalia
C. ensiformis
C. gladiata
Candida
C. debaryomyces
C. kefir
C. utilis
Canna
C. edulis
Caragana
C. arborescens
Carica
C. papaya
Cassia
C. hirsuta
C. laevigata
C. obtusifolia
C. occidentalis
C. tora
Casuarina
Catha
C. edulis
Catraria
C. islandica
Caulerpa
C. racemosa
Ceiba
C. acuminata
C. parviflora
C. pentandra
Ceramium
Chaetomorpha
Chanos
C. chanos
Chayote
Chlamydomucor
Chondrus
C. crispus
Cicer
C. arietum
C. cassia
C. zeylandicum
Cladonia
C. rangifer
Cladosiphon
Claviceps
C. purpurea
Clostridium
C. botulinum
Cnidoscolus
C. palmeri
Cocos
Codium
C. edule
C. reediae
Combretum
C. apiculatum
Corchorus
C. capsularis
C. olitorius
Cordyline
C. australis
C. fruticosa
C. terminalis
Correa
C. speciosa
Corynocarpus
C. laevigata
Cossidae
Cruciferae
Cryptococcus
Cucurmis
C. naudianus
Curcumia
C. agustifolia
Cyamopis
Cymbopogon
C. citratus
Cynara
C. carbunculus
Cyperus
Cyrtosperma
C. chamissona
Cytoseira

D

Decapterus
Dictyopteris
D. australis
D. plagiogramma
Dioscorus
D. bulbifera
Diospyrus
Dolichos
D. lignosus
Durio
D. zibethinus
Durvillea
D. antarctica
D. potatorum

E

Ecklonia
Eisenia
Elaeis
Eleocharis
Eleusine
Endisme
Endomycopsis
E. burtonii
E. fibuliger
Ensete
E. ventricosum
Enteromorpha
Eragrostis
E. tef
Erythrina
E. variegata
Espalter
E. gariepina
Eucalyptus
E. johnsoni
Eucheuma
Eugenia
Euryhale
E. ferox
Eutrema
E. wasabi
Evernaria
E. prunasta

F

Faba
F. vulgaris
Fabaceae
Ferrobacillus
Ferula
F. assafoetida
Ficus
F. carica
Flavobacterium
Fomes
Fortunella
Frankia
Fucus
Fugu
Fusarium
F. gramineum
Froseum
Fuscus

G

Garcinia
G. cambogia
Gelidium
G. amansii
Geotrichum
G. candidum
Giardia
Glycine
G. max
Glycorrhiza
G. echinata
G. glabra
Gracilaria
G. verrucosa
Grateloupia
Grewia
Guaiacum
G. officinale
Gyrophora
G. esculenta

H

Hansenula
H. subpelliculosa
Hansula
H. anomala
Helix
H. aspersa
Hepialidae
Heterozostera
Hibiscus
H. sabdariffa
Hiizikia
H. fusiforme

I

Ilex
I. paraguayensis
Imperata
I. africana
Indigofera
Inga
Inocarpus
I. edulis
I. fagifer
Irdae
I. edulis
Isatis
I. tinctora

K

Kermes
K. vermilio
Kloeckera
K. ariculata
Kluyveromyces
K. fragilis

L

Lab-lab
L. purpureus
Lactobacillus
L. acidophilus
L. brevis
L. buchneri
L. bulgaricus
L. casei
L. caucasicus
L. cellobiosus
L. cucurmeris
L. cucurmeris fermenti
L. delbruckii

L. fermentati
L. fermentum
L. gasseri
L. helormensis
L. helvecticus
L. jugurt
L. kefir
L. lactis
L. lindneri
L. pentoaceticus
L. plantarum
L. reuteri
L. salivarius
Lagenaria
L. vulgaris
Laminaria
L. japonica
L. saccharina
Larix
L. laricina
Lathysis
L. sativus
Laurencia
L. pinnatifida
Lawsonia
Lecanora
L. esculenta
Lens
L. culinaris
Lentinus
Leucaena
L. glauca
L. leucocephala
Leuconostoc
L. citrovorim
L. cremornis
L. dextranium
L. m. cremornis
L. m. dextranium
L. mesenteroides
Locusta
Locusta bruchos
Luffa
L. cylindrica
Lupinus
L. alba

M

Macrocystis
Macrotermes
Macrotyla
M. uniflorum
Macrozamia
Madhuca
M. indica
Mallotus
M. philippensis
Manikara
M. hexandra
Maranta
Medicago
M. sativa
Melia
M. azedarachta
Metschnikowia
M. pulcherrima
Micrococcus
Miscanthus
M. sacchariflorus
Monascus
M. purpureus
Monilia
Monostroma
Monstera
M. deliciosa
Moringa
Mucor
M. meihei
Mucuna
M. pruriens
M. urens
M. utilis
Mycoderma
M. aceti
Myriopterum

N

Nemacystus
Nereocystis
Neurospora
N. intermedia
Nopalea
Nopales
Nostoc

O

Ophiocephalus
Opuntia
Oryzopsis
O. hymenoides
Oxalis

P

Pachyrhizus
P. erosus
Pandanus
P. tectorius
Panicum
P. burgu
P. palmifolium
Parkia
P. africana
P. biglobosa
P. clappertoniana
P. filicoidea
P. uniglobosa
Parmelia
Pediococcus
P. acidilacti
P. cerevisiae
P. halophilus
P. pentosaccus
Peltiger
Penicillium
Pentaclethra
P. macrophylla
Perilla
P. frutescens
Phaseolus
Ph. acutifolius
Ph. coccineus
Ph. lunatus
Ph. vulgaris
Phoenix
P. dactylis
P. reclinata
P. sylvestris
Phragmites
Pinguicola
P. vulgaris
Pinus
P. monophylla
P. pinea
Pisum
P. sativum
Plevetia
Podbralla
Polygonum
Porphyra
P. capensis
P. laciniata
Posidonia
Prosopis
Psilocybe
Psophocarpus
P. tetragonolobus
Psuedomonas
P. putida
Pueria
P. lobata
Puffinus
P. tenuirostris

R

Ramalina
Raphanus
R. longipinnatus
R. sativus
Rhizobium
R. melilot
Rhizoporus
R. oligosporus
Rhodotorula
R. glutinus
R. minuta
R. mucilaginosa
Ricinodendron
R. rautanenii
Roccella
R. montagneiss
R. tinctora
Rosa
R. centifolia
R. damascena
Rubia
R. tinctorum

S

Saccharomyces
S. cerevisiae
S. rouxii
Sagittaria
Salmonella
Sanseveria
Sargassum
Scenedesmus
S. quadricauda
Scirpus
Sclerocarya
S. birrea
Scomberomorus
Sesamum
S. indicum
S. orientale
Sesbania
S. grandiflora
S. sesban
Shorea
S. robusta
Solanaceae
Solanum
S. ajanhuivi
S. curtilobum
S. giganticum
S. juzepczuchii
S. tuberosum
Sphaerotrichia
Sphenostylis
S. stenocarpa
Spirulina
S. platensis
Sporobolomyces
Staphylococcus
Stichopis
Stolephorus
Streptococcus
S. cremornis
S. diacetilactus
S. faecalis
S. faecium
S. kefir
S. lactis
S. lactis diacetilactus
S. lactis lactis
S. raffinolactis
S. thermophilus
Swartzia
S. madagascariensis
Synapsis
S. alba

T

Tacca
Tamarindus
T.indica
Tectonia
Terfezia
T. pfeilli
Termitomyces
T. letestui
Thiobacillus
Tinocladia
Torulaspora
T. tosei
Torulopsis
T. bacillaria
T. etchellsii
T. nodaensis
T. sphaetica
T. versetilis
Trichrus
Tylosema
T. esculentum
Typha
T. angustifolia

U

Ulva
U. lactuca
Umbilicaria
Undaria
U. pinnatifida
Usnea
Ustilago
U. maydis

V

Vigna
V. acontifolia
V. angularis
V. mungo
V. radiata
V.umbellata
V. unguiculata
V. vexillata
Voandezeia
V. volvacae

W

Woroninella
W. psophocarpi

X

Xanthium
X. sibirium

Z

Zanthoxylum
Z. planispinum
Z. simulans
Zeusero
Z. coffea
Ziziphus
Zostera
Zygosaccharomyces

REFERENCES

Only the main references are listed here; I have annotated those of special interest for others contemplating further work. The problem with many such books is that they are expensive, or rare or both.

Although this book will also be limited, especially in specialised areas such as the cheeses, it will also remain open for detailed ground contributions from readers in the permaculture network. While I thank those friends who have sent short references, I may remind them that a note on the source should always be attached. My main hope is that many of you will find value in this book for your own household and your good work. It is the sort of book I wish I had ten years ago.

ACHARYA, Gyanu, 1991

Domestic Skills of Nepal

Unfinished manuscript. Tagari Publications, 31 Rulla Road, Sisters Creek, Tasmania 7325, Australia

AL AZHARIAJAHN, Dr. Samia, 1981

Traditional Water Purification in Tropical Developing Countries. German agency for technical cooperation (GTZ) Eschborn, Germany

A truly encyclopaedic study of village water purification processes.

ALEXOPOULOS, C.J., 1962

Introductory Mycology. John Wiley and Sons, New York

ANANTHANARAYAN, R. and. JAYARAM PANIKER, C.K. 3rd edition, 1986

Textbook of Microbiology. Orient Longman, Hyderabad, India

Essentially on medical microbiology, but in that area very comprehensive.

ANDRASSY, Stella, 1978

The Solar Dryer Book. Earth Books, New York

ARASAKI, Seibin and Teruko, 1983

Vegetables from the Sea. Japan Publications Inc. Tokyo

An essential gourmet reference for those who enjoy Japanese marine food—dozens of recipes.

AUBERT, Claude, 1986

Les Aliments Fermentés Traditionnels. Terre Vivant, 6 Rue Saulnier 75009, Paris

A very good overview of common ferment processes in many countries; in French.

AUBERT, Claude, 1983

Onze Questions-cles. Terre Vivant, Paris

AUMEERUDDY, Y. and PINGLO, F. 1989

Phytopractices in Tropical Regions. UNESCO, Institute of Botany, 7, Place de Fontenoy 57352 Paris 07 SP, France

AYYANGAR, N.R. et alia, 1977

A Course in Industrial Chemistry. Orient Longman, Hyderabad, India

Used mainly for sugar processing.

BECKER, Peter, 1975

Trails & Tribes in Southern Africa. Hart-Davis, MacGibbon, London

BEETON, Mrs I. M., First published 1861

Mrs Beeton's Book of Household Management. Ward Lock, London

BRISTOW, W.S., 1953, Proc. Nutr. Soc. 12 & 44

An Impressive List of Insects Known to be Appreciated by Man (not seen)

BRYCE-SMITH, D. and HODGKINSON, L., 1986

The Zinc Solution. Arrow Books Ltd. England

BUMGARNER, Marlene Anne, 1977

The Book of Whole Grains. St Martins Press, New York

CAMPBELL, I. and DUFFUS, J.H. 1988

Yeast: A Practical Approach. I.R.L. Press, Oxford; Oxford, Washington DC

CAMPBELL-PLATT, Geoffrey, 1987

Fermented Foods of the World. A Dictionary and Guide. Butterworths, Kent, UK

A useful compendium for cross referencing local names and for concise recipes; with many wines, ferments, and sausage recipes.

CAMPBELL, R. and MACDONALD R.M. (eds), 1989

Microbial Inoculation of Crop Plants. IRL Press, Oxford University Press; Oxford, New York

CHRISTIANSON, I.G. et alia, 1981 (Arasaki, 1983; and Fortner, 1978)

Seaweeds of Australia. A.H. & A.N. Reed, Sydney

Good for identification only.

COUFFIGNAL, Huguette, 1979

The Peoples Cookbook. Pan Books (Macmillan Publisher).

A practical guide to practical recipes.

CREASY, Rosalind, 1982

The Complete Book of Edible Landscaping. Sierra Club Books, San Francisco

An essential book on growing a great variety of foods, with some good recipes.

DAVID, Elizabeth, 1979

English Bread and Yeast Cookery. Penguin Books, UK

DAVIDSON, Alan, 1978

Seafood of South East Asia. Macmillan, London

DAVIDSON, Sir Stanley et alia, 1979

Human Nutrition and Dietetics. Churchill Livingstone, London, New York

A good overview of the role of amino acids, minerals, and vitamins in diet, with data on metabolic and nutritional imbalances. Basically a medical text.

DAVIES, Flyndon, and LAW, Barry A., 1984

Advances in the Microbiology and Biochemistry of Cheese and Fermented Milk. Elsevier Applied Science Publishers, London and New York

DUKE, James, 1981

Handbook of Legumes of World Economic Importance. Plenam Press, New York

Good data on protein content, and some ferment recipes.

FACCIOLA, Stephen, 1990

Corrnucopia, A Source Book of Edible Plants. Kampong Publications, 1870 Sunset Drive, Vista, CA 92084, USA

One of the best plant compendia of modern times, including fungi, algae, bacteria, extensive notes, cultivar lists and Asian species.

FAIRFAX, Kay, 1986

Homemade. Methuen Haynes, Sydney, Australia

FORTNER, Heather J., 1978

The Limu Eater. University of Hawaii. Sea Grant Program.

Hawaiian seaweed recipes—mainly from fresh algae.

GAJUREL, C.L. and VAIDYA, K.K., 1984

Traditional Arts and Crafts of Nepal. S. Chand and Co. Ltd. Ram Nagar, New Delhi 110055

Covers food, art, crafts, pottery, and so on; a useful book.

FREUCHEN, Peter, 1961

Peter Freuchen's Book of the Eskimos. Fawcett Premier, New York

GAST, M., and SIGAUT, F., 1979

Les Techniques de Conservation des Grains à Long Terme. Centre National de la Rechereche Scientifique, Paris

GARRETT, S.D., 1963

Soil Fungi and Soil Fertility. Pergamon Press, Oxford

GELFAND, Michael, 1971

Diet and Tradition in an African Culture. E. and S. Livingstone, Edinburgh and London

Useful for African foods.

GRIGSON, Jane, 1973

Fish Cookery. Penguin

GRIGSON, Sophie, 1991

Sophie Grigson's Ingredients Book. Pyramid Books, London

An excellent overview of ingredients, well illustrated. A good book to add to your kitchen.

HALE, P. R., and WILLIAMS, B. D., 1977

Liklik Buk: A Rural Development Handbook Catalogue for Papua New Guinea. Wantok Publications, PO Box 1982, Boroko, PNG

A model of useful data on every aspect of food, health, and village technology.

HARRIS, D. R. and HILLMAN, G. C., 1989

Foraging and Farming: The Evolution of Plant Exploitation. Unwin Hyman, London

HEDGE, M.V., et alia, 1978

A Course in Industrial Chemistry. Part III. Orient Longman, Bombay. Chapter 3 on fermentation, pp.124-166.

HESSE, Zara G., 1973

Southwestern Indian Recipe Book. The Filter Press, Palmer Lake, Colorado

Food of Apache, Papago, Pima, Pueblo, and Navajo.

HESSELTINE, C.W. and WANG HWA, L., 1986

Indigenous Fermented Food of Non-Western Origin. J. Cramer, Berlin

An excellent source for data on home culture of yeasts, moulds, and specific ferments.

HETZEL, Basil S., 1989

The Story of Iodine Deficiency. Oxford University Press, New York, Tokyo

ISRAEL, R. and SLAY, Reny Mia, 2nd edition., 1975

Homesteaders Handbook. Self published (out of print).

A very useful assembly of food and energy techniques.

JAFFREY, Madhur, 1983

Eastern Vegetarian Cooking. Jonathan Cape, London

JUMELLE, Henri, (undated, 1850s?)

Les Cultures Colonial. J.B. Bailliere et Fils, Paris

KANDER, Mrs. Simon, 1945

The Settlement Cook Book. Milwaukee, Wisconsin

A very sound cook book. It is now reprinted.

KAY, Daisy E., 1979

Food Legumes. Tropical Products Digest N°3. 56/62 Gray's Inn Road, London WC IX 8LU

KOSIKOWSKI, Frank V., 1977

Cheese and Fermented Milk Foods. F.V. Kosikowski and Associates, PO Box 139, Brooktendale, New York 14817-139

A very comprehensive book on milk products and cheeses for both professionals and home or village use, written by a widely travelled expert. A classic.

LAPPÉ, Frances Moore, 1975

Diet for a Small Planet. Ballantine Books, New York

MABEY, David, 1984

The Little Mustard Book. Judy Piatkus Pubs. Ltd. London

Everything you wanted to know about mustard.

MADLENER, Judith C., 1977

The Sea Vegetable Book. Clarkson N. Potter Inc. New York

MAJUPURIA, Indra, 1986

Joys of Nepalese Cooking. Tiptop Offset Press, Delhi

MARSH, Philip, 1980

Oral Microbiology. Nelson, London

MICROBIOLOGICAL PROCESSES: *Promising Technologies for Developing Countries.*

National Academy of Sciences, 1979. (USA) Washington DC

A very useful overview of the potential microbiological processes, readable and comprehensive.

Maori Cookbook (undated, ca 1980?) Glenfield College Home and School Association, New Zealand Consolidated Press Ltd., Auckland

Recipes noted as Maori or New Caledonia are taken from this useful book.

MOLLISON, Bill, 1988

PERMACULTURE: A Designers' Manual. Tagari Publications, 31 Rulla Road, Sisters Creek, Tasmania 7325 Australia

A full treatment of the techniques and design of sustainable land-use and settlement.

MOLLISON, Bill with SLAY, Reny Mia, 1991

Introduction to Permaculture. Tagari Publications, 31 Rulla Road, Sisters Creek, Tasmania 7325, Australia

An introductory text on the subject of sustainable land and settlement design.

MONZERT, Leonard, 1866

The Independent Liquorist, or, The Art of Manufacturing and Preparing All Kinds of Cordials, Syrups, Bitters, Wines. Dick & Fitzgerald

MOWAT, Farley, 1973

Tundra: Selections from the Great Accounts of Arctic Land Voyages. McClelland and Stewart, Toronto

MURPHY, Edith Van Allen, 1959

Indian Uses of Native Plants. Mendocino County Hert. Soc., 243 Bush St., Fort Bragg, CA 95437

OCHSE, U., 1931 (English edition)

Vegetables of the South East Indies (Indonesia). Archipel Drukkerij Buitenzorg, Java

Notes on food preparation.

OLANNA, Melvin and Karen, 1989

Shishmaref Eskimo Cookbook. Franklin Press, Seattle, USA

Basic cold-area meat ferment, seals, walrus.

PARKINSON, Susan and LAMBERT, J., 1983

The New Handbook of South Pacific Nutrition. UNICEF, PO Box 2351, Government Building, Suva, Fiji

PARKINSON, Susan, et alia, 1991

Storing and Preserving Pacific Island Foods. South Pacific Commission Nutrition Project, Suva. Fiji

PETARD, Paul, 1986

Plantes Utiles de Polynesia Francaise et Raan Tahiti. Editions Haere Po No. Tahiti

RICHIE, Donald, 1985

A Taste of Japan. Kodensha International. Tokyo

RODEN, Claudia

A Book of Middle Eastern Food. Penguin UK

RODY, Nancy, (undated. ca. 1988)

Breadfruit Bread and Papaya Pie. Printed anonymously in Hawaii

This book, crammed with recipes, sidesteps ferment, as do many western books.

ROMERA, Philomena, 1970

New Mexican Dishes. Self published.

ROSE. A.H., (ed.), 1982

Fermented Foods. (Economic Microbiology Vol. 7). Academic Press Inc. London, New York

A sound overview by many authors and contributors of the principles involved in ferment, and a range of products of fermented food.

RYDER, M.L., 1983

Sheep and Man. Duckworth and Co., London

SAONO, S. et alia (eds.), 1986

A Concise Handbook of Indigenous Fermented Foods in the ASCA Countries.

Association for Science Cooperation, Government Printer, Canberra, Australia

A very useful listing of recipes and flow charts, micro-organisms, analyses.

SCHAAL, E. and CROWLEY, B.M., 1989

The Great Aussie Sausage Book. Matchbooks, South Melbourne

SHURTLEFF, William and AOYAGI, Akiko, 1983

The Book of Miso. Ten Speed Press, PO Box 7123, Berkeley, CA 94707

A model of sound work on a specific subject; a definitive book. The same authors have several books on ferment products, of equal thoroughness.

SPENCER, J.F.T. and SPENCER, D.M.,(eds.), 1990

Yeast Technology. Springer-Verlag, Berlin, New York

SRIVASTAVA, C.B.L., 1985

A Textbook of Fishery Science and Indian Fisheries. Kitab Mahal 15, Thornbill Road, Allahabad, India

STAMETS, Paul and CHILTON, J.S., 1983

The Mushroom Cultivator. Agarikon Press, Olympia, Washington, USA

Probably the essential reference for the home and commercial mushroom grower.

STEINBECK, Helmut, 1981

Mushrooms in the Garden. Eugen Ulmer Press, Stuttgart

Ideal for those of us who would like to "naturalise" edible mushrooms on logs or below specific tree hosts.

STEINKRAUS, Keith H., 1989

Industrialization of Indigenous Fermented Foods. Marcel Dekker, Inc; New York, Basel

Taking key foods, Steinkraus reviews their history and subsequent developments, and the potential for future work. An acknowledged classic; a range of contributors.

STEPHENSON, Winifred A., 1968

Seaweed in Agriculture and Horticulture. Bargyla Rateaver, Pauma Valley, CA 92061

SUBBA RAO, N.S., 1982

Advances in Agricultural Microbiology. Butterworth Scientific, London, Singapore

Deals with a wide range of pathogens, processes, utilisation of wastes and so on.

TANIKAWA, Eiichi, 1985

Marine Products in Japan. Koseisha Koseikaku Co. Ltd., Tokyo

The essential reference for the processing of marine products and ferment in Japan.

TREDGOLD, Margaret H., 1986

Food Plants of Zimbabwe. Mambo Press, Senga Road, Zimbabwe

Margaret Tredgold not only describes the plants, but gives a great deal of data on traditional food preparation, omitted by many authors on like subjects.

VEERESH, G.K. and RAJAGOPAL, D.,1983

Applied Soil Biology and Ecology. Oxford and IBH Publishing Co, New Delhi

WANG, Daniell I.C. et alia, 1979

Fermentation and Enzyme Technology. John Wiley and Sons, New York

WARD, O.P., 1989

Fermentation Biotechnology. Open University Press, Milton Keynes

WILLIS, J.C., 1909

Agriculture in the Tropics; An Elementary Treatise. University Press, Cambridge

WOLFE, Jennifer A., 1987

The Potato in the Human Diet. Cambridge University Press, Cambridge

The Wealth of India.

A Dictionary of Indian Raw Materials and Industrial Products. (Ten volumes) Council of Scientific and Industrial Research, Government Printer, New Delhi

Would that every country had such a mine of information, mainly on plants but including animals, minerals, and above all a variety of processing methods for those materials, analyses of ammo acids and essential oils in foods, and specific uses. A veritable goldmine of data.

INDEX

G

H

I

J

K

L

M

N

O

P

R

S